UNIVERSAL CREATIVITY and INDIVIDUATION of the BIOSPHERE: A CONSTRUCTIVISM EPIC

DONALD PRIBOR

University of Toledo

KENDALL/HUNT PUBLISHING COMPANY
4050 Westmark Drive   Dubuque, Iowa 52002

Contents

UNIT I

AN EXPANDED THERMODYNAMIC UNDERSTANDING OF UNIVERSAL CREATIVITY   1

Chapter 1	AN EXPANDED THERMODYNAMIC UNDERSTANDING OF PHYSICAL CREATIVITY	3
Chapter 2	SYSTEMS THEORY OF CREATIVITY AS EXEMPLIFIED BY THE EMERGENCE OF LIFE	41
Chapter 3	THE FOUR CREATIVE LEARNING STYLES OF SCIENTIFIC THINKING	55
Chapter 4	HOMEOSTATSIS: ONE DEFINING CHARACTERISTIC OF LIFE	83

UNIT II

BIOLOGICAL INDIVIDUATION AS CONTROL OF STRESS   93

Chapter 5	CELL AS THE FUNDAMENTAL UNIT OF LIFE	95
Chapter 6	CELL AS A CHEMICAL MACHINE THAT REPRODUCES ITSELF	135
Chapter 7	HUMAN HOMEOSTATIC BODY SYSTEMS	181
Chapter 8	NEURO-ENDOCRINE-IMMUNE SYSTEM	219
Chapter 9	NEURO-PSYCHIC HOMEOSTASIS	243
Chapter 10	ECOSYSTEM AS A HOMEOSTATIC MACHINE	275

UNIT III

INDIVIDUATION OF THE NON-HUMAN BIOSPHERE   293

Chapter 11	CORE IDEAS OF GENETICS	295
Chapter 12	EPIGENESIS: DEVELOPMENTAL INDIVIDUATION	331
Chapter 13	BIOLOGICAL THEORY OF EVOLUTION	377
Chapter 14	CORE MUTUALITIES OF NON-HUMAN INDIVIDUATION	389

Preface

As the title suggests, the core ideas of this textbook are creativity, individuation, constructivism, and epic, a kind of narrative. Of these four ideas constructivism for many readers may be the most unfamiliar. Constructivism presumes that Reality is mysterious, that is, unknowable, but Reality is interpretable. This assumption leads to the practice that each human and each human society does not discover structure in nature, but each can interpret one's individual or socially objectified collective experience of Reality. Though not usually explicitly stated this assumption was adopted by modern scientific communities that emerged and expanded in Europe between 1500–1700 A.D. From 1700 to the present this scientific constructivism focused on creating theories primarily for the purpose of solving practical problems. The mutual influence alongside of the mutual antagonism of scientific constructivism and religious or philosophic subjective belief in a creator God led to the emergence of democracy in the American colonies in the 1740s. The core idea of this democracy that was formalized by the Declaration of Independence and the Constitution (that includes the Bill of Rights) is a political version of narrative constructivism. According to this constructivism principle, a democratic government must not only allow but guarantee that each citizen has the right to choose to construct for himself an interpretation of one's subjective experiences that may be represented as the existence or non-existence of a creator God. This personal constructivism also may include the name and description of God as provided by any particular religion.

In the twentieth century, especially after World War II, the scientific, technological, and business communities have consciously or unconsciously collaborated to produce a radical, individualistic perspective that overshadows and undermines the original spiritual vision of the "fathers" of American democracy. Especially at the end of the cold war in late 1991, globalization and a more inclusive understanding of democracy have emerged and now are challenging the restrictive vision of the patriarchal, creator God perspective of the white male "fathers" of the American Revolution. In order to overcome this challenge we need a higher level of constructivism that provides a foundation for this more inclusive, non-patriarchal understanding of democracy. This new type of constructivism also must transcend the opposition between radical individualism of capitalistic democracy and the spiritual visions of diverse religions and philosophies that may assert or deny the existence of a creator God.

Based on a deep analysis of the operation of an ideal machine (the Carnot ideal heat machine), Chapter 1 describes a narrative, scientific constructivism that is a subjective, metaphorical conceptual understanding of any scientific theory. As a result of metaphorical concepts, abstract ideas in science can be understood in relation to one's personal experiences. This way of understanding science does not oppose but rather can complement other types of narrative constructivism such as found in the humanities and in Eastern as well as Western philosophical-religious perspectives. Collaboration among diverse types narrative constructivism can bring a rebirth to the spiritual vision of the "fathers" in a postmodern democracy suitable for globalization. Each culture can integrate its unique narrative perspectives with the science-based, capitalistic democracy. What makes this possible is that narrative, scientific knowledge always can be reduced to scientific constructivism which is totally objective and correspondingly suitable for solving concrete problems.

At the same time a narrative scientific approach to teaching facilitates the student to understand any abstract, new idea in terms of her own personal experiences. This understanding involves memorization coupled with creative insights which remain even after the memory of details is hazy or lost. As a result, a brief review can bring back all that was once very familiar, and sometimes this rebirth of understanding is deeper and more personal. Creative understanding of ideas provides "hooks" for retaining a large number of facts that a person may need in a particular situation. Learning becomes much more efficient when it involves creative understanding of many ideas from which one selects a few that will be further developed and enhanced by factual information that one needs to know. This relieves the burden of memorizing many facts that are not understood in terms of personal experiences and that may be irrelevant to one's needs.

The narrative, scientific perspective of this textbook implies that quality of understanding is more important than the quantity of information that is memorized. With this in mind I never would teach all the information in this text even to a class of well-prepared, highly motivated, science majors. I think science majors could benefit from reading this text in that they are deprived of creative learning as a result of the great quantity of information they must memorize in introductory science courses. Nevertheless, this book primarily is written for a one semester or one quarter survey of biology course for non-science majors. How much material in this book one requires students to read depends on the length of the course and the preparation, quality, discipline, and motivation of the students in the course. For most survey courses, even those at the most prestigious schools, the first 20 pages in Chapter 1 dealing with fundamental ideas in physics can be passed over, but the ideas in the second half of Chapter 1 are very important for understanding the rest of the book. In my experience of teaching this material, at first students are dismayed at the prospect of attempting any creative learning, and this is more especially the case with regard to the abstract ideas of physics and of constructivism. I accommodate them by giving many metaphors, examples, and personal stories to facilitate them internalizing these ideas, and I proceed at a slow pace. Many students give up and stop coming to class, but then some of these come back and eventually begin to catch on. When this happens, they not only understand the abstract ideas at some level, they realize a new dimension in themselves. They realize that they can creatively learn ideas and then sometimes immediately apply these ideas to concrete situations in their lives. Moreover, they begin to apply this way of learning to other classes. In the process of understanding aspects of the individuation of the biosphere, they, themselves, individuate. When this happens even to only a small extent, the resulting personal transformation provides the basis for more extensive learning later in life. To me, anyway, this qualitative change is more important than memorizing a great quantity of information.

I would like to make one final observation/suggestion. Many chapters, especially Chapters 5, 6, and 12 contain some technical details that I provide for completeness and for the student who may be interested in the subject of that chapter. For most students it is not necessary to learn these details.

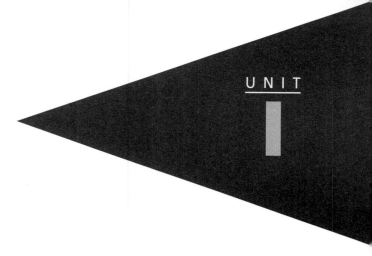

An Expanded Thermodynamic Understanding of Universal Creativity

An Expanded Thermodynamic Understanding of Physical Creativity

Thermodynamics depends upon the theory of the operation of any machine which, in turn, depends upon the generalized concept of energy and the related ideas of work and heat. All of these latter ideas depend on Newton's theories of motion involving contact forces and non-contact forces, for example, gravitational forces. Therefore, this chapter begins with Newton's theories which then are expanded to a generalized concept of energy related to work and heat processes. Understanding the functioning of any machine contradicts core ideas of the expanded Newtonian mechanics. In particular, machine dynamics introduces the necessity of explicit, subjective insights that Newtonian mechanics excludes, for example, entropy is a measure of chaos and the irreversibility of time as one subjectively experiences events contrasted with reversible, measurable time of Newtonian mechanics. A higher level thermodynamic perspective transcends this contradiction by modifying and subordinating mechanistic dynamics to machine dynamics. This chapter concludes with an extension of thermodynamic, subjective insights to propose fundamental laws governing physical creativity. *Note: On page 17 there is a section entitled: "INTELLECTUAL SHORTCUT FOR PHYSICS" consisting of two paragraphs that can take the place of pp. 1–16 of this chapter.*

General Idea of Motion

Our experience of motion is that of one object changing its spatial relation to some other reference object during some time interval. Furthermore, the motion of one object may "influence" (Aristotle would say "cause a change in") the motion of another object such as a bowling ball knocking over stationary pins or a baseball bat reversing the direction of motion of an oncoming baseball. Thus, our experience of motion is somehow involved with our intuition of changing spatial relationships, time intervals, and interactions (causality?) among objects. We now come to a crucial question: can we humans rationally comprehend motion or at least construct a rational model that approximately comprehends motion? The major Greek philosophers, Medieval thinkers, seventeenth century scientists, and perhaps most modern scientists would answer yes.

I exhort all who would hear me out that we answer NO to this question based on the following value commitment. By denying rational comprehension, we allow ourselves to participate in motion, marvel at it, be in awe of it much like the baby paying attention to without comprehending some of the activity occurring around him. The value commitment is to keeping participatory consciousness coequal with control consciousness so that the appropriateness of poetry, myth, pure subjectivity, or rational analysis depends on context rather than on the "object" of our consciousness. This denial of rational closure allows for the possibility of integrating science with the humanities and with our experienced subjectivity. At the same time we still can pursue ever greater control over nature according to the following analogy. The theologian can never rationally comprehend God but nevertheless can describe aspects of God which can guide how humans interact with one another and with nature. In like manner, the scientist may admit that he can never rationally comprehend motion but nevertheless can describe aspects of motion with ever greater clarity and generality. These descriptions will give us some control over natural phenomena.

Newtonian Idealized Motion

Even though Newton brought to maturity the modern scientific way of describing nature, he still was born into and influenced by a Medieval culture which analyzed motion according to an Aristotelian perspective. In this perspective, a material thing exists and expresses its existence by moving toward some goal, the thing's final cause, and sometimes by being the efficient cause for the motion of some other material object. We could say that the movement present as a property of one object caused a second object to move toward some goal. Newton sought a quantitative description of causal interactions of this sort. To achieve this he would have to create a quantitative model which he then could apply to his or anyone else's observations of interacting moving objects. As is true of all human creativity, he had to imagine a new kind of vision which nevertheless was built upon the creative imaginations of others. The new vision would be like a myth or fairy tale. But instead of saying, "in the beginning, out of the void came" or "once upon a time," he started with: "suppose we consider material objects existing in an on-going duration which is the same for all of them and in absolute space which extends indefinitely in three dimensions."

This choice allowed Newton to use the mathematical and physical ideas of other thinkers extending from his near contemporaries such as Rene Descartes and Isac Barrow to Galileo, and all the way back to the Greek mathematicians. At this point a story of motion almost begins to tell itself. Absolute space has the properties described by Euclidian geometry, and absolute time is linear and can be represented and measured by a numbered straight line in the same way as each of the three dimensions of space. We then can locate an object in space and define any motion it would express as the change of its position in space during some time interval. The distance the object traveled would represent its change in position. Because both distance and time can be measured, all continuous motions could be represented by one of many possible types of continuous functions in which each type makes distance a function of time.

For example, take the simplest type of motion in which an object moves in a straight line at a constant speed. This type of motion could be represented by a function such as $D = 2t$ as shown in figure 1.1 and table 1.1. The angle that this straight line makes with the x axis would uniquely specify the direction of the motion, and the formula, $2t$, specifies the magnitude of distance traveled during the motion.

A somewhat more complex motion, which we often may observe, was described by Galileo. It is the free fall of a body toward the earth in which speed changes at a constant rate. This motion can be represented by a family of three functions: (1) $D(t) = 1/2\ gt^2$; (2) the direction of the motion is always perpendicular to the surface of the earth; so all we need to worry about is the magnitude of the change in velocity which is given by the function, derived from the calculus and represented as $D'(t) = v = gt$; (3) the rate of change of velocity with respect to time is analogous to velocity. Just as velocity is the rate of change of distance with respect to time and is calculated by finding the derivative, $D'(t) = v$, so also *acceleration* $= a$ is the rate of change of velocity with respect to time and is calculated by finding the derivative of the velocity function (a concept in calculus), e.g., $v'(t) = a = gt^0 =$ the constant, g, in this example, see figure 1.2 and table 1.2. In the more general case, motion is not in

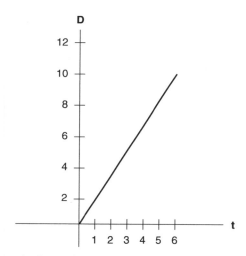

FIGURE 1.1   Distance as a function of time.

Table 1.1   Velocity as a function of time

D	0	2	4	6	8	10	12
t	0	1	2	3	4	5	6

Table 1.2   Acceleration as a function of time

Time t	Distance D	Velocity V	Acceleration A
0	0	0	0
1	16	32	32
2	64	64	32
3	144	96	32
4	256	128	32
5	400	160	32
6	576	192	32

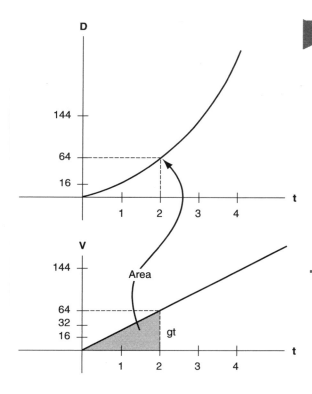

FIGURE 1.2

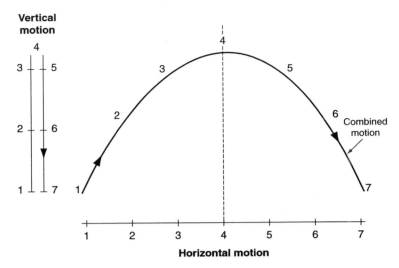

FIGURE 1.3

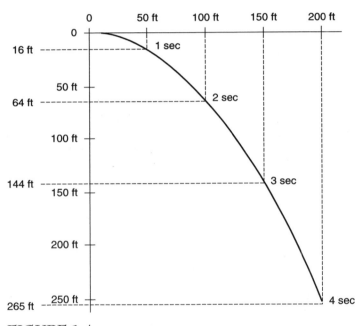

FIGURE 1.4

a straight line but rather is some curve indicating a continuous change in direction with respect to time. Galileo's analysis of the motion of a projectile such as a bullet provided an insight for how one can describe non-linear motion. He demonstrated that the path of a projectile where one can ignore air resistance is that of a parabola, see figure 1.3. For example, consider the somewhat simpler case of firing a cannon at the edge of a cliff where the velocity of the cannon ball as it leaves the cannon is 50 ft/sec in the horizontal direction. Galileo assumed that if the cannon were lying flat on a horizontal plain both the magnitude and the direction of this velocity would remain constant, that is, the ball would move in the horizontal direction 50 ft after 1 sec, 100 ft after 2 sec, 150 ft after 3 sec, 200 ft after 4 sec, and so on. From a previous section we already know that if the ball merely was dropped from the edge of the cliff, it would move in the downward direction according to the function $D(t) = 1/2 \, gt^2$, that is, 16 ft after 1 sec, 64 ft after 2 sec, 144 ft after 3 sec, and 265 ft after 4 sec, see table 1.2. As shown in fig. 1.4 the actual path of the cannon ball fired at the edge of the cliff would be that of a

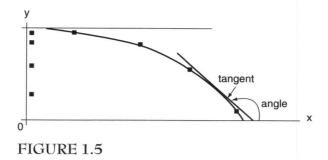

FIGURE 1.5

half parabola which at each point is the resultant of the independent movements in the horizontal direction and in the downward direction. The tangent to any point on this parabola as shown in figure 1.5 would represent the velocity of motion at that point. The angle that this line makes with the x axis uniquely defines the direction of the velocity at that point. The magnitude of the velocity would be given by the rate of change of distance in the horizontal direction which always will be 50 ft/sec and the rate of change of distance in the downward direction which will be gt.

Newton generalized Galileo's insight and proposed that any motion in a plane can be represented as the resultant of two functions: $D_x(t)$ = distance moved in the (+) or (–) direction along the x axis as a function of time and $D_y(t)$ = distance moved in the up (+) or down (–) direction along the y axis as a function of time. From these functions one can derive two new functions that represent the magnitude and direction of the velocity as a function of time. Likewise from these velocity functions one can derive two functions that represent the magnitude and direction of acceleration as a function of time.

Having imagined describing motion in this way, Newton may well have asked the question of how much motion does an object have at a particular point in time? This quantity of motion must have a goal which in this case would be given by its direction. The quantity also must describe to what extent the object can change the motion of some other object. In the imaginary world of mathematics, velocity quantifies the motion of a point at a particular moment in time, but in the empirical world of experience some new factor comes into play. For example, consider two balls of the same diameter, each traveling at 80 miles/hr in a straight line toward me. If the one object is a tennis ball, when it hits me, this would smart a bit but the ball would not knock me over. If the second ball was made of marble, it would knock me over upon impact. Thus, though both balls have the same velocity, the marble ball has a greater quantity of motion by virtue of being able to change my motion to a greater extent than the tennis ball is able to do.

Of course we know that the marble ball weighs more than the tennis ball. Newton assumed that weight is a measure of "quantity of matter," which he called *mass*. Mass is something like distance in that both are assumed to be continuous and can be measured. That is, just as we define a standard unit of length and then compare any other distance to this standard length, so also we can define a standard unit of mass and compare any other mass to this standard mass. Having created the physical concept of mass, Newton could choose between the two simplest ways of defining quantity of motion: either it is mass + velocity or it is mass × velocity. In either case motion always is an object's change of position in space. But if quantity of motion is defined as mass + velocity, then when velocity = 0, the object would still be considered to have a quantity of motion = mass + 0 = mass even though the object is not changing its position in space. This is a contradiction; therefore, instead Newton defined quantity of motion called *momentum* = p as p = mass × velocity = Mv.

At this stage in Newton's theory causality easily can be visualized as one object transferring some or all of its momentum to some other object. Thus, during any causal interaction, momentum is conserved. Actually it was Rene Descartes who first proposed this idea and later Christian Huygens generalized it to the law of conservation of momentum. In his book, *Physics for Poets*, Robert March nicely illustrates the law as follows. Suppose we let m = meter = unit of length and kg = kilogram = 1000 grams = unit of mass. Then suppose that for two objects that are free to move on a "frictionless" surface, the one having 3kg mass is stationary and the other having 2kg mass is moving toward the

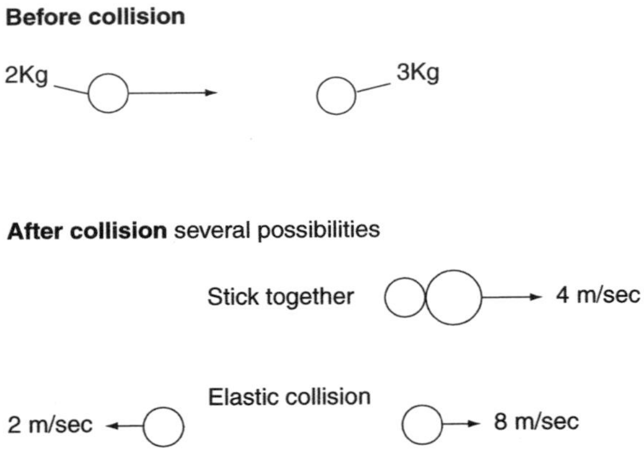

Before collision

After collision several possibilities

Stick together 4 m/sec

Elastic collision

2 m/sec 8 m/sec

FIGURE 1.6

first with a constant velocity of 10m/sec. Before any interaction all momentum is in the 2kg object and is equal to (p = Mv) 2kg × 10m/sec = 20 kg-m/sec. After the collision, one of two outcomes may occur, see figure 1.6. The two objects may stick together in which case their combined mass would be 5kg and their new velocity would be 4m/sec which would produce the momentum 5kg × 4m/sec = 20 kg-m/sec. The two objects may undergo what physicists call an *elastic* collision in which they "bounce off one another" and start moving in opposite directions. Huygens is the one who realized that we must take into account the direction as well as the speed in the law conservation of momentum. He assumed that motions in opposite directions cancel each other; so if we take motion to the right as positive momentum, then motion to the left will produce negative momentum. After an elastic collision the 2kg object may be moving 2m/sec to the left and the 3kg object moving 8m/sec to the right. If we don't account for direction, the total momentum would be p_1 = 2kg × 2m/sec + p_2 = 3kg × 8m/sec = (4 + 24) kg-m/sec = 28 kg-m/sec. However, in taking direction into account, the net momentum would be − 4kg-m/sec + 24kg-m/sec = 20kg-m/sec.

Transfer of momentum describes in a general way what happens during efficient causality of motion. Newton was in possession of the calculus which enabled him to propose a measure of the quantity of "efficient causality of motion" that is occuring at a point in space. At that time it was reasonable to assume that mass = quantity of matter does not change during any change in motion. Thus, when one object changes the momentum = p of another object, the mass remains constant and only the velocity changes. Acceleration defines the change of velocity with respect to time at a moment in time and at a point in space; so the quantity of efficient causality of motion, which Newton called *force*, is equal to mass times $\Delta v/\Delta t$ = mass × acceleration. **f = ma.** Note: Δ means change.

Newton referred to efficient causality of motion as *action*. Force, then, is a measure of action. When one object increases its momentum in a positive direction, it has expressed action which is represented by the measure called force. That is, just as area is a measure associated with a space in a plane surrounded by straight and/or curved lines, so force is a measure applied to action. An object which loses momentum or even reverses its momentum during a causal interaction expresses a *reaction*. According to the law of conservation of momentum, the momentum gained by one object must always equal to the momentum lost by the other. Thus, the reaction must be equal but in the opposite direction to the action. Newton expressed this as "for every action there is an equal and opposite reaction." For every force we can measure, there is another force we could measure which is of the same magnitude but in the opposite direction.

It is important to note that with respect to this aspect of Newton's theory, force is *not* a cause of motion, but it is a measure of action (or reaction) which is the process of an object being caused to change its motion. As far as Newton was concerned, when an object is moving with a constant velocity, it is not causally interacting with any other object. If and when such an interaction does occur, there will be an action (change in momentum) that is measured by force. This was in direct conflict with Aristotle and the prevalent common sense notion that if a body is moving, something is causing it to move. Newton expressed this radical departure from Aristotelian physics as his first law of motion, the so-called law of inertia: a body at rest will tend to remain at rest, and a body in constant velocity (with respect to magnitude and direction) will tend to remain in this state of motion.

It will be helpful to summarize the key features of Newton's theory so far:

1. Material objects exist in an ongoing duration which is the same for all of them and in absolute space which extends indefinitely in three dimensions.

2. The measures length, area, volume, and time may be rigorously defined and assigned to the corresponding aspects of a space and duration.

3. Most fundamentally, motion is the change of position of an object in space during some interval of time. Some motions are continuous and may be represented by a family of continuous mathematical relationships of distance as a function of time and its derived functions of velocity and acceleration.

4. The quantity of motion an object has at a particular point in space and moment in time is momentum defined as $p = mv$.

5. When one object causes another to change its motion, there may or may not be a transfer of momentum from one object to another, but in either case, the law of conservation of momentum always is followed.

6. The process of one object causing another object to change its motion (change its momentum) is called action, and force, defined as $f = ma$, is a measure of action.

7. For every action there is an equal and opposite reaction; the force that is a measure of a reaction is equal in magnitude and opposite in direction to the force that is a measure of the action associated with this reaction.

8. Law of inertia: a body at rest will tend to remain at rest and a body in constant velocity (with respect to magnitude and direction) will tend to remain in this state of motion.

Newton's Law of Gravity

A second unsigned preface to Copernicus's book *On the Revolutions of the Celestial Spheres,* emphasized that the Copernican system was not presented for debate on its truth or falsity, but merely as another computing device for predicting the location of "heavenly bodies" in space. After 1609 when Galileo invented a much-improved telescope and began exploring the heavens with it, he and many others became convinced that this heliocentric, i.e., sun centered, model represented the heavens as it really is. Galileo then set about trying to understand how objects when thrown directly upward fall back to about the same spot on the earth even though during the interim the earth has moved many miles on its circuit around the sun. Kepler sought to simplify the Copernican system to a form that would show the celestial harmonies among the motions of all the heavenly bodies. Galileo remained convinced of the Aristotelian notions that the heavenly bodies must move in circular paths and that

> motion still means "local motion," a translation from one place to another, a motion to a fixed destination and not a motion that merely continues in some specified direction forever—save for circular motion. [Thus Galileo's law of inertia applies only to motion parallel to the surface of the earth.][1]

As a result, Galileo was never able to generalize his theories to describe the motion of heavenly bodies. Kepler also was too enmeshed in Aristotelian physics to apply the harmonies he discovered among the heavenly bodies to terrestrial physics. Newton used the ideas of Galileo and Kepler to construct a single physical theory that described motions of terrestrial and heavenly bodies.

---

[1]Bernard Cohen, *The Birth of a New Physics* (New York: Anchor Books, Doubleday & Co. 1960), p. 125.

A key insight for simplifying the Copernican system was the idea first proposed by Kepler in his three laws of planetary motion in which he states that planets move in elliptical rather than circular paths around the sun. Kepler's work was not taken seriously by scientists of his day and in fact his three laws were acquired by Newton secondhand ". . . very likely from Seth Ward's textbook on astronomy."[2] This is understandable because

> Kepler was a tortured mystic, who stumbled onto his great discoveries in a weird groping that has his most recent biographer [Arthur Koestler, *The Sleepwalkers,* Hutchinson & Co., London, 1959] to call him a "sleepwalker." Trying to prove one thing, he discovered another, and in his calculations he made error after error that canceled each other out.[3]

Nevertheless Kepler's three laws are accepted as true to this day.

1. Each planet travels in a plane with its orbit in the form of an ellipse with the sun located at one focus, see figure 1.7
2. "The area swept out by a line drawn from the sun to a planet is the same in equal time intervals."[4] See fig. 1.8.

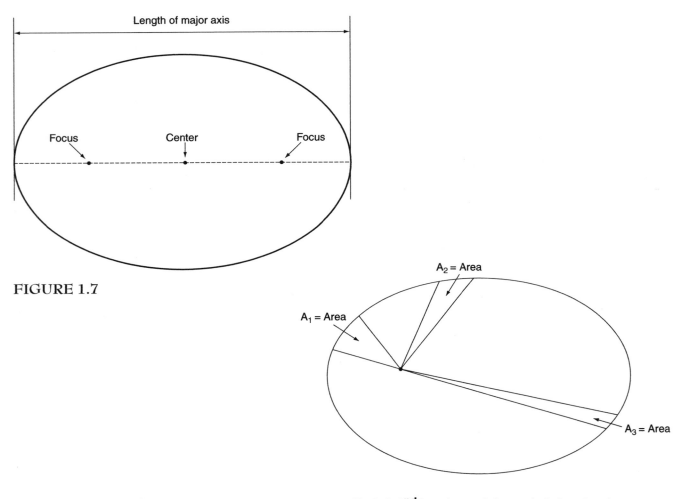

Length of major axis

Focus   Center   Focus

FIGURE 1.7

A_2 = Area

A_1 = Area

A_3 = Area

Kepler's 2**nd** Law: in equal time periods $A_1 = A_2 = A_3$

FIGURE 1.8

[2]Ibid., p. 134.

[3]Ibid., p. 135.

[4]Robert H. March, *Physics for Poets* (Chicago: Contemporary Books, Inc., Reprint. Originally published: 2nd ed. New York: McGraw-Hill, © 1978), p. 48.

3. ". . . the squares of times of revolution of any two planets around the sun (earth included) are proportional to the cubes of their mean distances from the sun."[5] OR "The square of the length of each planet's year is proportional to the cube of the major axis of its orbit. Fig. [1.7] shows the *major axis,* the largest dimension of the ellipse. This law implies that the outer planets move more slowly in their orbits than the inner ones, and thus the length of the year increases more rapidly than the size of the orbit."[6]

Newton used Descartes' analytical geometry, Kepler's three laws, the calculus, and his laws of motion involving contact force to formulate a law to describe the motion of planets about the sun. This limited law of gravity that only applied to "celestial bodies," implied that a gravitational force is "causing" a planet to accelerate toward the sun. The velocity of the planet that always is perpendicular to the line of force interaction between the planet and the sun is what keeps the planet circulating around the sun rather than actually moving toward it. Newton also used his "law of gravity" to describe the circulation of moon(s) around planets. In particular he accurately described the motion of the earth's moon around the earth. Like the planets the moon is continuously accelerating toward the earth. In effect, the moon is *falling* toward the earth, but because of a velocity perpendicular to this "falling" represented as a force interaction, the moon continues to circulate around the earth. This description of the moon circulating around the earth served as a metaphor and analogy for a more general idea. This metaphorical conceptual knowing led Newton to propose that any free falling body toward the earth results from the gravitational force interaction between the earth and the falling body.

Newton was able to prove that the mass of a solid object may for purposes of calculation be considered to be concentrated at its center. Then any free falling body is like the earth's moon (and like any planet circulating around the sun) according to the following analogy: the moon accelerates toward the earth and this force interaction is directly proportional to $1/r^2$ where r is the distance between the center of mass of the moon and the center of mass of the earth. In like manner, a free falling body accelerates toward the center of the earth and the force interaction is directly proportional to $1/r^2$ where r equals the radius of the earth. For bodies close to the earth's surface, r is the same for all of them no matter how high up they are since these differences in height are insignificant compared to the radius of the earth. Thus, all free falling bodies have the same acceleration toward the earth no matter what their weight is. This means that this acceleration should be equal to Galileo's constant, g, and weight is the gravitational force between the solid object and the earth, i.e., $F = mass \times g$. This is why weight is a measure of mass, but of course this is only true if one stays close to the surface of the earth. Weight depends on the distance, r, from the center of the earth; mass is independent of distance. Thus, humans in a space shuttle are sufficiently far from the center of the earth for the gravitational force interaction to be very small. In effect, the humans are "weightless," but they have the same mass as they did while on earth.

> The bold idea that the falling body is governed by the same law as the cellestial bodies had been conceived by Newton in 1666. At the age of twenty three he was in possession of the whole theory which enabled him to make the calculation. But with the numerical values then at his disposal his result differed by about 20 per cent from the value of *g;* it was disastrous. . . . We may imagine the suspense with which Newton, in 1682, was awaiting the results of degree measurements which Picard had begun in 1679. Legend has it that, when Newton learned of the corrected radius [of the earth], he was too excited to insert it into his formulas to see whether now his theory would be verified and that he asked a friend to do it for him. The result was thoroughly satisfactory, and only then did Newton publish his work containing his entire theory.[7]

The idea that any free falling body is like the moon falling toward the earth led Newton to propose an even more general law. According to this *universal law of gravity,* any two mass objects anywhere in

---

[5]Cohen, *loc. cit.,* p. 143.

[6]March, *loc. cit.,* p. 49.

[7]Ibid., pp. 164–165.

the universe continuously undergo a gravitational force interaction that is directly proportional to $1/r^2$ where r is the distance between the two mass objects. Thus, in general, let m_1 and m_2 be two masses anywhere in the universe, on earth or in the heavens that are separated by a distance, r, then there is an action at a distance measured by the force, F, as follows

$$F = G\, m_1 m_2 / r^2 \text{ (G is a universal constant)}$$

> The motion of the freely falling body showed that terrestrial objects of any shape or form are attracted by the earth in accordance with the law of gravitation. But it was a mere hypothesis that any body, no matter of what size or shape, having its specific gravitational factor, imparts an acceleration to any other body, like the sun or the planets. . . . It was not until 1789 that Cavendish demonstrated with his torsion balance that two lead balls do exert a gravitational effect on each other.[8]

The formulation of the universal law of gravity was a culmination of a grand synthesis, usually recognized as one of the greatest achievements in the history of human thought:

> For him [Newton] to invent the needed tool [the calculus] and by its aid to reduce the major phenomena of the whole universe of matter to a single mathematical law [law of gravity], involved his endowment with a degree of all the qualities essential to the scientific mind—pre-eminently the quality of mathematical imagination—that has probably never been equaled. Newton enjoys the remarkable distinction of having become an authority paralleled only by Aristotle to an age characterized through and through by rebellion against authority.[9]

However, the law of gravity makes the idea of force problematic. Force is a measure of action which represents one object causing another object to change its motion. Our common sense view of this physical causality is that it must involve contact to allow a transfer of momentum from one object to another, and that it is finite in having a beginning and an end in space and in time. Newton's force of gravity denies all of these common sense properties of physical causality: (1) there is action at a distance rather than contact action; (2) sometimes there is no net transfer of momentum such as when planets circulate around the sun; (3) there is no beginning (other than at the "creation" of the universe) or end to this action at a distance since all bodies in the universe always have and always will continue to attract one another; (4) this "action at a distance" is present everywhere in space; and (5) the force of gravity operates in all time intervals.

Machine Causality

The Newtonian synthesis still had its roots in the Greek and Medieval tradition of contemplative reason. His theory enabled one to rationally comprehend the universe, but it did not lead to many practical applications.

> For at least a century after the publication of the *Principia*, [in which Newton published his famous law] calculations of the motions of the planets remained the most important application and confirmation of both Newton's laws of mechanics and his law of gravity. It was not until well into the nineteenth century that the design of machinery, for example, ceased to be a trial-and-error process, and Newton's laws of motion received their first practical application. . . . Newtonian mechanics took force as its starting point and went on from there. It offered no ready answer to the question that the pioneers of the industrial revolution confronted daily: What does it take to generate a force? Engineers and inventors wanted to create motion where none had existed before, and all Newton had to offer them was the assurance that however they managed to do it, an equal and opposite motion would inevitably arise in the process. This was some help, but not much.[10]

---

[8]Ibid., pp. 168–169.

[9]Edwin Arthur Burtt, *The Metaphysical Foundations of Modern Physical Science* (Garden City, N.Y.: Doubleday & Co., The Anchor Books ed., 1954), p. 207.

[10]March, *loc. cit.*, pp. 52–54.

Furthermore, what started out as an attempt to rationally understand physical causality ended up with precise, symmetrical, and aesthetically pleasing mathematical relationships that posed serious philosophical problems. How can physical causality occur at a distance, and what does it mean to say that presumably finite causal interactions occur everywhere in space and during all time intervals?

If we change our focus from physical causality (mechanical causality) to machine causality, at the very least we will make Newtonian physics more practical and perhaps overcome some of its philosophical difficulties. The evolution of the emergence and further differentiation of human control consciousness is closely linked to the evolution of machine causality. Hominids (the first "proto-humans") evolved a network of traits which led to a nuclear family structure. This gave the hominids a survival advantage over all other animals even though as individuals, hominids were weaker and less efficient in doing things than some other types of animals. In the nuclear family the large task of staying alive is divided into component tasks such as the mother feeding the children at the home base while the father gathers and brings food to the home base. By virtue of having the two forelimbs with hands differentiated for grasping and carrying and having a pelvis and the two hind limbs differentiated for walking, the father is able to be a "carrying machine" coordinated into the goal of survival of the nuclear family.

The task of the carrying machine is to move food from one place to the home base. In some manner, the father acquires a net quantity of motion, for example, the resultant of complex motions of fingers, hands, arms, legs, and body, and some of this quantity of motion shows up in accomplishing the task of the carrying machine. "Something happened" in the distance between the place where food is growing and the home base. In general, any machine takes in a quantity of motion and converts some of it into accomplishing a large task that always may be divided into one or more small tasks. Each small task "contains" a quantity of motion. To make Newtonian physics practical we need to define quantity of motion in such a way that it can be a measure of any small task. Then we could calculate the efficiency of any machine, which would be the quantity of motion of the large task divided by the total quantity of motion we added to the machine to accomplish the large task. If this ratio comes out to be one, then the machine is 100% efficient; all the motion put into the machine shows up as motion in the accomplished task. We usually experience machines to be very much less than 100% efficient. At any rate if we know the quantity of motion of the task and the efficiency of the machine, we know the minimum amount of motion we somehow must be able to add to the machine.

Work Resulting from Kinetic Energy

The simplest type of machine causality would be one in which a body starts out at rest and accelerates to some final velocity at a constant rate. This is the free fall motion described by Galileo in which distance is a function of time, i.e., $D(t) = 1/2\, gt^2$, and velocity is a function of time, i.e., $v(t) = gt$, and the constant acceleration is g. We can use Newton's concept of force to analyze free fall motion as a result of machine causality. By the end of a free fall an object has acquired a quantity of motion which enabled it to change from being at rest to some final velocity after traveling a certain distance. At every point in this distance interval, the body underwent a constant change in momentum which we call force, $f = mg$. Thus, the net change in motion in this interval is the sum of all the changes in motion at each point in the interval, i.e. the sum of all the forces in the interval which is the area of the rectangle shown in figure 1.9. This area is called *work* which is the process of an object changing its quantity of motion as a result of a continuous force (action) in a distance interval.

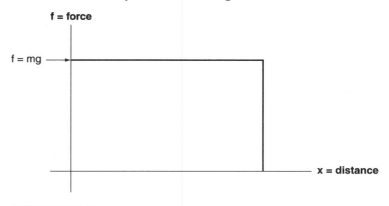

FIGURE 1.9

In free fall the work process is represented as W = force × distance. In this case the change in quantity of motion which is a measure of work done can easily be evaluated. [f = mg and the distance traveled during free fall is D(t) = 1/2gt and the velocity of the body just as it hits the ground is D′(t) = gt = v. By substituting these values into the work equation we obtain:

W = mg × 1/2gt² = 1/2mg²t² = 1/2m(gt)² = 1/2mv².] As indicated in the parentheses, the work accomplished is equal to 1/2mv². This final quantity of motion that the free falling body has acquired by the time it hits the ground is a new measure of motion called *kinetic energy: KE = 1/2mv²*.

Generalization of the Concept of Kinetic Energy

As a body falls to the ground it acquires K.E. as a result of "action at a distance" of the force of gravity acting at each point in the distance the body falls. The idea of machine causality representing work done implies that there must be a transfer of quantity of motion. Where does this quantity of motion come from and once the body comes to rest on the ground, where does it go to? One way out of this dilemma is to say that the body at rest (no motion) at h distance above the ground somehow has a potential quantity of motion called *potential energy = PE. P.E. is defined as the KE an object can acquire as a result of some force, in this case, the force of gravity, "acting at a distance.* At the same time we know that under some circumstances such as hitting a drum, K.E. is converted into vibrations that we eventually perceive as sound. Also K.E. is associated with temperature changes, for example, when we expand and contract a rubber band several times we can feel that it is warmer. As we will see later in this chapter, a change in temperature is associated with what we will call heat energy. Thus, we can give a consistent though not philosophically satisfying description of free fall in terms of machine causality as follows. The P.E. that a body has when at a height above the ground is continuously converted into K.E. during its fall to the ground. Just as the body hits the ground all the P.E. is in the form of K.E. After hitting the ground, the K.E. is converted into sound energy and heat energy.

The idea of P.E. representing "action at a distance" may be somewhat more philosophically satisfying if understood in the following way. P.E. represents what we may call a "potential" which in turn represents a certain organization pattern. A disruption of any organization pattern always leads to two consequences: (1) some aspect of the organization pattern disappears forever and will show up again only as a consequence of something else recreating the original pattern sometimes a

▶ **Table 1.3   Contrast Mechanical vs. Machine Causality**

Mechanical Causality	Machine Causality
1. A body at rest tends to stay at rest; a body with constant speed in a straight line will continue that way unless acted upon by some other body.	1. A *net* transfer of motion has occurred in the interval Δx, if and only if the magnitude of the final velocity does not equal the initial velocity.
2. Force = f = ma is a measure of the interaction at a point in space; it is a vector quantity in that it has direction.	2. Power = P = Energy/Δt is a measure of the interaction in a distance interval; it is a scalar quantity in that it does not have direction.
3. For every action there is an equal and opposite reaction.	3. When two objects interact over a distance interval, the work done *by* one object is equal but opposite in sign to the work done *on* the other object.
4. Conservation of momentum: in any interaction at a point in space, the momentum lost by one body is equal to the momentum gained by another body; $m_1v_1 = m_2v_2$.	4. Conservation of E = mechanical energy. In any interaction in a distance interval: K.E. + P.E. = a constant value.

machine may do this, and (2) the "amount" of organizational pattern that is lost is converted into KE (in the idealized Newtonian world that does not consider heat processes). Thus, P.E. = potential, only is a measure of various types of organization patterns. Science studies the machine causality of various types of organization patterns in nature. This provides us with the basis for designing machines and for interfering with patterns in nature in order to accomplish a particular goal. However, the quantitative measure of a pattern, for example, P.E., is not identical to the pattern; there are other ways of understanding aspects of it which include poetry, myth, metaphysics, or even our own subjective feeling awareness of it.

Heat Interactions

Heat

Heat is associated with the primal experience of one thing "making another thing warmer." Primitive people as well as modern couples today huddle together to keep one another warm with their body heat. We huddle around a fire and are "warmed by its heat." However, a major change with regard to heat came in 1770 with James Watt's invention of the modern steam engine. Heat associated with our subjective experiences of hotness now showed the intriguing possibility of producing mechanical energy. Heat was thought of as a kind of substance that leaves an object colder as it flows out of it, and makes an object warmer as it flows into it. Temperature, an objective measure of the hotness or coldness of an object is determined by the relative mechanical expansion or contraction of a simple (ideal) gas such as helium. Thus, temperature not only provides a way of measuring heat, it suggests how heat can produce mechanical energy, for example, mechanical expansion of a gas. Heat may be defined as an exchange (of something) between two objects that results in at least one object changing its temperature. Usually as one object decreases its temperature, the second object increases its temperature. However, there are situations in which an object loosing heat decreases its temperature to liquid water at $100°$ C; instead of the water increasing its temperature, it changes from a liquid to a gas.

Temperature

Temperature, like all other measures, is to be understood metaphorically. We experience a stable mixture of pure liquid water and ice as being cold in contrast to a stable mixture of pure liquid water and water vapor as being hot. We also note that any pure substance such as liquid mercury will expand when heat is added to it or contract when heat is withdrawn from it. We now have a way of measuring temperature. We take a small spherical reservoir containing mercury connected to a uniform glass cylinder and let it be our "temperature ruler" called a thermometer. We note the height of the mercury column when it is immersed in the liquid-ice mixture and when it is immersed in the liquid-vapor mixture. We then divide the difference in mercury column heights into 100 units where each division represents a unit called a degree of temperature on the centigrade scale. Using the expansion or contraction of pure helium, it is possible to calibrate other thermometers that can measure temperatures far below water's freezing point and far above its boiling point. Thus, just as we compare the mass of any particular object to a standard unit of mass; so also, we compare the temperature of any particular object to a standard unit of temperature. In the world of scientific magic, it is not necessary to understand the deeper meanings of nature. We still can invent formulas and construct machines that will give us some mastery over her.

Heat Is Similar to Work

Work like heat also is associated with a temperature change. If we feel an object that has undergone a work interaction several times, such as a hammer or a rubber band stretched and contracted several times, or a paper clip bent and straightened out several times or our hands rubbed against one another several times, we experience an increase in hotness. At the end of an earlier section we concluded that kinetic energy can be converted into "an organization pattern" whose measure is called potential energy. We now propose that some of the K.E. involved in a work interaction becomes involved in a heat interaction which results in a change in temperature. This implies that some of the change in K.E. of the work interaction is converted into a "non-mechanical" form of energy we may call *thermal* (temperature) *energy*. Just as K.E. of an object is its capacity to do work, that is, enter into a work interaction; so the *thermal energy* of an object is its capacity to do heat, that is, enter into a heat interaction. Work leads to a change in motion and heat leads to a change in temperature. "Actual" energy underlies both of these changes. When it shows up as work, it is called a change in K.E.; when it shows up as heat, it is called a change in thermal energy.

Thermal energy may be measured and metaphorically understood in the following way. Different pure substances have different capacities for increasing or decreasing temperature during a heat interaction. It takes more thermal energy to raise by one degree the temperature of liquid water than it does to raise the temperature of ethyl alcohol. In terms of our everyday experience, on a cold day an iron fence extracts thermal energy from one's hand much more quickly than a wooden fence. A standard unit of thermal energy, the *calorie*, is taken to be the quantity of energy required to raise a unit mass (one gram) of pure liquid water one degree (centigrade) in temperature. Then just as units of KE, for example, joules, are used to measure work units of thermal energy, calories are used to measure heat.

Heat now may be defined in a way analogous to work. Just as work is the interaction resulting in a transfer of kinetic energy, heat is the interaction resulting in the transfer of thermal energy. Let it also be noted that *neither work nor heat is a form of energy*.[11] An object is characterized by having a certain K.E., but it never may be said to have a certain work; rather, work equals the amount of K.E. it receives or gives off to some other object. Work equals *transferred* K.E. In like manner, an object is characterized by a certain amount of thermal energy, but it never may be said to have a certain heat; rather heat equals the amount of thermal energy it receives or gives off to some other object. Heat equals *transferred* thermal energy.

In summary, Table 1.4 contrasts work and heat processes.

Potential

Any object or system in the universe is embedded in a network of interactions. All measurable changes in a system (or object) result from a net potential between the system and its immediate environment. A *potential* consists of a heat potential and/or a work potential. A *heat potential* consists of a

---

[11]See: William F. Luder, *A Different Approach to Thermodynamics* (New York: Reinhod Pub. Corp., 1967) "Because in general terms, the *energy* of a system is its capacity to do *work* the two words are closely related and frequently confused." p. 5. "A student may avoid confusion between the ideas of work and energy by stressing that the energy of a system is its capacity to do work. . . . Although the system has no work in it, it has the capacity for doing work. . . . Work *is not* a property of the system; energy is a property of a system." p. 6. "The subject of heat has been one of the most confused in all science; it still is for most people . . . according to the thermodynamic definition of heat, the sun as a complete system has no heat in it. No system has heat in it. Heat is not a property of a system any more than work is. *Heat is energy transferred between a system and its surroundings because of a temperature difference between them.*" pp. 6–7.

▶ **Table 1.4**

Work	Heat
1. is not a property of a system; is not equal to K.E. which is a property of a system	1. is not a property of a system; is not equal to thermal energy which is a property of a system
2. is *transferred* kinetic energy	2. is *transferred* thermal energy
3. results in change of position	3. results in change in temperature

difference in temperature which may produce heat, the transfer of thermal energy. A *work potential* consists of K.E. or potential energy (P.E.) or their equivalents such as a difference in pressure or concentration of a substance that produces a net movement of the substance from one place to another.

Intellectural Shortcut for Physics

The ideas of mass, quantity of motion, and force provided the foundation for Newtonian mechanics. This system was modified by introducing the ideas of temperature analogous to mass density and heat, the transfer of thermal energy, analogous to work, the transfer of kinetic energy. Einstein fundamentally transformed this system with his special and general theories of relativity. The intellectual shortcut described next is a thermodynamic, metaphorical synthesis of these diverse mechanical ideas.

Energy is the core idea for understanding physical changes. Energy has two aspects: (1) potential and (2) flux. *Potential,* when applied to a system or applied to a system's environment, represents Order and the possibility of generating a type of event. *Flux* is the way energy as potential manifests itself. Energy flux leads to one or both of two kinds of change: (1) motion, that is, change of location in space and (2) change in temperature. Energy flux leading to motion is called *work* and energy flux leading to change in temperature is called *heat.* Potential representing Order "points to" sequential, reversible changes of structure in which time is reversible and Order is conserved. That is, at any moment during these changes Order never is lost. Rather the disappearance of Order associated with one event is coupled with the emergence of a new and equal degree (quantity) of Order of the sequential next event. For example, in the ideal seesaw, as one person goes down, the other person goes up; the quantity of going down equals the quantity of going up. The conservation of Order means that there is no Chaos.

Flux may be metaphorically thought of as the transfer of a quantity of energy that leads to motion or as the transfer of *thermal energy* that leads to a change in temperature. When we imagine these transfers of energy to occur "infinitely slowly," then flux also involves reversible processes in which Order is conserved. However, energy flux of finite human experience implies: (1) time is irreversible, (2) time based events are irreversible, (3) Order is not conserved which means there always is a loss of some degree (or quantity) of Order and a corresponding gain of degree (or quantity) of Chaos, and (4) flux may involve only heat or heat and work, but it never only involves work.

The fluxes associated with the operation of any machines we shall describe are continuous. There are fluxes in nature and in mechanical systems that are discontinuous. Descriptions of these systems lead to profound insights that will not be talked about in this text. As a result of the metaphorical conceptual assumptions described in the above two paragraphs, one may skip reading pages 1–16 of this chapter and still understand the idea of a machine and the profound, universal principles associated with the operation of a machine.

Machines

Characteristics of Any Machine

As pointed out in previous sections, our ideas of force, especially action at a distance, and energy, especially potential energy, are rather nebulous. But in assigning measures to these notions we can treat them as if they really exist. In particular we can describe energy as like a substance that can be transformed and/or transferred from one place to another until it results in a particular type of motion, a task, which we desire to happen. We can devise formulas which enable us to create machines that take in energy and transform it into the type of work we wish to accomplish. The machine is our magic wand, for as Eliade notes:

> modern science [is] the secular version of the alchemist's dream, for latent within the dream is "the pathetic program of the industrial societies whose aim is the total transmutation of Nature,

its transformation into 'energy.'" The sacred aspect of the art became, for the dominant culture [of Europe], ineffective and ultimately meaningless. In other words, the domination of nature always lurked as a possibility within the Hermetic tradition, but was not seen as separable from its esoteric framework until the Renaissance. In that eventual separation lay the world view of modernity: the technological, . . . , as a logos.[12]

At the core of the operation of any machine is one or a sequence of energy couplers. An *energy coupler* is a part of a machine that goes through a two step cycle: (1) the coupler receives energy and goes to a particular higher energy state; (2) the coupler gives off energy and goes to a lower energy state wherein (a) some of the energy given off always accomplishes a particular type of work and (b) the lower energy state is ready or can be modified to repeat step 1. A human using a hammer to pound a nail into a block of wood is a "nail pounding machine," and the hammer is the energy coupler. The hammer receives energy from arm muscles and rises above the nail and then acquires kinetic energy in its downward movement toward the nail. The hammer is at its highest energy level just before hitting the nail. Upon hitting the nail some of the K.E. of the hammer accomplishes the task of moving the nail into the block of wood. The hammer then is ready to start a new cycle.

In general, machines exhibit three characteristics: (1) Energy flows from an energy source into the machine, through it, and out again as work. (2) Incoming energy causes an energy coupler to go from a low to a high energy level; then in going from a high to a low energy level, the coupler accomplishes some task. (3) A large task is accomplished by the sum of many small tasks each accomplished by one cycle of an energy coupler. For example, in a one-cylinder motorbike, the combination of gasoline and oxygen injected into the cylinder has chemical potential energy. Just after the spark plug initiates the chemical reaction between gasoline and oxygen, the products of this reaction (CO_2 and H_2O gases) have high kinetic energy. As these expand, the K.E. decreases as the piston in the cylinder rises. When the valve opens allowing the gases to escape through the exhaust pipe, the piston falls and becomes ready to start a new cycle. The rise and fall of the piston is converted into one rotation of the wheel of the motorbike. Thus, each cycle of the piston accomplishes the small task of one rotation of the wheel. The addition of many of these cycles accomplishes the large task of the bike moving from one place to another.

Fundamental Energy Limitation of Any Machine

The most fundamental limitation of any machine is that it functions only in accord to the heat and work potentials it can express and continually re-create. The machine must be in a context in which it continually receives energy and therefore continually re-creates a potential which then is focused on accomplishing a specific task. The best, that is, the most efficient, machine imaginable is the ideal heat machine first described by Sadi Carnot in 1824.

Ideal Heat Machine

The ideal heat machine only operates when its energy coupler is a connecting link between a heat energy source and a heat energy "sink" at a lower temperature. The energy coupler is a cylinder filled with an ideal gas, for example, helium and a piston. The ideal gas goes through four phases during one cycle of the energy coupler. In this description it will be convenient to define task of the machine as equaling task work a + task work b.

Phase 1.   The transfer of heat energy from the heat source to the ideal gas causes it to expand but in such a way that there is no temperature change and no increase in internal energy represented by the letter U; i.e., $\Delta T = 0$ and $\Delta U = 0$ (Δ represents change). Thus, q representing the transfer of heat energy *into* the ideal gas equals expansion work. Only some of the expansion work accomplishes part of the task of the machine; the remaining expansion work increases the environment's potential to cause the gas to contract. Let us call this remaining expansion work, *environment contraction potential*. Then, *expansion work* = task work a + environment contraction potential.

---

[12]Berman, loc. cit., p. 99. [quote from Eliade: Eliade, *Forge and Crucible*, pp. 172–73. Cf. Brown, *Life Against Death*, p. 258.]

Phase 2.   The internal energy of the ideal gas decreases causing further gas expansion, but this happens in such a way that there is no change in temperature, that is, $\Delta T = 0$. Thus, the decrease of internal energy does not show up as heat; rather it is equal to further expansion work. This work, in turn, equals task work b + further increase in the environment's potential to cause the ideal gas to contract. Thus, *further expansion work* = task work b + *further environmental contraction potential*.

Phase 3.   The heat energy that entered in phase 1 now is transferred to the heat sink in such a way that there is no temperature change or increase in internal energy, i.e., $\Delta T = 0$ and $\Delta U = 0$. This transfer of q away from the ideal gas causes the gas to contract; q = heat energy transferred *away from* the ideal gas equals *contraction work*. The environment contraction potential generated in phase 1 coupled with the transfer of heat energy away from the ideal gas causes the gas to contract by the same amount that it expanded in phase 1, i.e., q into the gas = q out of the gas and expansion in phase 1 = contraction in phase 3.

Phase 4.   The further environment contraction potential generated in phase 2 causes the gas to contract and increase its internal energy, but this happens in such a way that there is no change in temperature, $\Delta T = 0$. That is, the increase in internal energy shows up as contraction work; none of the increase shows up as heat. The increase of internal energy in phase 4 equals the decrease of internal energy in phase 2.

Generalizations about the Ideal Heat Machine

The generalizations about the ideal heat machine apply to any machine no matter how complex it is and how far from ideal. In any real machine, i.e., non-ideal machine, any energy transfer shows up as work *plus heat*. This is the "non-ideal" aspect of real machines. As we shall see when discussing the second law of thermodynamics, the energy transfers that show up as heat influences the efficiency of the machine. The more heat that is expressed, the less efficient is the machine.

Seven Generalizations from the Functioning of an Ideal Heat Machine

One: Machine Efficiency.   Even the ideal heat machine is not 100% efficient where efficiency is defined as task work divided by total energy input to the machine for a single cycle of the energy coupler. If energy input equals task work, the machine is 100% efficient. But some energy input always shows up as environment contraction potential which, of course, does not contribute to the task work. Thus, task work is less than the energy input; how much less this is determines the inefficiency of the machine. For any real machine the inefficiency is even greater because some of the energy input shows up as heat rather than as task work or environment contraction potential.

Two: Machine Taoism.   An energy coupler cycle is seen to consist of two component processes. These two processes are radically different but mutually dependent on one another. Phases 1 and 2 comprise the *action process* where the energy coupler accomplishes task work and creates an environment contraction potential. Phases 3 and 4 comprise the *active-passive process* where the environment contraction potential created in phases 1 and 2 *prepares* the energy coupler to start a new cycle. Even though the two sub-processes of the cycle of an energy coupler are radically different, their mutual dependence is represented by the reversible changes that occur within them. The expansion of gas that occurs in phase 1 equals the contraction of gas that occurs in phase 3. Likewise the "further expansion" of gas that occurs in phase 2 equals the "further contraction" of gas that occurs in phase 4. Also, the decrease of internal energy that occurs in phase 2 equals the increase of internal energy that occurs in phase 4.

Nature's Cycles.   At the core of machine function is the cyclic activity of an energy coupler. Taoism as expressed in the *I Ching* describes the same idea as the core of all activity in the universe.

> The Chinese philosophers saw reality, whose ultimate essence they called Tao, as a process of continual flow and change. In their view all phenomena we observe participate in this cosmic process and are thus intrinsically dynamic. The principal characteristic of the Tao is the cyclical nature of its ceaseless motion; all developments in nature . . . show cyclical patterns. The Chinese gave this

idea of cyclical patterns a definite structure by introducing the polar opposites yin and yang, the two poles that set the limits for the cycles of change: "The yang having reached its climax retreats in favor of the yin; the yin having reached its climax retreats in favor of the yang."[13]

An energy coupler cycles between high and low energy levels. During phase 2 an energy coupler, the gas in an ideal heat machine, goes from a higher to a lower energy level and accomplishes work on the environment; $\Delta U = w$. In the Taoism perspective this is the yang process which having reached its climax now must give way to a yin process. In phases 4, the energy coupler goes from a lower to a higher energy level. This is the yin process which having reached its climax now must give way to a yang process.

A yin-yang cycle as described . . . "by Manfred Parkert in his comprehensive study of Chinese medicine,"[14] is explicitly analogous to the cycle of the energy coupler of an ideal heat machine.

According to Parkert, yin corresponds to all that is *contractive*, responsive, and conservative [phases 3 & 4], whereas yang implies all that is *expansive*, aggressive, and demanding [phases 1 & 2].[15]

In the heat machine the ideal gas is made ready to do work on the environment by contracting during phases 3 and 4, the yin process; then the gas does work on the environment by expanding during phases 1 and 2, the yang process.

The Direction of Harmony.   The ideal heat machine only is able to function by allowing the flow of thermal energy from high to low. Using the metaphor of energy as a substance we say that thermal energy flows into the machine during phase 1, but some thermal energy must flow out of the machine in phase 3 in order for the cycle to start again. The machine functions only as a result of being interposed between two heat reservoirs, one that adds heat energy and the other that takes away heat energy. If the machine is not open to receive thermal energy, it will not do work. If the machine does not give up thermal energy, it will not be prepared to receive thermal energy. The machine functions by being in harmony with a heat potential in nature. The machine merely couples the flow of thermal energy down a heat potential to accomplishing work.

The heat potential is an intrinsic aspect of the universe that becomes manifest in the functioning of a machine. This thermodynamic insight is in accord with an ancient Chinese recognition that:

activity—"the constant flow of transformation and change," as Chuang Tzu called it—is an essential aspect of the universe. Change, in this view, does not occur as a consequence of some force but is a natural tendency [a potenial that spontaneously may disappear in producing work or heat], innate in all things and situations. The universe is engaged in ceaseless motion and activity, in a continual cosmic process that the Chinese called Tao—the Way.[16]

The machine coupling to a heat potential in order to accomplish work is what Taoist philosophy means by *wu wei*.

In the West the term is usually interpreted as referring to passivity. This is quite wrong. What the Chinese mean by *wu wei* is not abstaining from activity but abstaining from a certain kind of activity, activity that is out of harmony with the ongoing cosmic process. . . . [this is in accord with] a quotation from Chuang Tzu: "Nonaction does not mean doing nothing and keeping silent. Let everything be allowed to do what it naturally does, so that its nature will be satisfied." If one refrains from acting contrary to nature or as Needham says, from "going against the grain of things," one is in harmony with the Tao and thus one's actions will be successful [the task will be accomplished]. This is the meaning of Lao Tzu's seemingly puzzling statement: "By nonaction everything can be done."[17]

---

[13]Fritjof Capra. *The Turning Point* (New York: Bantam ed., 1983). p. 35; quote in passage taken from: Wang Ch'ung, quoted in F. Capra, *The Tao of Physics* (Berkeley: Shambhala, 1975), p. 106.

[14]Ibid., p. 35: Manfred Parkert, the Theoretical Foundations of Chinese Medicine (Cambridge, Mass.: MIT press, 1974), p. 9.

[15]Ibid., p. 36.

[16]Ibid., p. 37; quote in passage taken form quote in Fritjof Capra. *The Tao of Physics* (Berkeley: Shambhala, 1974, p. 114).

[17]Ibid., p. 37.

The flow of thermal energy down a heat potential has direction; it always is from high to low. The great heat potential of the universe is the measurable manifestation of time, and time is the existential unfolding of finite being. The universe is like a great water wheel coupled to the directional flow of water, which is time. All things are in harmony with the Tao when "they go with the flow."

Machine Harmony.   The Tao of an energy coupler. Let Y represent a type of energy coupler that only accomplishes a small task in going from high to a low energy level; then let Y^h = the energy coupler at the higher energy level and let Y_l = the energy coupler at the lower energy level. Then the energy coupler cycle may be represented as follows:

YIN
work done on Y

YANG
work done by Y = task

Wu Wei of an Energy Coupler.   One way of defining *wu wei* operationally is to discover a particular work or heat potential in the universe and an energy coupler that can convert this potential into a desired task. Then one can connect a machine containing the energy coupler to the potential in such a way that: (1) the *YANG* process of the energy coupler cycle will accomplish the desired task and then immediately give way to the *YIN* process; (2) the *YIN* process of the cycle will re-create the machine potential for accomplishing the desired task and then immediately give way to the *YANG* process.

Wu Wei of Directional Harmony of a Machine.   The directional harmony in the universe may be thought of as a network of work and heat potentials. According to the 2nd law of thermodynamics, any machine task is accomplished as a result of a *net* decrease in the work and/or heat potentials in the universe. Machine *wu wei* is the balanced YIN/YANG cycle of an energy cycle (see Table 1.5) may be summarized as follows:

YIN   individual initiates the preparatory action plus chooses openness (= *wu wei*) to allow potentials already present in any particular situation to complete the preparation for accomplishing a specific task.

YANG   individual initiates focused action and chooses openness (= *wu wei*) to allow potentials created in the YIN phase to accomplish a specific task.

Our Enchanted Cosmos.   Though we can measure heat and work potentials, we cannot rationally comprehend them. We still need qualitative as well as quantitative metaphors. Energy which underlies all observable events in nature is like water in a river that flows down stream. If we acknowledge our need for qualitative metaphors, we can come home again to an enchanted cosmos which sustains us in existence. The universe is alive; all actions are spontaneous. If an event appears non-spontaneous, for example, a potential increases rather than decreases, then some other spontaneous event gave birth to it. Some things rise while other things fall, but underlying all cycles of change is the unfolding of time manifested as the directional interactions of all finite beings.

Three: Subjective Idea of Machine Mutuality Rather than Objective Idea of Machine Autonomy. Functioning of the ideal heat machine in one sense supports but in a holistic perspective opposes the mechanistic claim that any system may be totally understood in terms of the interactions among its parts. On the one hand, the ideal machine converts the energy flux of a potential into a sequence of autonomous cycles where each cycle consists of a Yang phase that produces an elemental task work and a Yin phase that produces a potential in an energy coupler to produce the Yang phase. In this view, the machine appears to have the intrinsic power to convert the energy related to a potential, for example, potential energy, into a sequence of autonomous work tasks that taken together equals the work goal of the machine. However, this view breaks down in two ways. Just as the idea of child implies parent and likewise parent implies child, so also machine work and energy flux imply one another. The machine is not an autonomous entity that converts energy into machine work. Likewise an energy flux may produce some work, but it does not produce machine work. Rather, an

▶ Table 1.5   Tao of Human Action in Relation to the Carnot Ideal Heat Machine

YIN	3. *preparatory action* = use energy received by the human machine to position oneself to participate in action of YANG.	*Phase 3:* thermal energy received from heat source in phase 1 is transferred to heat sink; $\Delta T = 0$ $\Delta U = 0$ $-q_{rev} = +w$ = ideal gas is compressed
	4. *active passivity* allows potential of any situation (this potential includes the internal energy of the human machine) to complete the preparation for another **YANG** process task	*Phase 4:* potential in environment does work on the ideal gas increasing its internal energy, but without changing its temperature. $\Delta T = 0$ $\Delta U = U_{final} - U_{initial} = +w$ = work done on gas by environment = gas compressed.
YANG	1. *action* = take in energy (food) and use it to accomplish a specific task	*Phase 1:* thermal energy from heat source is transferred into the gas and immediately converted into the work of gas expansion. $\Delta T = 0$ $\Delta U = 0$ $+q_{rev} = -w$ = gas expands = does work on environment
	2. *active focus* on allowing a task potential created in the YIN phase to be converted into the specific task	*Phase 2:* some of the internal energy of the gas is converted into the work of further expansion of the gas but in such a way that the temperature is unchanged. $\Delta T = 0$ $\Delta U = U_{final} - U_{initial} = -w$ = work done by gas on the environment = gas expands

autonomous energy flux and an autonomous machine must enter a collaborative mutuality in order to produce machine work. A paraphrase of what a Taoist would say is that once this collaboration is established, the work task will occur on its own accord. Humans using machines cannot make things happen. All they can do is create the collaboration between any energy flux and an appropriate energy coupler; then the task will occur without further human effort. Likewise, just as *response* exists only in relation to a *stimulus* and vice versa, in each work cycle, the Yang phase exists only in relation to a Yin phase which legitimately is called *Yin* only if it leads to a Yang phase. Thus, in relation to the operation of a machine, the idea of autonomous parts must be supplemented with the idea of mutuality and with the subjective ideas of irreversible time and events.

The ideal heat machine—the prototype for all other machines—in itself cannot accomplish anything. Rather *this machine collaborates with an aspect of its environment.* The energy coupler, in this case, the ideal gas, is oriented so that it can convert the flow of heat energy from a "heat source" to a "heat sink" to accomplish a specific task. Neither the heat energy flow nor the autonomous machine can accomplish the desired task work, but the collaboration of these two factors can. This insight is a special case of the postmodern perspective that the meaning of a thing or event is partially determined by context. The meaning of any machine partially depends on the environmental work and/or heat potentials it exploits. Moreover, the flow of energy down an isolated potential always is from higher to lower energy levels; never the reverse. Of course, this energy flow occurs over time. Thus, time associated with any machine is irreversible just as humans' subjective experience of time is irreversible. Mechanistic time, on the other hand, is reversible just as distance is. Thus, machine dynamics implies irreversibility of time and correspondingly also implies some degree of subjectivity in contrast to the dogmatic objectivity of the mechanistic perspective (Pribor, 1999).[18]

Four: Mutuality of Order and Chaos.   The ideal heat machine introduces another subjective insight which is the idea of *mutuality of Order and Chaos*. In so many words in 1865 Clausius saw this mutuality in relation to phases 2 and 4 of one cycle of an ideal heat machine. The decrease of internal energy in phase 2 means that the ideal gas now has less ability to do work. Clausius called this change an increase of entropy where *entropy* is a measure of a system's *inability to do work*. Correspondingly, the increase of internal energy in phase 4 means that the ideal gas has regained a greater ability to do work, and this change is represented by a decrease in entropy. The ability to do work may be subjectively understood as degree of organization represented by the word *Order*. The decrease of the ability to do work is a decrease of Order which we may represent as *Chaos*. Thus, entropy changes in the universe involve an increase of entropy where Order goes to some degree of Chaos or a decrease of entropy where Chaos goes to some degree of Order. Thus, Order and Chaos always imply one another for all changes in the universe; there is a mutuality between Order and Chaos.

At a more general level, the physicist studies physical potentials and the expression of these potentials. With respect to Newtonian mechanics, mass and the related idea of momentum are potentials in that they have meaning only in relation to an interaction called *force* which directs the expression of the potential. The expressed potential is a change in the quantity of motion represented either as a change in momentum or a change in kinetic energy. Mass or momentum representing a potential is static—this idea is expressed in Newton's law of inertia—and force representing a change of velocity is non-static. Thus mass as a potential represents Order that expresses itself by a force interaction producing an event that represents a new order. Understood in this way force is the ambiguous transformation of an old order into a new order. In one perspective, force is the loss of an old order, e.g., a decrease in momentum, and thus the mutuality of mass as a potential and force is the mutuality of order and loss of order, which is chaos. In the other equally valid perspective, the chaos of a decrease in momentum implies the creation of a new order, e.g., the loss of momentum of one object shows up as an increase in momentum of another object. Thus, there also is the mutuality of chaos and creativity.

Energy represents a potential for expression of an event that is work or heat or work and heat. Energy as a potential is order and *flux* is the transformation of energy into work and/or heat. Flux is analogous to force and thus represents chaos. Perhaps an easier way of seeing this is in relation to a pencil falling off a table to the floor. The pencil on the table has potential energy due to the force of gravity. As the pencil falls toward the floor, it progressively converts potential energy into kinetic energy. This conversion is the flux of the flux of the potential energy which is ambiguous. From one point of view potential energy is lost and of course simultaneously the potential disappears which means order becomes loss of order, i.e., chaos. From another point of view, the chaos of the loss of potential equals to the creation of new order, e.g., during the free fall this new order is the emergence of kinetic energy. Thus, the simultaneity of these two points of view produces two mutualities: (1) the mutuality of order and chaos and (2) the mutuality of chaos and creativity.

The functioning of any machine points to a more radical mutuality of order and chaos. An ideal machine converts a large potential into the sum of autonomous work cycles. This is analogous to breaking up a ten foot-long stick into ten one foot-long pieces, but from the point of potential to do task work, this analogy breaks down. Without a machine the large potential can be converted totally into work (when the conversion is done infinitely slowly). But with an ideal machine the large potential is converted into a specific task work which is less than the total amount of work that the potential

---

[18]Pribor, D. 1999. "Transcending subjectivity double binds of science." Proceedings of the Institute for Liberal Studies, vol. 10, pp. 27–37.

could express. This is because each of the work cycles produced by the ideal machine consists of a Yang phase that produces a small unit of task work plus a Yin phase that produces work to prepare the energy coupler for another Yang phase. Thus, when a large potential is converted into machine work, there is an overall loss of order and therefore an increase in chaos. The order of the large potential is greater than the order of the machine work. Of course a non-ideal machine generates even greater chaos in that some of the order of the large potential is converted into heat.

Five: Hierarchal Mutuality of Chaos and Machine Creativity.   The mutuality of order and chaos associated with an ideal machine points to another subjective, profound insight about nature. The order of a potential in nature and the order of machine work may be represented by the same mathematical-physical symbols, i.e., units of energy, but the orders are *qualitatively* different. The autonomous potential in nature is not a potential to accomplish a particular task; it has no machine purpose. It only becomes a "task potential" when it collaborates with an appropriate energy coupler. This collaboration is not determined by the physical laws that specify the potential in nature nor by the laws that specify the operation of an energy coupler. There are an indefinite number of ways the collaboration can occur resulting in the efficiency of the machine varying from low to high. Csanyi makes an equivalent evaluation of this situation in stating that there is no algorithm for building the most efficient machine (Csanyi, 1989, pp. 8–10).[19] Machine work results from machine order which in turn contains within its definition the subjective idea of collaboration between a flux and an energy coupler. An autonomous potential in nature does not contain or in any way refer to this idea of collaboration. Therefore, order of a potential in nature and order of machine work are related but fundamentally different. Another way of stating this difference is that a machine has an organization pattern that can convert flux into machine work. Energy flux drives the *translation* of machine organization into machine work organization.

The orders associated with energy flux and machine work organization respectively are different but are fundamentally related by a hierarchal transformation. A machine transforms the order of a potential into a machine work organization by dividing a potential into a sum of machine work cycles. An "old order" of a potential goes into chaos which is the flux mirroring this order. The flux collaborating with the machine organization creates a "new order" which is the machine work organization. During the process the old order is incorporated into the new order which is thereby said to be at a higher level of order. That is, a potential in nature is incorporated into a machine work organization, but as stated earlier, the potential in nature does not "contain" machine work organization. This is what a two-level hierarchal system, AB, is. A incorporates B but B does not contain A. Even though the phases of an ideal heat machine are reversible, each work cycle is irreversible. This is because heat flow equal to flux is unidirectional; it only flows from high to low. It would take another machine exploiting a different potential in nature to make the flux occur in the opposite direction. This unidirectional flow in collaboration with a machine generates a two-level hierarchal arrangement in nature.

The above insight about hierarchal orders may be generalized to a statement about hierarchal mutuality of chaos and machine creativity. Machine creativity is the process of a machine work organization emerging from the chaos of a flux associated with an autonomous potential in nature. The overall process is Order, Chaos, New Order that includes a modified old order. That is, Order represented by a potential goes to Chaos represented by flux. The flux is converted into work cycles each generating an element of machine work. The sum of these elements equals machine work organization which sometimes may be represented by a new autonomous potential in nature which is less than the old potential from which it came. There is an overall increase in entropy in nature as a result of this machine creativity. The increase in entropy represents the transformation of some order into chaos in nature, but it also represents the "machine creations" of a new order.

Six: Machine Creativity Involves a Narrative Interpretation of an Historical Process.   If we focus only on the mathematical description of machine creativity, then machine work organization is not fundamentally different from order of a potential, and therefore, this creative process does not generate a

---

[19]Csanya, V. 1989. *Evolutionary systems and society.* Durham, N.C.: Duke University Press.

hierarchy of orders. However, if we focus on an interpretation of what is happening during machine creativity, then we see that this is an historical process that produces a new two-level hierarchy of orders. During this historical process, new order emerges from chaos. Thus, there is a mutuality of chaos and creativity. The idea of this mutuality is stated independent of Prigogine and Stengers' mathematical, thermodynamic theory of order emerging from chaos.

Description of an historical process is a story interpretation of events. The story interpretation or "narrative knowing" uses metaphorical concepts that relate to some objective pattern in nature to one's subjective experiences of nature. The subjective aspect makes metaphorical concepts ambiguous sometimes leading to contradictory interpretations. This, I believe, is because the experience of reality is at times paradoxical. Of course if one assumes modern that science is and must always be only utilitarian, then any metaphorical conceptual interpretation of nature must be replaced by a logical, conceptual model that has been tested for validity. However, *modern science is not only useful; it also provides valid metaphorical interpretations of nature*. In partial agreement and partial disagreement with this last statement, Bohm takes a non-hierarchal view of sciences description of nature that is utilitarian and provides metaphorical insight about reality. Bohm proposed that entropy can be defined in an objective way with no recourse to subjective insight. This definition is possible because, ". . . according to the metaphor that *chaos is order* [italics mine], an increase in entropy has to be understood in a different way, that is, in terms of a kind of change of order." (Bohm, 1987, p. 138).[20] Bohm's argument for this rests on a probabilistic model of thermodynamics. If thermodynamic events, such as machine creativity, can be adequately *interpreted* by the theory of probability, then I agree with Bohm. But this probabilistic interpretation is a subtle form of reductionism as also employed by system scientists. That is, all higher levels of order are reduced to a single level of order that may be adequately interpreted by a probabilistic model. Bohm proposes a transformation of science in which a *knowable* but not ever completely *known* implicate order generates matter that is a sub-total explicate order. In a dialogue with Weber, Bohm explains:

> I am not trying to deduce life and consciousness from physics, but rather to see matter as a part of a relatively independent sub-total which includes life. Leaving out life, we get inanimate matter, leaving out consciousness we get life; . . . I don't call this a hierarchy but rather a series of levels of abstractions, which is somewhat different, since if you abstract something you [don't] call it lower or higher (as in a hierarchial system), but merely different. (Wilber, 1982, p. 191)[21]

Because I reject any form of reductionism as an adequate interpretation of reality, I propose that any probabilistic interpretation of thermodynamics is a *legitimate* but not an *adequate* way of describing nature. First of all, just as no one ever has observed an atom, likewise no one ever has observed a random event. Both ideas (like all theoretical ideas in science) were created to provide a way of interpreting things—events that we do observe. *Random event* is an idea related to *probability* and as such cannot be rigorously defined independent of the idea of probability. A random event is an outcome of a trial that can produce two or more possible outcomes. Further, if the trial is repeated many times under the exact same conditions and the frequency of each possible outcome approaches a specific ratio between zero and one, then this limit frequency is called the probability of the outcome. Each event produced by such a trial is called a random event. For many situations we can know ahead of time all the various finite number of possible, autonomous outcomes, and then it may be useful to assign a probability equal to $1/n$ for each possible outcome. More generally, any measurement is a trial which has a non-denumerable infinite "number" of possible outcomes (this is so because measurement of a continuous quantity will never be exact and any continuous quantity is non-denumerably, infinitely divisible). Then we cannot assign probabilities to all the possible outcomes ahead of time. What we can do is invent a mathematical entity called a *probability distribution function* such as the Gaussian function and superimpose this function onto the measuring process. That is, we make it represent probability mass associated with all possible subsets that together contain all the possible autonomous outcomes of the

---

[20]Bohm, D. and Peat, F. D. 1987. *Science, order and creativity.* New York: Bantam Books.

[21]Wilber, K. editor, 1982. *The holographic paradigm and other paradoxes.* Boulder: Shambhala Pub., Inc.

measuring process. Then, often we can convert this distribution function into a frequency distribution function where each frequency is interpreted to represent the probability of one of the possible outcomes of the measurement process. For example, many measurement processes can be represented by the so-called "bell-shaped" frequency function (curve) which is derived from the Gaussian probability distribution function.

What the systems scientists do is assume that the universe they are describing is a probabilistic universe or in the case of Bohm that the explicate order of matter is adequately described by a probabilistic model. Such a universe often may have a non-denumerable, infinite "number" of possible manifestations at any moment in its evolution. Some or all of these manifestations are said to be random. First of all, by interpreting the universe in this way, systems theorists limit themselves to a mechanistic analysis; i.e., the universe at that moment is a "whole" that may be considered to be a set of autonomous possibilities from which only one will be manifested. Probability theory presupposes that all elementary random events are autonomous and the collection of (the set of) all these possible events completely represents the whole event that is being studied, e.g., in this case the moment in the evolution of the universe that is being observed. In other words, each moment in the evolution (history) of the universe is identical to a set of non-denumerable, infinite "number" of autonomous, possible manifestations of the evolving universe. This, of course, is another form of subtle mechanistic reductionism.

Systems scientists and in an analogous way, Bohm, describe creative evolution as a sequence of transformations each from a lower level to a higher level of organization. However, this *systems concept of organization level* reduces our tacit, subjective understanding of qualitatively different levels of organization in nature to a quantitative, objective representation of levels of organization.

For example, we may represent the development of an embryo to a fetus just before birth as the increase in *length* from a few millimeters to 16–20 inches. This is an objective valid representation of development, but it ignores the subjectively obvious, non-quantifiable, organizational changes that occur during this development. In like manner, a systems science description of this development would represent it as a large decrease of entropy. This is an objective, valid representation of fetal development, but it also ignores the subjectively obvious, non-quantifiable, organizational changes that occur during this process.

Wilber's critique of systems science brings in the importance of analogical thinking in various types of knowing.

> . . . the claim of the evolutionary systems theories (from Bertalanffy to Laszlo to Jantsch) is that, although *none* of the levels can be reduced to any other, general laws or regularities of dynamic patterns can be found that are the same in all three realms [matter, life, and mind]. These are called "homolog laws" and not "analog laws," which means they are the *same* laws and not just *similar* laws.
>
> I [Ken Wilber] agree with that position (as far as it goes), and in chapter 2 [of Wilber's book] we outlined twenty tenets, or "homologue laws," that are characteristic of holons *wherever* they appear. So far, so good.
>
> But these tenets have to be of such a general nature that they will apply to all three general realms, and that means, in essence, that they apply basically to the realm . . . [of matter], since . . . [matter] is the only thing that all three realms have in common (Wilber, 1995, pp. 114–115).[22]

Wilber's point is that if these "homolog laws" are not supplemented by other ways of knowing especially as applied to life (and to mind), then this in effect is a subtle way of reducing life to non-life. However, I argue that science uses "homolog laws" only when this way of knowing is reduced to utilitarian reductionism or utilitarian positivism. At a deeper level of understanding, utilitarian science emerges from a "primordial science" that requires metaphorical concepts and analogies for expressing meaning. Thus, at this deeper level of understanding, science does propose "analogical laws." Nevertheless, Wilber is correct in accusing systems scientists of being subtle reductionists

---

[22]Wilber, K. 1995. *Sex, ecology, spirituality (the spirit of evolution)*. Boston: Shambhala Pub., Inc.

because, for the most part, most of them reject or, at least, do not acknowledge the metaphorical foundation of science. To put it differently, they do not understand or recognize that the conceptual, utilitarian models emerge from the *more fundamental narrative understanding,* which uses metaphors and analogies, for understanding the cosmos.

Seven: Machine Creativity Points to a New, Higher Narrative Way of Understanding Science.   Machine creativity may be described from two different, hierarchal perspectives. The metaphorical conceptual narrative perspective not only interprets machine creativity but provides a way to have empathetic, participatory understanding of nature. This partially subjective knowing cannot guide objective action, but it can be reduced to logical, conceptual knowing, e.g., mathematics and measurement, which is utilitarian. The metaphorical narrative knowing is ambiguous not only in content but also in level of knowing. On the one hand, it is a primordial kind of knowing that can guide the formation of a rational hypothesis. On the other hand, when a hypothesis becomes a valid theory, then a metaphorical, narrative understanding of it is a higher level of knowing. It is higher for several reasons. The metaphorical conceptual understanding of a rational theory gives one the openness to see possible connections with other areas in science and with non-science areas of knowledge. This type of knowing provides ordinary language as a way to communicate with thinkers in diverse disciplines; i.e., it partially overcomes the fragmentation of knowledge in postmodernism. This type of knowing also actuates the spiritual-mystical aspect of being human. The mutuality of the two levels of knowing is what Heidegger refers to as the mutuality of calculative thinking and meditative thinking. In the introduction to Heidegger's book, *Discourse on thinking,* John Anderson summarizes Heidegger's view of this mutuality as follows. Meditative thinking

> Requires two attributes . . . two strands which man can take, and which he calls *releasement toward things* and *openness to the mystery* . . .
>
> He also drives home the importance of such thinking to man's very being, claiming indeed, that even the ultimate meaning of the calculative thinking of modern science and its humanly significant applications are discerned in and through meditative thinking. But fundamentally Heidegger is urging his hearers and readers toward a kind of transmutation of themselves, toward a commitment which will enable them to pass out of their bondage to what is clear and evident but shallow, on to what is ultimate, however, obscure and difficult that may be (Heidegger, 1966, pp. 12–13).[23]

Machine Cosmology

Mechanical potential energy due to the force of gravity is a mystery. We cannot visualize how two masses such as the sun and the earth that are separated by empty space can effect one another's motion, but Newton's laws enable us to measure this interaction. Potential energy is not a "substance," but it is convenient to think of it as such; each body is thought of as possessing potential energy. In like manner how the temperature of the sun can increase the temperature of the earth is a mystery, but we can describe radiation of light from the sun through empty space to the earth. We also can measure the effects of this radiation in terms of increase in temperature and in terms of motion. Again it is convenient to think of this interaction as an energy "substance" traveling from the sun to the earth. Other types of radiations such as X rays and radio waves as well as light are thought of as "possessing" energy. Thus, a great diversity of events in the universe are metaphorically related; they all can be described as work or heat events which in turn may be measured in units of kinetic energy (Joules). Energy is the unifying idea amidst the diversity of observed events, and in particular it underlies both thermo (heat) and dynamic (motion) events. If we agree to think of it as a substance, then we can formulate a fundamental law of the universe, the so-called first law of thermodynamics: during any event in the universe, energy is neither created nor destroyed; it is transferred from one place to another, and it may exist in any one of many different forms. Nevertheless,

---

[23]Heidegger, M. 1966. Anderson, J. & Freund, F. H. ed. *Discourse on Thinking.* New York: Harper Torchbook, Harper & Row Pub.

energy is not a substance. It is the mysterious, underlying thing-process of all observable events that can be measured.

Starting with the metaphor of energy as a substance we can go on to think of any object in the universe as a machine. Any system (or object) is embedded in a network of work and heat potentials. A system does not create or "cause" work or heat to be done; rather it merely transfers energy to and from its environment. These energy transfers are what we call work and heat. Moreover, though a heat process may occur alone, the converse is not true; every work process always is associated with heat. The insights of machine cosmology can be formulated in terms of thermodynamics.

Thermodynamics

This model connects the theory of heat (thermo comes from the Greek word meaning *heat*) with *dynamics,* the theory of mechanical energy changes. Thus, thermodynamics exemplifies the way unification of science occurs: two apparently different kinds of phenomena each described by a very different theory are seen to be fundamentally related. The core ideas of thermodynamics are implicit in two statements, the first and second laws of thermodynamics. What also is very interesting is that the first law represents the high point of scientific mechanistic thinking, but the second law is the first serious challenge to scientific mechanism. Let us first understand the first law and then explain how William Clausius, the first to formulate the second law, radically challenged total scientific objectivity.

First Law of Thermodynamics

The first law often is stated as: during any event in the universe, energy is neither created nor destroyed; it is transferred from one place to another, and it may exist in any one of many different forms. However, a more precise formulation of this law depends on three ideas:

1. *Events* in nature may be understood in relation to a *system* interacting with its *environment.* In thermodynamics the boundary between the system and its environment must be clearly specified.
2. Any system interacts with its environment in *only* two ways: (1) heat process and (2) work process.
3. Energy is like a substance and so it can be transferred into or out of a system via heat or work or both.

Let U = total internal energy of a system where this energy may be present in many different forms; and let q = heat and w = work and ΔU = change of internal energy of a system. Then the first law of thermodynamics states:

$$\Delta U = q + w$$

that is, the change in internal energy of a system is equal to the amount of energy transferred into or out of the system via q plus the amount of energy transferred into or out of the system via w.

Entropy

In 1865 Clausius invented a measure of the inability of any system in the universe to convert the energy "within it" or the energy "it receives" into work. Clausius called this measure, entropy; the higher the entropy, the greater the inability of the system to do work. The concept of entropy reminds us of the fallacy of thinking of energy as a substance in an autonomous system. If this metaphor were universally valid rather than merely a convenient way of thinking, then we could compare a machine doing work to filling up a glass with water from a pitcher of water. We simply empty the pitcher and fill up one or more glasses with water. All of the water, i.e., energy, in the pitcher, the machine, is transferred to the work of filling glasses with water. Thomson's first statement of the second law of thermodynamics and what we described as the limitation of any machine is that even the best machine represented by the ideal Carnot heat engine is less than 100% efficient. The energy in it or the energy it receives cannot be totally converted into the work of some desired task.

Second Law of Thermodynamics

Stated as a Limitation of the First Law.   (1) According to the first law of thermodynamics, $\Delta U = q + w$ where sometimes $\Delta U = q$ or $\Delta U = w$. According to the second law, sometimes $\Delta U = q$ but ΔU may never equal w. In the ideal heat machine where all processes are reversible, $\Delta U = +w$ (during phase 4) and $\Delta U = -w$ during phase 2, but in any real machine, all processes are irreversible because some of the internal energy decrease or increase shows up as q, thermal energy lost to the environment.

(2) No real machine is 100% efficient. Even in an ideal machine, some of the incoming energy must be lost to the environment as thermal energy rather than work so that the energy coupler can start a new cycle. Then too some of the incoming that ideally could contribute to the maximum work done by the machine is lost as thermal energy to the environment because the machine processes are irreversible.

Stated in Terms of Final Causality.   (1) Any process in nature is spontaneous or is the result of a spontaneous process. A process is spontaneous if a heat or work potential becomes zero or decreases to the lowest possible value the environment will allow. (2) All real processes in nature occur in a finite time and express the directionality of time; underlying measured time, which is reversible, is the real time of our subjective experience which is irreversible.

(3) Every system has a characteristic property called entropy which represents its inability to express spontaneous change—the higher the entropy, the lower the ability to be spontaneous. In an isolated system, the entropy will increase to the maximum value the environment will allow. Another way to say this is that entropy is a measure of the degree of Chaos of a system. The higher the entropy, the greater the degree of Chaos and the lower the degree of Order. This means that the system is less able to do work, that is, is less able to spontaneously change. In an isolated system, the degree of Chaos will increase to the maximum value the environment will allow. That is, if no thermal energy is transferred into or out of the system and if the environment allows the system to do work on it due to a work potential equal to the possible change in internal energy of the system, then the entropy will increase to a maximum value. In the process of the system becoming more disorganized (increase in Chaos), it loses its potential to do work on the environment. If the system is open, i.e., allows matter or thermal energy to move into or out of the system, then the net entropy change of the system and its environment always increases. If the entropy of the system decreases in an irreversible manner, there always will be an even greater increase in the entropy of the environment. For example, when the ideal gas in a Carnot heat machine is reversibly compressed by the environment in phase 4, the decrease in entropy of the gas is offset by an equal increase in entropy of the environment. When the gas compression is irreversible, then there is an even greater increase in entropy of the environment due to the thermal energy passing into the environment during this irreversible process.

Stated in Terms of Indeterminacy.   It is possible to reformulate "classical thermodynamics" and in particular, the second law in terms of the modern theory of probability which includes the intuitively understandable complementarity of order versus randomness. Then, every system has some degree of randomness and correspondingly, some degree of order, and every system in nature tends to increase its randomness and correspondingly decrease its order.

Physical Mutualities

Law 1: No Event Is Autonomous

Rather every event is mutually related to some other event. For example, Newton's laws of motion which implies the conservation of momentum make up one aspect of the foundation of modern science. Momentum is the quantity of directional motion of an object at some point in space; that is, momentum = mass × velocity. Then, according to the conservation of momentum, if some object increases its momentum at a point in space, some other object must have decreased its momentum by an equal amount. The event of increasing momentum is never autonomous but always simultaneously involves another event during which momentum decreases. Likewise, the conservation of energy, another aspect of the foundation of modern science, involves mutuality of events. Energy represents the potential quantity of motion plus the actual quantity of motion an object has at some point in space. Then, according to the conservation principle, if some object increases its energy at a point in space, some other object must have decreased its energy by an equal amount. Again, as is the case with momentum events, there are no autonomous energy events. As our ideas for describing the universe become more complex, the mutuality of events becomes even more obvious and undeniable.

The scientific revolution (1500–1700) was the first movement of Western thinking toward the Buddha's insight of dependent co-origination. Physics then and now abandons the Aristotelian ideas of being, substance, and causation. Force which equals mass times acceleration (F = ma) of some body at a point in space and moment in time is not a cause of motion as many textbooks erroneously proclaim. Force, always and everywhere, is an interaction between two things. This interaction is quantitatively defined in terms of the quantity of mass of one object involved in the force interaction times the rate of change of the velocity, which is something like speed, of the object. This definition of force, in turn, depends upon the mutualities of: (1) conservation of mass, (2) conservation of momentum, and (3) forces of action and reaction and on what may be called the suchness of inertia.

Newton proposed that every material body in the universe has a quantity of inertia called *mass* that expresses itself in a way that I call *dependent co-origination with respect to motion*. If there were only one mass-object in the universe or if it were possible to totally isolate one mass-object from all others, then two statements eternally describe its state of motion. If the mass-object is not moving, that is, velocity = 0, it will never move. If the mass-object is moving at some constant velocity, then it always will move at that constant velocity. The mass-object is *such-that-it-is* which is a state of motion independent of any notion of causality or it being a type of substance, that is, the mass-object is not an entity with an essence that determines (causes) its existence to manifest as a particular state of motion. The independent mass-object simply is a particular state of motion (velocity = 0 or velocity not equal to 0). There is no "why" and it has no purpose. But Newton proposed—and all physicists to this day agree—that all mass-objects are interdependent. This means that mass-objects in physical contact with one another influence one another's motion *not* by causation but by the mutualities described in laws of conservation such as the conservation of momentum. Newton's so-called three laws of motion specify this influence during physical contact. When mass-objects are separated, they influence one another's motion in a way as specified by Newton's law of universal gravitational force. All the interdependent mass-objects are interdependent in their motion. Yet, each mass-object is equally in itself, of itself, without one being prior to the other. Also there is nothing whatsoever more substantial or more real which grounds the interdependence of all mass-objects. As a group, scientists did not come to a full realization of this last statement until the late 1700s. Pierre Laplace (1749–1827) gave expression to this realization with his pronouncement: "We no longer need the God hypothesis."

Law 2: Mutuality of Order and Chaos

The order of the universe may be represented by potential events. A potential event may be expressed by some mutuality as described in law 1. For example, Newton's law of universal gravitation quantitatively specifies the potential for any two objects in the universe to move toward one another; the potential is called a force of gravity. Likewise, the most fundamental understanding of energy is that it is a potential for some energy event to occur. In physics there only are two types of energy events. One is *work* which quantitatively specifies motion of an object in an interval, and the other is *heat* which quantitatively specifies the transfer of thermal energy that results in the rise of temperature at some one or two points in space. Correspondingly, the universe is described in terms of work potentials (which include potentials for changing momentum) and heat potentials. Though the modern theory of quantum mechanics and Einstein's general theory of relativity are based on assumptions that contradict one another, both theories acknowledge the validity of the second law of thermodynamics. This law may be understood in terms of potential events wherein every potential event at a point in space and/or moment in time will tend to expresses itself as some mutuality as

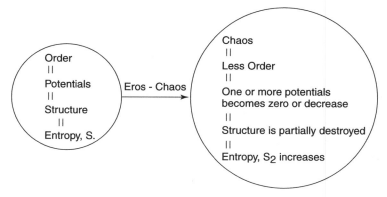

FIGURE 1.10

described by law 1. When the potential event does express itself, the original potential event disappears, e.g., when a rock falls to the ground, it loses its potential to fall to the ground. Likewise, when a bottle of milk loses thermal energy so as to come to the same temperature as that of a refrigerator, it loses its potential to decrease its temperature in these circumstances. Thus, Order represented by a potential is also a potential for a loss of Order which we refer to as Chaos, and as Order decreases, Chaos simultaneously always increases to the same extent.

One of the original formulations of the second law of thermodynamics spells out this mutuality in terms of *entropy*. Entropy is a concept created to represent the quantitative increase or decrease of any potential event. Entropy is defined in such a way that when the quantity of the potential event increases, e.g., greater force of gravity or greater temperature difference between two objects, entropy decreases and Order increases. Conversely, when the quantity of the potential event decreases, i.e., an increase in Chaos, the entropy increases. According to the second law of thermodynamics, the entropy of any object or system in the universe tends to increase; i.e., the isolated potential event associated with that object tends to decrease and correspondingly, a decrease in Order is simultaneously replaced by an increase in Chaos. Thus, the second law of thermodynamics really is a statement about the mutuality of Order and Chaos anywhere in the universe, see fig.1.10.
All isolated systems in the universe tend to change:

1. From Order to Chaos
2. From lower Entropy to higher Entropy
3. From networks of potentials to one or more potentials decreasing (sometimes to zero)
4. From nonequilibrium to equilibrium
5. From structure to less structure
6. From maintained separation to union = gross harmony

Machine Creativity, Law 3: Mutuality of Chaos and Creativity

At the most elementary level as described by physics, *creativity* is the emergence of a new potential event = new Order. The mutuality of Chaos and Creativity means that New Order always is coupled with some Chaos resulting from the decrease or disappearance of some potential event representing Old Order.

For example, any machine has at least one energy coupler. If the machine has more than one, the interconnection among two or more couplers may be thought of as a single complex energy coupler. An energy coupler takes in energy and focuses it on accomplishing a particular task defined as the creation of a new potential event = a New Order in the universe. For example, a hammer takes in energy from the human arm and hand muscles and focuses it on pounding a nail into a block of wood—perhaps connecting one piece of wood with another as in a cabinet. This outcome represents a New Order that like any order in the universe will tend to break down into chaos. The energy coupler goes through one or more (usually many) cycles where each cycle consists of a YANG phase and a YIN phase. In the YANG phase the energy coupler starts out as a particular potential event represented by a particular value of entropy. When the potential event is expressed, entropy increases as the quantity of the potential event decreases. The mutuality of energy events—Law 1— shows up as a decrease of the energy of the energy coupler which equals to the increase of energy of one or more other objects. When this Order of the energy coupler is lost and replaced by Chaos, some New Order = the task of the machine is *created* and so we have Chaos-Creativity mutuality. In the YIN phase the energy coupler receives energy from some other objects in the universe. As a result, the energy coupler decreases its entropy as it increases its quantity of potential event = increase in Order. The mutuality of energy events—Law 1—shows up as an increase in energy of the energy coupler which equals the decrease of energy of some other objects in the universe. As the quantity of potential event of other objects in the universe decreases (entropy increases) and thus go to Chaos, the quantity potential event of the energy coupler increases, i.e., it is re-Created. Thus, the YIN phase also expresses Chaos-Creativity mutuality, see figure 1.11.

Every machine expresses *machine creativity* as a result of a mutuality between the machine and its environment. In the YIN phase Order in the environment goes to Chaos in such a way that the quantity of the potential event of the energy coupler, i.e., energy coupler Order, is re-Created. Then, in the YANG phase, the energy coupler Order goes to Chaos in such a way that a task is accomplished. The task is the New Order that is Created by the YANG phase. Machine creativity also may be represented by a mutuality between Logos = Order = potential event and Eros = drive to express a potential event. The word *Eros* comes from the word *erotic* which relates to the human sexual drive in its various manifestations. The sex drive is a metaphor for the tendency of any potential event in the universe to be expressed. Correspondingly, it seems fitting to use the word *Eros* to represent this tendency for any potential event to be expressed in the universe. Beginning with this terminology I use machine creativity to introduce new fundamental terms that will be useful in other situations. During the YIN phase, environmental Order (Logos) goes to Chaos and the drive to do this is called *Eros-chaos.* Eros-chaos is coupled with, i.e., is mutual with, re-Creating Order (Logos) of the energy coupler and the drive to do this is called *Eros-order.* In the YANG phase the energy coupler Order (Logos) goes to Chaos and again the drive to do this is Eros-chaos. The Eros-chaos is coupled with creating

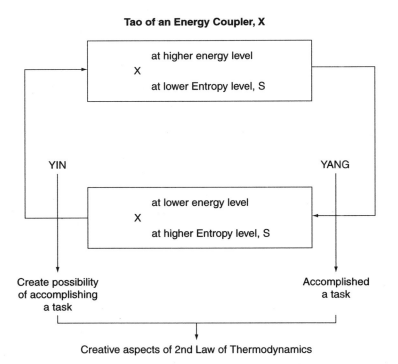

FIGURE 1.11

task Order (task Logos) and the drive to do this Eros-order, see figure 1.12. The mutuality of Chaos and Creativity may be represented as the mutuality of Eros-chaos and Eros-order. Generalization 4 and 5 about the ideal heat machine provides another way of stating this law. An energy flux is Eros-chaos where Order goes to Chaos as energy is transferred from one place to another. The collaboration between an energy coupler and flux produces potentials that represent Eros-order. Eros-order in each cycle of a machine creates machine order = continuous cycles + work task for each cycle. Thus, the mutuality of Eros-chaos and Eros-order generates the mutuality of flux and machine order.

Law 4: Hierarchal Mutuality of Chaos and Creativity

Let us call "order in a particular system" *Logos*, then the loss of Order may be called *Chaotic Logos*. Eros-chaos drives some Order (Logos) in the universe to Chaotic Logos where the chaos makes potential events available to be expressed which before the chaos could not be expressed. Eros-order coordinates the expression of one or more of these potential events with remaining Order of the chaotic Logos to produce a New Order that includes aspects of the old order. Thus, Eros-chaos is mutual with Eros-order in such a way that a modified Old Order is incorporated into a New Order. The New Order expresses *emergent properties* which have the following four aspects: (1) these properties only can be described in terms of the whole higher order pattern. (2) These properties cannot be understood in terms of (reduced to) interactions among components; i.e., the properties cannot be understood in a gross, mechanistic way. (3) These properties cannot (or could not) be predicted to occur from knowing the components of the new pattern. (4) The properties are the expression of imminent potentials of a set of apparently autonomous entities that interact in a context that actualizes the imminent potentials. One can know the existence of these imminent potentials only as a result of knowing their manifestation that shows up as emergent properties. One can know the context for the expression of these imminent potentials (and thus count on their expression when the

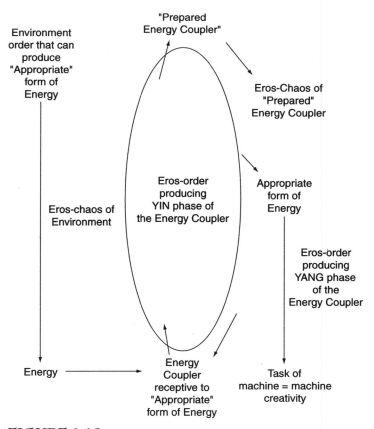

FIGURE 1.12

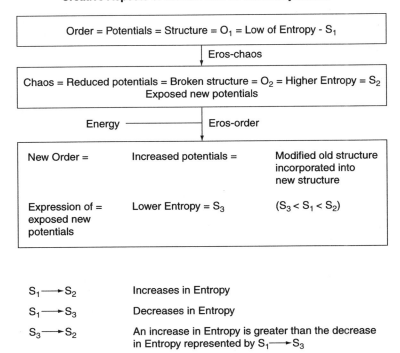

FIGURE 1.13

appropriate context is set up as we do in chemistry, for example), but one never can know the *how-why* the context leads to their expression. Knowing the *how-why* implies a mechanistic description, and the expression of emergent properties is a non-mechanistic and indeed, a non-rational process. The overall process of *hierarchal mutuality* is: Old Order, Chaos, New Order expressing emergent properties and incorporating aspects of the old order, see figure 1.13.

For example, under certain conditions, such as very high temperatures, the Order in a population of hydrogen gas molecules and oxygen gas molecules goes to Chaos = Chaotic Logos. The hydrogen molecules each consisting of two hydrogen atoms chemically bonded to one another, break up into hydrogen atoms. Likewise, oxygen molecules, each consisting of two oxygen atoms chemically bonded to one another, break up into oxygen atoms. Then, the one negative electron orbiting the positive nucleus (a proton) of each hydrogen atom breaks away from the hydrogen atom producing chaotic hydrogen Logos = positive hydrogen atom (= positive hydrogen ion). Each of the free oxygen atoms has a potential to receive into its structure, i.e., its orbital system, two electrons. The now available electrons that broke away from the hydrogen atoms go to the oxygen atoms thus producing chaotic oxygen Logos = negative oxygen atom with two negative sites (= negative oxygen ions). The resulting chaotic hydrogen and oxygen Logos makes available new potential events. Eros-order leads to the expression of one of these new potential events. Namely, two positive hydrogen atoms (ions) are received by the two negative sites of each negative oxygen atom (ion). The resulting chemical bonding between each negative oxygen atom and two positive hydrogen atoms produces a water molecule, H_2O, see figure 1.14. The population of water molecules has emergent properties that could not be predicted from knowing the properties of hydrogen gas or oxygen gas or a mixture of hydrogen and oxygen gases.

Law 5: Physical Individuation Is an Order, Chaos, New Order Process.

Individualism is an acquired degree of uniqueness of an entity that enables it to enter into new types of interactions. For example, each electron (fundamental unit of negative charge) in a cloud of electrons and each proton (positive charge) in a cloud of protons has no uniqueness. Under the right con-

Hierarchal Mutuality of Chaos and Creativity

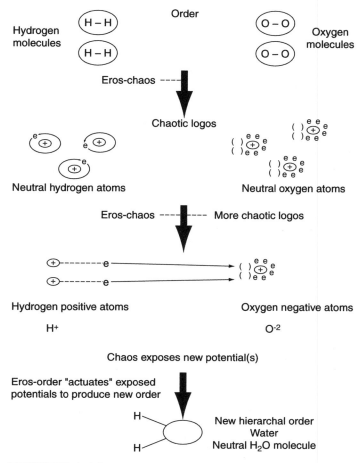

FIGURE 1.14

ditions a particular electron may break away from its population and a particular proton may break away from its population. Now the more unique electron and the more unique proton have the possibility of interacting with one another. If and when they do so, the electron will begin to orbit the proton thus producing a hydrogen atom. That is, the electron and the proton each acquired a degree of uniqueness expressed as their mutual interaction in forming the hydrogen atom. The overall process involves breaking away from the low level order of a population to acquire autonomy and then entering into a mutual interaction to form a new structure a new order. This process of individuation involving going to a higher level of individualism and autonomy exemplifies the pattern of transformations that gave (and gives) rise to all levels of organization in the universe. Things achieve some degree of separateness and autonomy by breaking from one level of order and thus going into one corresponding level of chaos. The separated, autonomous things in chaos then may mutually interact to produce a new level of order. The overall transformation process may be represented by the sequence: Old Order, fragmentation producing Chaos, from which emerges a New Order which includes the modified fragmentations of the Old Order. More briefly the pattern is: Order, Chaos, New Order. The characteristic of the universe that allows for transformations of this sort are patterns of mutuality, laws of mutuality. These laws of mutuality interrelate all "parts" of the universe so that parts of the universe only are relatively autonomous.

Tearout #1

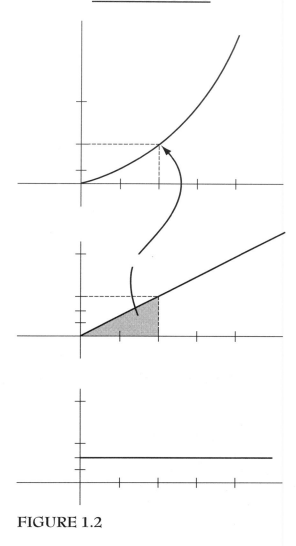

FIGURE 1.2

Directions:

1. Using figure 1.2 as a reference, fill in the appropriate labels.
2. Using figure 1.2 as a reference, indicate where there is a "higher energy level" and lower entropy and where there is a lower energy level and higher entropy.

Tearout #2

Hierarchal Mutuality of Chaos and Creativity

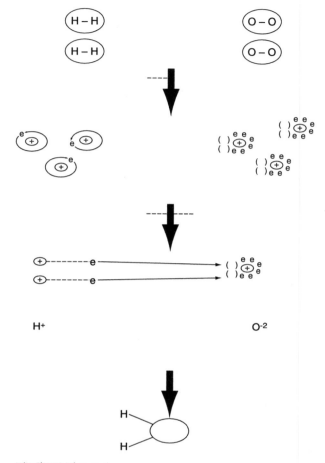

FIGURE 1.14

Directions:

1. Using figure 1.14 as a reference, fill in the appropriate labels.

Systems Theory of Creativity as Exemplified by the Emergence of Life

Evolutionary Transformation of Physics

In 1924 Edwin Hubble demonstrated that there are many galaxies besides the one that includes our sun. We now know that there are several hundred thousand million galaxies each containing several hundred thousand million stars! Then in 1929, Hubble reported that wherever one looks through a telescope, distant galaxies are seen to be moving rapidly away from us; i.e., the universe is expanding.

> This discovery that the universe is expanding was one of the great intellectual revolutions of the twentieth century. . . . This behavior of the universe could have been predicted from Newton's theory of gravity at any time in the nineteenth, the eighteenth, or even the late seventeenth centuries. Yet so strong was the belief in a static universe that it persisted into the early twentieth century. Even Einstein . . . was so sure that the universe had to be static that he modified his . . . [General theory of relativity] to make this possible . . . [he] introduced a new "antigravity" force, which unlike other forces, did not come from any particular source, but was built into the very fabric of space-time.[1]

Based on the work in the 1920s of the Russian physicist and mathematician Alexander Friedmann, there are three possible models of the uniformly expanding universe discovered by Hubble: (1) the universe expands and recollapses so that space is bent in on itself like the surface of the earth; (2) the universe is expanding so rapidly that the gravitational attraction can never stop it and instead of space bending in on itself, it is bent the other way like the surface of a saddle, and (3) the universe is expanding only just fast enough to avoid collapse and space is flat but similar to the second model in that it also is infinite. All of these models are based on Einstein's general theory of relativity and

> . . . have the feature that at some time in the past (between ten and twenty thousand million years ago) the distance between neighboring galaxies must have been zero. At that time, which we call the *big bang*, the density of the universe and the curvature of space-time would have been infinite.[2]

---

[1]Stephen W. Hawking, *A Brief History of Time* (New York: Bantam Books, 1988) pp. 39–40.

[2]Ibid., p. 46.

From the late 1940s until 1970 there were a number of attempts to explain Hubble's data and Friedmann's models without assuming that there had been a big bang. The most popular of these was the so-called steady state theory. In 1965 the British mathematician and physicist Roger Penrose used Einstein's theory to show that an aging star which is collapsing under its own gravity is trapped in a region whose surface eventually shrinks to zero size implying also zero volume; so the density of matter and the curvature of space-time become infinite. We call these regions in space-time *black holes* because any matter radiation such as light is bent back on itself by the force of gravity so no light can get out of this region in space.

> In 1965 I [Stephen Hawking] read about Penrose's theorem that any body under going gravitational collapse must eventually form a singularity. I soon realized that if one reversed the direction of time in Penrose's theorem, so that the collapse became an expansion, the conditions of his theorem would still hold, provided the universe were roughly like a Friedmann model on large scales at the present time. Penrose's theorem had shown that any collapsing star must end in a singularity; the time-reversed argument showed that any Friedmann-like expanding universe must have begun with a singularity. . . . During the next few years I developed new mathematical techniques to remove [technical problems with Penrose's theorem] . . . The final result was a joint paper by Penrose and myself in 1970, which at last proved that there must have been a big bang singularity provided only that general relativity is correct and the universe contains as much matter as we observe.[3]

Thus, the universe has a history that like any history can be represented by a story. Any story is an expression of causality interacting with chance events to produce a plot. The plot of the universe is a drama involving transformations at all levels of organization within it.

Evolution of Non-Living Matter

Origin of Elements and Some Molecules During Stellar Evolution

> In the *big bang model* of the evolution of the universe, an extremely dense and hot *ylem* expands and cools, giving rise to stars by collapse of local supercritical masses of mixtures of hydrogen and helium (~25 percent helium). During collapse, half the gravitational potential energy of a supercritical mass is converted into kinetic energy, causing the gas temperature to rise; the other half is radiated into space. As the temperature soars, ionization strips electrons from hydrogen atoms, leaving protons.[4]

Hydrogen nuclei, that is, protons, fuse to form helium nuclei with less mass than the sum of protons that fused to form them. As predicted by Einstein's special theory of relativity, this mass difference shows up as an enormous amount of energy that drives more complex fusions that produce lithium, boron, and beryllium nuclei. Further collapse of these forming stars due to their own gravity, that is, the forming protogalaxies, produce the so-called "red giant stars" within which fusion of subatomic particles (alpha particles) and various atomic nuclei produce carbon, oxygen, and neon nuclei. These types of fusions stabilize the red giants so that no further gravitational collapse occurs.

Later energy from reactions among alpha particles and from further collapse of the core of stars lead to a great increase in temperature which leads to supernova explosions that spread hot "star dust" into space that, in turn initiate formation of new stars. In these smaller stars a new set of nuclear reactions called *carbon-nitrogen cycle (CN)* begin to occur. The cycle regenerates carbon nuclei and produces oxygen, neon, magnesium, silicon, and sulfur nuclei. Light energy released in the process of photodisintegration during a second series of supernova explosions and star births leads to nuclear reactions that produce heavier nuclei including iron. In a third series of star births our own galaxy, the Milky Way including our own Sun, was "created."

---

[3]Ibid., pp. 49–50.

[4]Fox, R. F. *Energy and the Evolution of Life* (New York: W.H. Freeman and Co.), p. 8.

In certain areas of the galaxy huge clouds of [star] dust [from supernova explosions] assembled. . . . [this] is the place of very active chemical processes (J. M. Greenberg, 1984: "Chemical Evolution in Space." *Origins of Life* 14: 25–36.) Most of the dust formed from tiny silicate grains surrounded by a mantle of various simple compounds created by photochemical reactions . . . water, carbon monoxide, carbon dioxide, ammonia, various cyanides, simple sugars, amino acids, etc. (W. M. Irvine and H. Jjalmarson, 1984: "The Cultural Composition of Interstellar Molecular Clouds." *Origins of Life:* 14: 15–23). The amount of the organic material thus formed is huge, an estimated 0.1 percent of the Milky Way's total mass.

. . . The densest clouds are the most active regions of ongoing star formation. . . . The stellar clouds occasionally undergo a gravitational collapse in which huge masses of dust form high-density protostars. The condensed material gets hot very quickly, and in certain stages nuclear fusion reactions start with hydrogen burning. . . . In most cases the birth of a new star is accompanied by the formation of planets that also originate from condensed stardust. First, the protostar is surrounded with dust particles, and when the nuclear burning of the star begins, the dust is heavily irradiated by the star's radiation. Some of the organic substances formed evaporate and are destroyed; some others undergo further photochemical reactions, probably resulting in rather complex organic compounds. In the distant dust local condensation may begin, and meteorites and comets of various sizes (from centimeters to several kilometers in diameter) are formed. Planets are created by the further condensation of meteorites. It is most important that the new planets are subjected to the shower of organic dust and meteorites for millions of years following their birth (A. H. Delsemme, 1984: "The Cometary Connection with Prebiotic Chemistry." *Origins of Life* 14: 51–60).

Enormous amounts of water and organic materials flow to the surface of the new planets. In other words, all basic compounds necessary for the origin of life are available in great quantities and high concentrations. Of course, life does not start on all planets. Planets too close to the central hot star lose their water and atmosphere very soon because of high temperatures, while others get too far from the central star and lack the radiant energy needed for starting life. Very probably Earth is in the "biogen zone," as is shown by the evolved biosphere (J. Oro, K. Rewwes, and D. Odom, 1982: "Criteria for the Emergence and Evolution of Life in the Solar System." *Origins of Life* 12: 285–305). We can deduce [summarize] from the above facts that the origin of life is not an unusually rare phenomenon but an essential part of the Universe's evolution.[5]

Evolution of the Earth and Formation of Small Molecules

Prevailing models of the evolution of the earth envisage an early stage of accretion about 4.5 billion years ago. Around 4 billion years ago the planet was so hot that the iron-group elements (Cobalt, Nickel, and Iron) melted and, because of their great density, passed through the lighter silicate rock to form the core of the earth. This event is called the *iron catastrophe.* The surface then began to cool and develop a solid crust, which was very thin relative to the earth's radius. The iron-nickel core solidified under great pressure. Current opinion is that the solid iron-nickel core is surrounded by a region of liquid iron and nickel; together these regions account for roughly half the earth's volume. Floating on this liquid iron and nickel is a molten mantle, primarily of silicates, perhaps 2000 km thick. The solid siliceous surface crust is only about 100 km thick, about as thick as an eggshell compared with the diameter of the egg.

Outgassing through the crust built an atmosphere rich in H_2O, N_2, CO_2, and CO. By 3.6 billion years ago, the oceans had begun to form; and great crystal plates, solidified out of the lighter material and floating on the molten interior, were being shaped and moved on the planet's surface by an on-going process called *plate tectonics.* On this earth with its nascent oceans, temperatures were about $3 \times 10^2 K$ and densities were between 1 and $10 g/cm^3$.

Because light gases such as hydrogen and helium had escaped from the atmosphere, the environment was composed of hydrogen in molecules more complex than H_2 and of heavier elements.[6]

[5]Csanyi, Vilmos. *Evolutionary Systems and Society A General Theory of Life, Mind, and* Culture (Durham, NC: Duke Uni. Press, 1989), pp. 27–28.

[6]Ronald F. Fox. *Energy and the Evolution of Life.* (New York: W.H. Freeman & Co., 1988).

Elements formed during stellar nucleosynthesis combine into varieties of small molecules at the relatively low temperatures on a planetary crust. Energy fluxes, such as lighting, ultraviolet radiation, and volcanic heat, convert the most stable small molecules, such as CO_2, N_2, and H_2O, into combinations such as sugars and amino acids, which become the building blocks for life.[7]

Individuation of the Earth to a Biosphere

Origin of Life: A Uroboric Puzzle

The Uroboros Puzzle What is the fundamental problem in biology? Max Delbruck answers as follows: "Thus there is a clear case for the transition on Earth from no-life to life. How this happened is a fundamental, perhaps *the* fundamental question of biology." [in XIII the Nobel Conference, Gustavus Adolphus College, Oct (1977). *Nature of Life,* edited by William H. Heidcamp, Uni. Park Press, Baltimore, 1978] The problem of the initiation of life is beautifully embodied in the ancient idea of the uroboros. The uroboros, symbolized by a serpent with its tail in its mouth . . . , represents an entity that is self-generating and self-sustaining.

Life itself is a self-generating and self-sustaining system that has evolved into a state of being in which its origins are no longer discernible. Organisms use proteins in energy-transducing structures for obtaining the energy to make the proteins needed for energy transduction. The uroboros puzzle of the transition of no-life to life is: how could this have begun?[8]

Major biological organic molecules of life are large *polymers* composed of *monomers:* 1) proteins consist of many thousands of amino acids; 2) complex sugars consist of many simple sugars joined together, e.g., glycogen in animals and starch in plants are made up of glucose subunits; 3) nucleic acids consist of nucleotides joined together.

Uroboros Puzzle Restated.   Organisms have complex structures stemming from bonding among polymers.

. . . what organisms must do to achieve these structures involves energy. The energy requirement exists because biological polymers are surrounded by water, which tends to degrade them by hydrolysis. Organisms acquire this energy through the catalytic activity of enzymes, special proteins that the organisms manufacture from instructions coded in genes (DNA). This is the essence of the uroboros problem: to make polymers, polymers are needed. Alternatively, this may be expressed: to make the energy needed for synthesis of polymers, energy for synthesis is needed. The transition from monomers to polymers poses the first real difficulty.[9]

The environment of the earth before the emergence of life thermodynamically favored the formation of water as well as siliceous crystal rock (energy from the sun plus a high chemical potential [large negative free energy of formation] drives these reactions). In this environment polymers that form by chance tend to immediately degrade by *hydrolysis* into its constituent monomers. We know that living cells are able to take in and use chemical or light energy to activate monomers to form polymers more rapidly than hydrolysis degrades them; so polymers accumulate over time. Likewise, any *primordial form of life* must be able to do the same thing, but the problem is that already existing polymers are needed to activate monomers to form more polymers. Somehow what must emerge is a primitive Uroboric polymer system, i.e., a "self-begetting" polymer system.

Summary of Uroboric Puzzles

1. A biochemical version.
   a. in the non-living world, polymers that spontaneously form also break down more rapidly than they are formed

---

[7]Ibid., pp. 20–21.

[8]Ibid., pp. 3 and 5.

[9]Ibid., pp. 34–35.

    b. a living system consists of polymers that are "self-sustaining," i.e., the polymers capture energy from the environment and use this energy to make new polymers more rapidly than the spontaneous breakdown of newly formed polymers

    c. how can "non-self-sustaining" polymers become "self-sustaining" polymers; i.e., how can a living system evolve from the non-living world?

2. Two general ways of stating the Uroboric Puzzle with respect to the emergence of life:

    a. If any living organism has properties that a non-living system does not have, then how can a primitive living system evolve from non-living matter?

    b. Life is a self-generating and self-sustaining system that evolved from non-self-generating and non-self-sustaining systems. How can this occur?

3. Two human analogies for a Uroboric Puzzle:

    a. *Learning to ride a bike.*

      1) In order to ride a bike, one must have a "feel" for riding a bike.

      2) In order to have a "feel" for riding a bike, one must get it by riding a bike.

    b. *Learning to solve a type of mathematical problem*

      1) In order to solve a type of mathematical problem, one must have a "feel" for solving these kinds of problems

      2) In order to get a "feel" for solving these kinds of problems, one must experience solving these kinds of problems.

General Solution to *The Uroboric Puzzle:* Emergent Properties

Random events of energy available for bonding between monomers that by chance "bump" into one another leads to the beginning of monomers. Since hydrolysis is fairly slow, it is possible that a particular chain may grow to having several linked monomers. If such a chain were to have the new property of facilitating more monomers to be added, then the polymer would continue to grow faster than it decomposed into monomers; i.e., the polymer would be a "self-begetting macromolecule" because of its *emergent property* of catalyzing its own formation. With respect to this idea, R. Fox makes a remark that betrays his leanings toward reductionism:

> *Emergent properties* of systems can be predicted only with great difficulty from knowledge of the properties of a system's components. Thus the properties of macromolecules are difficult to predict from the behaviors of monomers or oligomers.[10]

The fact of the matter is that quantum mechanics and chemistry itself defeats the reductionism of classical mechanistic physics. We cannot at all predict the properties of hydrogen atoms from our knowledge of electrons and protons, and what is more, quantum mechanics dictates that we cannot even picture in our imagination an electron or a proton. Likewise we cannot at all predict the properties of a water molecule from our knowledge of hydrogen and oxygen atoms. Of course, macromolecules will have emergent properties that cannot be predicted from knowledge of their constituents.

    Any theory for the origin of life must propose the spontaneous formation of a complex unit, such as a macromolecule, that has emergent properties that enable it to form a higher level complex unit, for example, a population of protein-like molecules called proteinoids that have the emergent property of forming into microspheres. The microspheres, in turn, must be able by chance to spontaneously form a still higher level complex system, and this progressive formation of higher level systems must continue until one reaches a system with the minimum set of properties that causes it to be a primitive form of life. Fox describes the plausible emergence of such a system as follows:

> Once microspheres existed that were capable of making RNA and then small proteins on this RNA, a particular RNA sequence could, by chance, code for a short protein that would catalyze RNA copying (that is, would cause nucleotide polymerization on an existing RNA molecule). The second

---

[10]Ibid., p. 43.

copy of this RNA sequence would reproduce the original RNA. In this way, the model RNA could also act as a reproducible gene. This development could result in a system that is self-generating—the initiation of uroboros. This model thus provides a plausible scheme for the transition from no-life to life.[11]

Fox then goes on to further emphasize the importance of the progressive appearance of emergent properties:

> As evolution proceeded, slowly establishing an increasingly faithful genetic apparatus, the primordial uroboros would develop the ability to produce amino acid sequences that performed catalytic functions, which were needed for developing and refining metabolism and polymerization. The genetic apparatus seems to have made possible the *diversification of sequences* for diverse functions; it has not selected a limited number of sequences out of an astronomical number of combinatorial possibilities.
>
> What sets evolution, as a process, apart from other energy-driven processes are the very special properties of polymers, not just the energy input. The feedback effects of catalysis and regulation by proteins and of genetic memory of polynucleotides enable the energy flows to evolve complexity [organization superimposed onto this complexity]. A system without these emergent properties of biopolymers, even if it is energy-driven, will not have evolutionary capacity.[12]

Characteristics of Emergent Properties

1. These properties only can be described in terms of the whole higher order pattern
2. These properties cannot be understood in terms of (reduced to) interactions among components; i.e., the properties cannot be understood in a gross, mechanistic way.
3. These properties cannot (or could not) be predicted to occur from knowing the components of the new pattern.
4. The properties are the expression of imminent potentials of a set of apparently autonomous entities that interact in a *context* that actualizes the imminent potentials. One can know the existence of these imminent potentials *only* as a result of knowing their manifestation that shows up as emergent properties. One can know the context for the expression of these imminent potentials (and thus count on their expression when the appropriate context is set up, as, for example, we do in chemistry; i.e., put a mixture of hydrogen and oxygen gases at high pressure and high temperature. This context leads to the expression of the imminent potential in this mixture which is water molecules). But, one never can know the "how-why" the context leads to their expression. Knowing the *how-why* implies a mechanistic description, and the expression of emergent properties is a non-mechanistic, and indeed, a non-rational process.

Systems Theory of Creativity

The previous section described the apparent dichotomy between non-Life and Life in terms of the Uroboros—a snake eating its own tail. An ordinary snake is a linear creature with its head as its beginning and its tail as its ending, but a uroboros snake is paradoxical because its *beginning* is identical with its *ending*. In like manner, a seemingly common sense understanding of life on earth is that any living organism was begotten from another living organism; *life is self-begetting*. Another version of this uroboric cycle is that a chicken is an animal that develops from an egg, but a chicken only produces a chicken egg. How can one ever get a chicken without first having an egg and how can one ever get an egg without first having a chicken? One possible solution is that somehow either a chicken or an egg emerged from non-life; or more generally the chicken evolved from less complex life forms going all the way back to some primordial life form that evolved from non-life. This is not a

---

[11]Ibid., pp. 69–70.

[12]Ibid., p. 76.

logical paradox but it is an enigma. The beginning of a solution comes when we realize that all living organisms continuously use nonliving matter to maintain themselves alive and sometimes to produce other living organisms. Thus, life necessarily implies non-life matter, but non-life matter does not imply living organisms. Destroy all living organisms on earth; matter still remains and presumably will continue to do so until the end of time. Conversely, destroy all matter; all life also will disappear. Thus, the very presence of life on earth implies that there is a two-level organizational hierarchy in the biosphere: Life is the higher organization level that includes matter. Matter is the lower organizational level because it does not include Life but is capable of being incorporated into — used by—Life.

The *Big Bang* may be said to have initiated the differentiation of a material universe that at least on earth if not elsewhere further differentiated into a *Biosphere*. This implies that, at the very least, there came into being an undifferentiated "something" that first differentiated into a molecular network of interactions and then further differentiated into a life network of interactions. The life network incorporated the molecular network. Thus, non-life vs life is not an absolute logical dichotomy like non-existence vs existence. In order to avoid ultimate paradox, we have to make a choice between two options. The first option is that either matter (non-life) is a variation of a pattern we call life; or conversely, life is a variation of a pattern called matter. The second option is that matter and life are different forms—neither form can be reduced to some variation of the other—and yet matter and life are related as stages in a developmental-evolutionary process. Matter is the first stage that differentiates to a degree of complexity that then *transforms* to life, the second stage that includes the differentiated aspects of the first stage. This two-stage process is *developmental* because like an acorn becoming a tree or an egg becoming a chicken, it involves a progressive differentiation toward a goal. The two-stage process also is *evolutionary* meaning *creating* something new rather than merely actualizing a potential via differentiation. It is *evolutionary* because it involves a *transformation* from one stage to another which then via a developmental-evolutionary process "leads to" many diverse forms of life as a result of diverse transformations. Hereafter I will refer to developmental-evolution as (normal or natural) *holarchy*, a term which Wilber took from Koestler (in his book, *Ghost in the machine*). Wilber describes it as:

> . . . the sequential or stage-like unfolding of larger networks of increasing wholeness, with the larger or wider wholes being able to exert influence over the lower-order wholes.[13]

The systems theory of creativity stems from considering how this second stage, Life networks, emerged from the first stage. This is a paradox that may be stated in the form of a question. How can a system that is self-generating and self-sustaining evolve from—emerge from—a system that is non-self-generating and non-self-sustaining? The circular nature of this uroboric paradox is represented by the analogous question: which came first, the chicken or the egg? A chicken egg only comes from a chicken, but a chicken only comes from a chicken egg. A human analogy of such a circular paradox is a human learning how to drive a car. In order to drive a car, one first must have a "feel" for driving a car, but in order to have a "feel" for driving a car, one must get it by driving a car.

The systems solution to these types of paradoxes is a theory of creativity involving rejecting *gross reductionism* and proposing the interconnected ideas of random events, "self-organization," and emergent properties.

Rejection of Gross Reductionism

The new physics of the Enlightenment began with the mathematical description of objects moving in space, a process considered to be distinct from "life processes." A few scientists chose the first of the two options described earlier (that life is a variation of a pattern called non-life, that is, life is reduced to non-life) and created the first form of gross reductionism spawned by the Enlightenment. This formulation assumed that all things-events in nature were determined by the interactions among molecules and atoms which, in turn, were governed by the laws of Newtonian mechanics grounded on the idea of force = mass × acceleration. For this and the next form of gross reductionism, the laws of

---

[13]Ken Wilber. *Sex, Ecology, Spirituality The Spirit of Evolution.* (Boston: Shambhala, 1995) p. 22.

thermodynamics (especially the second law) were regarded as practical guidelines for building machines, but the laws did not help one understand ultimate reality. Then, Einstein's theories of relativity superceded Newtonian mechanics, and the ground for understanding ultimate reality was the idea of *field* described by Maxwell's and Einstein's field equations in relation to energy contained in fields. The formulation of modern quantum mechanics in the 1920s destroyed the basis of mechanistic reductionism in physics, but a new form of gross reductionism arose in biology.

Before the proposal in 1953 of the model of DNA structure (which implied how DNA could store and duplicate genetic information, see Chapter 6),

> . . . both biology and the social sciences . . . [had] been dominated by open or concealed vitalistic ideas, maintaining that there were particular laws governing biological or societal processes and that these laws were different in principle from the laws of physics and chemistry. Because of the many apparent weaknesses of vitalistic ideas, after a long and bitter fight a reductionist view . . . gained supremacy, at least in biology. According to this reductionist philosophy, all phenomena of biological or social systems can be explained by unified principles and ultimately reduced to chemical mechanisms in spite of their obvious hierarchical organization. The spectacular boom of molecular biology was regarded as proof of the reductionist view.
>
> Without diminishing the achievements of molecular biology, this proof can be questioned. . . .
>
> The antireductionist viewpoint was articulated most clearly by Polanyi (M. Polanyi, 1968: "life's Irreducible Structure." *Science* 160: 1308–1312), who wrote that "Mechanisms, whether man-made or morphological, are boundary conditions harnessing the laws of inanimate nature, being themselves irreducible to those laws. The pattern of organic bases in DNA, which functions as a genetic code, is a boundary condition irreducible to physics and chemistry. Further controlling principles of life may be presented as a hierarchy of boundary conditions extending, in the case of man, to consciousness and responsibility." . . .
>
> In my opinion, the interrelationships of components within the various organizational levels, which we may call *algorithms,* are specific only at a given level. In a general model, in which all organizational levels are embodied, all effects computed by the algorithms of the lower levels lead to random events in the event-space of the higher levels. The algorithms of the higher levels represent specific nonlogical constraints at the lower levels. This is clearly not a vitalistic standpoint, but from it follows the conclusion that models of the biological and societal systems are different in principle from models constructed for understanding the physical world.[14]

Csanyi then goes on to show how the second law of thermodynamics describes the boundary conditions for the operation of any machine, but the law does not give a set of directions, for example, an algorithm, for building the most efficient machine. As I described in Chapter 2, it is true that in comparing any two machines operating under identical conditions, the one with the lower entropy will be more efficient, and entropy is a measure of lack of complexity—the lower the entropy the higher the complexity. In general, complexity does not imply organization, but for machines, complexity does include organization so that the more highly organized machine also is more efficient. However, the quantitatively defined entropy concept does not give one an insight into the quality of the complexity. For example, entropy decrease does mean an increase in complexity, but one has no way of telling whether this increase is due to greater diversity of sub-units or some nonuniform distribution of sub-units or whether it also includes some organization pattern superimposed onto the collection of sub-units.

Csanyi describes the situation this way:

> . . . we cannot use these equations [derived from the two laws of thermodynamics] to predict the convertible useful portion of the total free energy of combustion, aside from the thermodynamic boundary as a maximal limit. We cannot use these equations or any other physical laws for designing a final perfect engine.
>
> The engine's structure, the shape and size of its parts and their arrangement [organization pattern], all are *special boundary conditions* that represent *non-holonomic constraints* on the

---

level of molecular kinetics (H. H. Pattee, 1977: "Dynamic and Linguistic Modes of Complex Systems," *Int. J. Gen. Syst.* 3: 259–266). Therefore, the combustion engine's design is an empirical science. Various designs are tried, and those that work best are used. We cannot calculate the perfect design from the kinetic equations. An engine's structure is in principle the *description* of its special constraints.

The space of the combustion reactions is bounded with metal walls, of which one (the piston) is movable and serves to transmit power [it is an energy coupler]. But the description of this structure is not based on the kinetics of the molecules; it is not reducible to kinetics. The combustion engine must be described on two different levels. The lower level includes the kinetic equations of gas molecules [the non-life matter], while the upper level includes the description of the specific constraints that originate from the engine's design [which is in the Mind of the human designer and in relation to this section is the life]. The working of the engine can be understood only by having both descriptions.

It is the same with a cell or a higher organism. At the molecular level these systems can be described by physical and chemical equations. But these equations do not show the constraints existing on higher levels; they do not show which *structures* constitute the boundary conditions for those physical and chemical laws.

This, of course, is not a denial of the causality principle—that the phenomena of the higher levels can be traced down to the lower levels as a *causal chain of events*. But an explanation based merely on causation is not comprehensive because it fails to describe the specific constraints existing on the higher level.[15]

Summary of Systems Science Theory of Creativity

From time to time the Order in the universe is disturbed by random events that produce local Chaos. As a result of one Order being destroyed, new structures become possible; that is, many new potential structures arise and conflict with one another. Sometimes the context of these conflicts is appropriate for some of these potentials to "self-organize" into a semi-stable new system. If this new system is "protected by circumstances" long enough to stabilize, it becomes the "rebirth from the death of the old order" and exhibits emergent properties. Besides not being present in the old order, these new properties cannot be predicted to occur from only knowing all about the old system. For example, under the right circumstances hydrogen and oxygen gases chemically combine to form water that has emergent properties. We cannot predict or deduce the presence of these properties of water from knowing the properties of hydrogen and oxygen gases.

Systems theorists reject gross reductionism such as that of molecular biologists, for example, Francis Crick[16] that claim to "explain" life and the human mind in terms of molecular interactions. Mechanistic science claims that physical laws that regulate the interactions among its autonomous parts determine the behavior of any system. In contrast, systems science is holistic.

. . . "everything is connected with everything else" describes a true state of affairs. . . . Scientific evidence of the patterns traced by evolution in the physical universe, in the living world, and even in the world of [human] history is . . . coalescing into the image of basic regularities that repeat and recur.[17]

Systems science describes "creative evolution" as a sequence of transformations each from a lower level to a higher level of organization.

The concept of organization level is understood here in the sense of a Chinese-box structural hierarchy of "boxes within boxes." On a given level of organization, systems on the lower level function as subsystems; on the next higher level of organization, systems jointly form suprasystems. Using this concept, we can readily appreciate that the products of evolution are distributed

[15]Ibid., pp. 10–11.

[16]Crick, F. (1995). *The astonishing hypothesis: the scientific search for the soul.* NY: Simon & Schuster.

[17]Laszlo, E. (1987) *Evolution. The grand synthesis.* Boston: Shambhala Pub., Inc., p. 5.

on multiple hierarchical levels. Several particles jointly constitute atomic nuclei, and nuclei surrounded by electron shells form the atoms of elements. Several atoms form simple chemical molecules, and more complex polymers are built from simpler molecules. Cells, in turn, are built from various kinds of macromolecules, organisms from cells, and ecologies and societies from populations and groups of individual organisms (Laszlo, 1987, p. 24).

Subtle Reductionism of Systems Science

Reduction to Autonomous, Potential, New Patterns.   While rejecting the all inclusiveness of mechanistic analysis, systems science proposes a subtle form of reductionism. Though the universe consists of interconnected systems, systems theorists reduce any individual system to a particular stable pattern plus several autonomous potentials to transform to new stable patterns. Any particular new pattern results from the partial breakup of the old pattern thus releasing components to reform into a new pattern or, more commonly, to interact with other systems to form a new pattern. An analogous example would be when two people divorce: each is released to find a new mate or join a commune or develop a new life centered on being single.

Laszlo describes these transformations to new patterns in the context of forming progressively higher levels of organization. When systems are in the right circumstances (i.e., third-state systems far from equilibrium with a high flux of energy passing through them), these systems

. . . can evolve through sequences of destabilizations and phases of chaos since they have multiple states of stability; when one steady state is fatally disturbed, other steady-state solutions remain accessible. . . .

The selection from among the set of dynamically functional alternative steady states is not predetermined. The new state is decided neither by initial conditions in the system nor by changes in the critical values of environmental parameters; when the dynamic system is fundamentally destabilized it acts indeterminately [thus, rejection of gross reductionism]. By an apparently random choice one among its possible numerous [autonomous!] internal fluctuations is amplified, then the fluctuation spreads with great rapidity [thus subtle reductionism because he assumes autonomous parts, i.e., autonomous fluctuations or potentials]. The thus "nucleated" fluctuation dominates the system's dynamic regime and determines the nature of its new steady state [the system in chaos "self-organizes" itself into a new pattern]. . . .

We cannot predict the precise evolutionary trajectory chosen by any system in the third state; all we can say is that the new steady state, if it is dynamically stable, assimilates the disturbance introduced by the destabilizing parameters within an open structure that is likely to be more dynamic and complex than the structure in the previous steady state.

It is through phases of chance and indeterminacy that evolution tends in the observed general direction of successive levels of organization, with growing dynamism and increasing complexity on each of the organizational levels (Laszlo, 1987, p. 36).

Tearout #1

Directions:

1. In the same form as represented in the text for learning to ride a bike, state an analogy of a Uroboric Puzzle with respect to learning *how to relate to other people in terms of feeling communications.*

2. Describe the solution to the above Human Uroboric Puzzle in terms of 1) trials to relate to people at a feeling level; 2) getting a little bit of "feel" for how to do this with each trial, and 3) the *emergence* of this new ability after many, e.g., 100, trials.

<u>Tearout #2</u>

Directions:

1. Describe what it means to say that a systems theory of creativity rejects *gross reductionism.*

2. Summarize the systems science theory of creativity in relation to the chemical reaction between O_2 gas and H_2 gas to produce water. (See figure 1.14 in Chapter 1).

The Four Creative Learning Styles of Scientific Thinking

Preparation for Scientific Thinking

Primitive Humans Were Embedded in Nature

Individual ancestor apes have an episodic awareness of other apes or other aspects of nature. The *episode* of being totally engaged with, that is, totally focused on, some object of awareness produces the episodic awareness. There is no awareness of a "self" being so engaged; rather the animal is totally embedded in the act of being aware of some object. As a result of this total embeddedness, there is no realization that the perceived object is only one of many parts of an outside world called *reality*. There only are objects of animal awareness. The lack of awareness of an objective "outside world" implies a lack of an inside world that generates a subjective self. Primitive human societies and 2-year-old children develop a "proto-subjective" self in that they are able to acknowledge their own feelings as distinct from those of others. Therefore, they are aware of their participatory embeddedness in a tribe or in a family, and at the same time they have a vague awareness of nature and of any other tribe or human as *other*. Anthropologists and developmental psychologists refer to this proto-subjective, participatory awareness as *participation mystique*. The "other" is ambiguous in that humans have a participatory affinity with it, and yet, they are vaguely aware of the actions they must take to keep this "other" from enticing them back to the total embeddedness in nature that characterizes ancestor apes. Eros-chaos (see end of Chapter 1) drives them to participatory affinity with the "other," and Eros-order drives them to separate from the "other" and exert some magical control over it.

Discovering One's Soul

The emergence of Mind, as occurs in the 4½-year-old child, enables humans to be aware of an inner subjective self that is episodically engaged with the "other." This awareness of an inner self is sometimes referred to as a human becoming aware of the source of its self-consciousness which is the *soul*. Now the ambiguity of the "other" is transformed into the ambiguity of the newly emerged Mind. The Mind is drawn outward toward engagement with the "other," and is drawn inward toward an inner, the *soul,* that generates its feeling engagements. The individual no longer always totally identifies with each feeling engagement it has. Rather the Mind turned inward becomes a *subjective ego* that, from time-to-time, "sees" that it *has* feelings and therefore can choose to express them or not express them. At the same time, the subjective ego is able to represent its outward engagements to itself. These imaginative representations result from developing a proto-conceptual language that spontaneously overflows into artistic expressions.

The subjective ego "sees" itself in different polar relations with the "other"; so it "knows" the "other" as an aspect of its relation to it. As a result, it sees itself as the center of all its outside engagements with "others." The mental ego does not yet have an external aspect. The ego does not "see" any particular "other" as a "part" of some grand, larger "Other." Thus, the subjective ego's knowledge of the "Other" is not of an object "out there" with a separate existence and characteristics that make it different from the knowing ego. As a result, humans at this stage of individuation can to some extent get along with members of the same tribe or of the same family by controlling the expression of their feelings. But these humans cannot imagine themselves as "parts" of a larger whole called society. To acknowledge oneself as a member of a society larger than the tribe or family, the human would have to give up its subjective egocentricity. The human would have to "die" to its imaginative, anthropomorphic way of knowing and be reborn to a new way of knowing that acknowledges the concepts of whole and parts.

However, with the further development of proto-language the subjective ego does recognize *different* polar relations vs. similar polar relations. The ideas of different vs. similar implies the conceptual categories of *same* vs *different.* This subjective ego's recognition of different vs. similar is a more differentiated, self-conscious recognition of diverse episodic feeling events. This suggests that the subjective ego is evolving toward a "proto-conceptual knowing" of whole vs. part and of same vs. different. Moreover, subjective egos have an anthropomorphic understanding of time and events. These humans tell stories in which there is an interconnected sequence of events. There is no conceptual understanding of individual events occurring in time. The vague sense of duration and sequential interconnection among episodic happenings implies that this *story knowing,* also called *mythical knowing,* is an individuation toward conceptual understanding of time, events and causation that connects one event with another. Thus, the subjective ego's anthropomorphic, proto-conceptual, polar knowing of "others" and the "Other" via myths or stories may be designated as *participatory, embedded knowing.* The transition to the next higher level of self-consciousness involves creating concepts to represent categories.

Becoming Civilized

Individuation from proto-conceptual, participatory embedded knowing to non-embedded, conceptual knowing is a mystery to each of us. In order to conceptually represent to ourselves what is going on while this individuation occurred in each of us, we needed to already have conceptual knowing. Not yet having this type of knowing, we could not realize what was happening while we were individuating toward it. But now as a result of having logical, conceptual knowing, the ego of each of us can separate itself from any anthropomorphic knowing and analyze itself as an object analogous to any object in nature. From this higher vantage point, it is possible to imagine the emergence of conceptual knowing corresponding to the first emergence of conceptual language.

Non-civilized humans experience aspects of the world episodically. These experiences are in some way represented and stored in the "knower." They are existential perceptions that can be associated with an arbitrary symbol, for example, a word. Individuation to Mind with a subjective ego corresponds to the "knower" "seeing" similarities among several perceptions which then become a domain of similarities in the mind. Individuation to the persona self in a civilized society enables the "knower" to conceptualize these similarities, which is to say that the knower creates and stores a pattern called a *concept.* Thereafter, each existential perception that relates to the same domain of experiences will be "seen" as exhibiting this pattern. Correspondingly, all these existential perceptions will be seen as identical. When this occurs, the knower sees each experience as an instance of a concept.

The individual, existential uniqueness of each experience is sacrificed for the sake of seeing patterns; i.e., conceptual knowing. At this point conceptual knowing still is a subjective process occurring within the individual knower.

However, growing up in a civilized society always leads to some of these subjective concepts being associated with a word in an artificial symbolic language. Many humans may have similar subjective concepts. The consensually agreed upon word associated with these similar, subjective concepts represents all of them as if they were identical thus producing *objective, conceptual, language knowing* of the world. This objective knowing within the context of civilized society leads each knower to see the world as a "whole" made up of relatively autonomous "parts." The parts are things or events that are conceptually known. This type of knowing conceptualizes concrete, physical experiences and produces sentences such as "I ate the apple," and "the book is on the table." Ideas in these types of sentences are called *literal concepts* because they are understood to be objective representations of the world as a whole made up of relatively autonomous parts that exist and interact with one another independent of any human knowing them. However, any concept emerges from self-conscious, episodic experiences each of which includes a subject-object polar interaction. These polar interactions are more real than the conceptual knowing derived from them. Put in another way, conceptual knowing produces the illusion that there is a world "out there" with a structure that is defined by interactions among the parts making up the world. This structure supposedly exists independent of humans and can be known by humans. This knowing process produces valid conceptual, language knowledge. This illusion is *valid* because it is necessary for one's day-to-day functioning in civilized society, but the conceptual knowing that supports this illusion is derived from human's more fundamental polar knowing of world. This polar knowing indicates that we do not know the "objective world in itself," but rather we conceptualize our polar knowing interactions with it. (This insight about the way humans know is a core insight of one interpretation of quantum mechanics that includes Heisenberg's uncertainty principle and a resolution of the particle-wave paradox.)

This first phase of the emergence of an *objective ego*—so named because it now understands the world by means of objective, conceptual, language knowing—is what Piaget calls *concrete operational thinking*. The drawings of my son, Andy, illustrate the transition from *participatory embedded knowing*—also called *non-conceptual mythical knowing*—to *conceptual mythical knowing*, which Piaget calls *concrete operational thinking*. Figs. 3.1a and 3.1b represent Andy's non-conceptual mythical thinking about his subjective world of experiences. Figure 3.1c is Andy's portrayal of his oldest brother, Don, who used to practice on the piano 5–6 hours each day. Fig. 3.1d includes a drawing of a "bird-man" who represents Matt, my third son, with the silly grin he used to have before he totally rejected this clown aspect of his early childhood. Figures 3.1e and 3.1f are drawings that Andy created several months later. They represent Andy's newly emerged conceptual, mythical understanding of the world corresponding to beginning to become civilized at the age of 7 years old. Now he could begin to understand *persona* expectations and the meaning of rules of behavior. He, therefore, was scolded or punished if he did not fulfill these expectations or if he did not obey these rules.

Metaphorical, Conceptual Knowing

The next stage of individuation of human knowing and of everyday language corresponding to that type of knowing is *metaphorical, conceptual knowing*. The evolution of classical Greek culture beginning with Homer in 700 B.C. and ending with the Presocratic sophists, e.g., Gorgias (485–380 B.C.) nicely illustrates the gradual emergence of this kind of knowing. I describe this evolution of human consciousness in the book, *The spiritual constructivism basis for postmodern democracy* (in press, winter 2005, Kendall/Hunt Pub. Co.) which I use for my course, Biology 1140, "The Biological Aspects of Human Consciousness." What follows only is a brief summary of this process. The non-Greek civilizations in the Middle East had the unacknowledged ambiguity of a mental ego with an external "face" and an internal "face." The *external ego* "saw" society as like a machine in which each autonomous part interacts with all the other parts in such a way that the machine accomplishes its tasks. Each civilized human has several *personas* as defined by the literally true myth accepted by a

FIGURE 3.1

particular society. The personas and codes of behavior define how each civilized person must interact with all other persons in his/her society. The *internal ego* looks inward to its *imaginative* representations of events, celebrates its participatory, embedded engagements with nature and other humans, and is spontaneously creative. For the sake of civilized, social harmony, all individuals are socialized to embrace the literal truth of a grand, overarching myth and its prescriptions. As a result, the external ego partially dominates the internal ego. The ambiguity of the mental subjective ego that is present with the first emergence of mind is transferred to a mind self characterized by a conflict between a dominating external ego and a rebellious internal ego.

Beginning in 700 B.C. some aristocratic Greeks moved away from accepting as literally true the Olympian gods and goddesses described by Homer's epic poems. This gradual humanistic individuation culminated in the Presocratic philosophers who constructed *metaphorical, conceptual stories* that attempted to describe humans in nature independent of their socially inherited, literally true mythologies. For the first time in recorded Western history, humans became the measure of all things. In this emerging humanistic view, Homer's Olympian gods and goddesses do not literally exist and determine the fate of humans. Rather these mythological, human-like beings represent metaphorical conceptualizations of powers and forces that humans experience in themselves.

According to the modern theory of metaphor as summarized by George Lakoff ("The Contemporary Theory of Metaphor" in Metaphor and Thought (2nd ed.), Ortony, A. (ed.), Cambridge: Cambridge University Press, 1992), metaphor emerges after the emergence of conceptual knowing and then profoundly enhances the power of ordinary language. Contrary to classical theories of language, metaphor is not a type of linguistic expression; rather it is a higher level of knowing than elementary conceptual knowing. George Lakoff defines metaphor as a general mapping across conceptual domains. That is, metaphor is a higher level of conceptual knowing in which one conceptualizes one mental domain of experience in terms of a different mental domain of experience that is "seen" to have in some ways the same internal structure.

Metaphorical knowing is the process of "seeing" similarities between two domains of experience, each represented by a concept. The knower starts out seeing that the two concepts, i.e., the two patterns, are quite different. Metaphorical knowing emerges when one sees similarities between the two patterns and then formulates a new higher level concept, the *metaphorical concept,* that represents the similarities between the "lower level concepts." The metaphorical concept then leads to new insights in the following way. While one sees similarities between concepts A and B, one may know a lot about A and not much about B. The metaphorical concept is the pattern: some structure relationship in A may be thought of as the same as some structure relationship in B. As a result of the metaphorical concept, one understands aspects of B in terms of what one knows about A. Lakoff formulates this idea as: *the structure of A is mapped onto B, where A is the source domain and B is the target domain.* However, this mapping is not an arbitrary, mathematical or prepositional mapping. Rather the mapping is the core of communication between some person, George, who already sees the similarities between A and B and another person, Don, who knows A but does not know much about B. By means of the metaphorical concept George gets Don to understand B in a new way, that is, he now understands B in terms of what he already knows about A.

For example, a core idea of modern biology is molecular communication, such as, hormone action on a cell that responds to the hormone in a particular way. One instance of this is the hormone, adrenalin, stimulating some muscle cells to contract. How can anyone visualize molecular communication when no one has ever seen a molecule let alone molecules communicating with each other? One can do so by means of the metaphorical concept, key-lock interaction. A, representing key-lock interaction is mapped onto B, representing hormone-cell interaction. A key is a source pattern that fits into an appropriate lock having a complementary receptor pattern; the key pattern is like one piece of a puzzle fitting into a complementary piece of a puzzle which is like the lock receptor pattern. When the key interacts with the lock, such as when the key fitting into the lock and then being turned to the right, something happens, e.g., the key-lock interaction in a car leads to the engine starting. In like manner, a hormone is like a key that fits into a lock which is like a receptor site in the membrane of a cell. Hormone-receptor site interactions in some muscle cells lead to muscle contraction. Lakoff would represent this metaphorical concept as: HORMONE-CELL INTERACTION is KEY-LOCK INTERAC-TION. It turns out that this type of metaphorical concept is valid for many areas of modern biology, for example, nerve cell interactions (neural transmissions), virus host interactions, and immunological interactions.

This metaphor involving key-lock interactions also applies to human communication and in particular to teaching. Using Lakoff's notation I propose that: TEACHING-LEARNING is KEY-LOCK INTERACTION. KEY representing conceptual information sent out by a teacher fits into and further interacts with a LOCK representing the conceptual domain of a learner that can recognize incoming information. When the learner "processes" the interaction between incoming information and the receptive mental domain, there emerges a new insight. This creative process is fundamentally mysterious, but nevertheless, the metaphorical concept does tell us something about what is going on. If the learner does not have the appropriate receptive mental domain, no new insight will emerge. Conceptual knowing becomes dynamic and potentially creative when one understands it as a mutuality between subjective, experiential knowing and objective, conceptual, language representation of experiential knowing. A dynamic concept involving this mutuality is malleable. It can "deform" its pattern so that it can interact with other seemingly contrary ideas to form a new, more powerful concept. However, if the concept is thought of as a rigid, totally objectively true or valid representation of an unchanging structure of the world, then it cannot "adapt" to incoming contrary concepts. The conceptual receptor of the mind will be like a lock that is not complementary to and not receptive to a key representing incoming new ideas. The learner with totally objective true or valid conceptual knowing only can accept compatible incoming ideas. In this situation the mind is like an incomplete puzzle. The only ideas that are received and processed are those that fit into the empty spaces. The present mental pattern is enhanced but not fundamentally changed. Thus, such a learner cannot create new insights.

The Presocratic "poetic philosophers" replaced the mythological beings of Homer's epic poems with more abstract, non-anthropological ideas such as Parmenides' metaphorical concept of *Being* and Heraclitus's metaphorical concept of *Logos*. However, these abstract metaphorical concepts are ambiguous. These ideas of Being and Logos point to an aspect of the external world that is eternal, non-changing contrasted with a human's subjective experiences of continuous change in nature and in himself—he/she grows old and dies. Thus, the ambiguity of the mental self looking outward as an *external, objective ego* and looking inward as an *internal, subjective ego* is transferred to the ambiguity of metaphorical, conceptual knowing.

The Presocratic stories allowed diverse interpretations including blatant contradictions such as all is eternal being versus all is continuous change. There was no way of eliminating these ambiguities or resolving the contradictions. The later Presocratic philosophers became rhetoricians in which the "truth" of any philosophical explanation depended on one's subjective power of persuasion. All "truths' became relative. This led to total skepticism that undermined the survival of the Greek city-state of Athens as it would any civilization. Athens was conquered by Alexander the Great during the later years of Aristotle (384–322 B.C.).

Individuation to Logical, Conceptual Thinking

Socrates (469–399 B.C.), Plato (427–347 B.C.), and Aristotle (384–322 B.C.) converted *metaphorical, conceptual thinking* into *logical, conceptual thinking*. Socrates's method of dialogue distilled metaphorical, concepts into pure logical concepts. Plato created the hierarchal chain of pure, eternal forms terminating in the ONE, metaphorically equivalent to the Hindu idea of Brahman or to the Judaeo-Christian-Islamic idea of a creator God. According to Plato, the ONE de-evolved down into the many that then evolve "upward" to the ONE. Aristotle invented the philosophical study of logical thinking. These heroes of classical Greek thought totally rejected conceptual, mythical thinking of the "childhood of civilization" (The Age of Myth which extended from 3300 to 750 B.C.), and the metaphorical, conceptual thinking of Presocratic philosophy. The ambiguity of metaphorical, conceptual knowing had evolved to a radical dichotomy within each autonomous, logical thinking person. In classical Greek culture virtually all of them were patriarchal, masculine males. Such individuals endure the Mind conflict of the external ego associated with control objectivity identified with masculine males that dominates the internal ego associated with participatory subjectivity identified with feminine females. Stereotypically, many aristocratic, classical Greek males idealized men and degraded women.

Mutuality between Metaphorical, Conceptual Thinking and Logical, Conceptual Thinking.   In the culture of Socrates-Plato-Aristotle, only a few Greek aristocrats individuated to logical, conceptual thinking. This classical Greek culture died and resurrected to the more inclusive Hellenic culture of the Roman Empire. In my book, *The Constructivism Basis for Postmodern Democracy*, I argue that the story of Jesus, the Christ, inspired individuation to a major new level of self-consciousness in the Western cultures and later in the Middle Eastern cultures. The conceptual mythology of the Old Testament was converted into the metaphorical, conceptual stories of the integrated "Old and New Testaments," the Catholic Bible. This provided the foundation for the emergence of the Catholic Church. The priest hierarchy of this church originally influenced by Plato [neo-Platonism adopted by St. Augustine of Hippo (354–430 A.D.)] and later by Aristotle [Aristotelian philosophy of nature adopted by St. Thomas Aquinas (1225?–1274 A.D.] totally rejected conceptual mythology, which we now call paganism. The followers of Jesus claimed that this human simultaneously also is the Son of God-the Father. The Jesus cult believed that Jesus died and resurrected as the Christ who then sent the Holy Spirit, the third person of the triune God, to guide humans toward salvation. These pronouncements resurrected the metaphorical, conceptual knowing rejected by Socrates. How else could the followers of Jesus make such claims about a man they experienced in this life on earth? The several death and resurrection myths of paganism were stories to describe generic human experiences; they were not descriptions of directly experienced historical events. Beginning with St. Paul of Tarsus, who was a Pharisee Jew and a Roman citizen familiar with Greek philosophy, the emerging hierarchy of the Catholic Church progressively constructed a logical, conceptual representation of "revealed truths" that only could be stated by means of metaphorical concepts.

Thus, the emergence of Catholic Christianity brought about the beginning of a mutuality between metaphorical, conceptual thinking and logical, conceptual thinking. The Christian metaphorical, conceptual thinking is similar to the subjectivity of Presocratic philosophy which was ambiguous and allowed for diverse interpretations. Was Jesus a man who lived and died or was he the eternal Son of God who only gave the appearance of being a man or was he similar to Heraclitus's idea of Logos implying both continuous change and eternal being; i.e., was Jesus simultaneously a man subject to change and death and eternal Son of God-the Father?

Regression to Exclusive Logical, Conceptual Thinking.   The logical, conceptual explanations, grounded in Platonic and Aristotelian thought, of "revealed truths" became dogmas of the Church. Salvation only came to those who made a blind leap of faith-commitment to these dogmas. The Catholic clergy had some logical, conceptual understanding of them in contrast to most Christians who clung to a literal understanding of them similar to the conceptual, mythical knowing of paganism. As a result, hundreds of thousands of pagans who converted to Christianity over the centuries did not have to individuate to a new, higher level of knowing. On the one hand, there always were the few who "fed" on the metaphorical, conceptual descriptions of these "truths" that only could be comprehended by direct mystical experiences of them. On the other hand, most educated clerics reduced these "truths" to logical, conceptual dogmas that the hierarchy imposed on the vast population of the "faithful." This set the stage for the emergence of the *patriarchal perspective*. All civilizations were and still are patriarchal, at least to some extent. In my view, only the Western and Middle Eastern cultures differentiated what I call the "patriarchal perspective."

Judaeo-Christian, Metaphorical, Conceptual Thinking Generated the Patriarchal Perspective

Definition of Patriarchy. (1) A form of social organization in which the father or the eldest male is recognized as the head of the family or tribe. Later with the emergence of civilization patriarchy came to mean (2) Government, rule, or domination through the male line.

Two Types of Rational Individualism. Evolution to the subordination of metaphorical, conceptual thinking to logical, conceptual thinking in Judaism and in Christianity led to the emergence of *rational individualism*. This new rational individuality presents each human (and each human society) with two new options. Firstly, he/she can choose to exert greater control of self-expression and of nature by differentiating *individualistic, control consciousness* manifested as *control individualism* with *control objectivity*. Secondly, he/she can choose to give himself/herself over to participatory awareness of self and of nature by differentiating *individualistic, participatory consciousness* manifested as *participatory individualism* with *participatory subjectivity*. Control consciousness involving reason opposes participatory consciousness involving feelings.

Control Objectivity vs. Participatory Subjectivity. This fundamental duality of Judaeo-Christian, rational, individuation includes many dualities that are summarized in Table 3.1.

▶ **Table 3.1 Participatory Subjectivity vs. Control Objectivity**

Subjectivity Via The primacy of metaphorical, conceptual knowing leads to the primacy of **Participatory Consciousness**	**Objectivity** Via Metaphorical-conceptual knowing that includes analogical knowing leads to the primacy of **Control Consciousness**
1. Subjective Knowing	1. Objective Knowing
2. Experiential Feelings	2. Rationalized Feelings
3. Open to new ways to approach a "solution" to a problem	3. Problem Solving via pre-established methods and guidelines
4. Situational Morality	4. Rule Defined Morality
5. Transformational Development	5. Adaptation Development
6. Expressive Individualism	6. Utilitarian Individualism
7. Receptive, Open	7. Socially determined perspective, Conservative
8. Oceanic—Explorative	8. Structured—focused
9. Gestalt-Integrative-Visionary	9. Mechanistic-Analytical
10. Perceptive-Engaged-Empathetic	10. Judgmental-Separate-Critical
11. Sharing Dialogue; Intimate Conversation	11. Discussion-Debate-Argumentative; Didactic Explanation
12. Unstructured Playful	12. Goal-oriented Play
13. Meaning determined by context	13. Meaning = Information independent of context
14. Non-verbal or Metaphorical	14. Conceptual & Literal or Analogical
15. Harmony	15. Order
16. Social cohesiveness via an "ethic of care"	16. Social cohesiveness via rule of law & order
17. Non-cultural-social but individual and personal	17. Non-individual and non-personal but cultural-social

▶ **Table 3.1   Participatory Subjectivity vs. Control Objectivity (*continued*)**

Subjectivity Via The primacy of metaphorical, conceptual knowing leads to the primacy of **Participatory Consciousness**	Objectivity Via Metaphorical-conceptual knowing that includes analogical knowing leads to the primacy of **Control Consciousness**
18. *Zen creativity* (dreaming, seeing creativity); the metaphor for this is Virgin birth of a god, e.g., the Christmas story and *passion creativity* the metaphor for this is: Life, Death, Rebirth process, that is, Order, Chaos, New Order that includes modified aspects of the old order	18. Maintain homeostasis rather than be creative or be open to a transformation
19. Communal—Interconnected	19. Autonomous within Hierarchal Community Structure
20. Unconditional love & feeling bonding	20. Respect based on merit & bonding via a social contract
21. Self-centeredness (ego centeredness) causes feeling love to become self-centered love, possessive love, narcissistic love	21. Self-centeredness (ego centeredness) causes the control oriented person to be 1) addicted to progress, 2) the Faustian person always seeking, 3) the idealist always pursuing a non-ending quest, 4) the tragic lover whose romantic love only can be consummated by death
22. Empathetic love that leads to joy and nurturing of another but also may lead to blocking the awareness of the need for "self-transformation" or the nurtured person's need for "self-transformation"	22. Commitment to autonomy which prevents receiving or giving empathetic love
23. Leadership style that empowers others and/or facilitates others; the "followers" are encouraged to feel good about themselves and only secondarily to admire the leader	23. style that seeks to dominate, command, and employ strategies for charismatic inspiration of others to follow the leader's ideas or programs

24. Employ strategies for *transcending conflicts*
  1. Assert one's position and then seek ways of achieving harmony that transcends the conflict; one's position may be modified but it is not compromised or totally abandoned
  2. Seek inner harmony (peace) in the context of external chaos or external defeat
	24. Employ strategies for *resolving conflicts*
	  1. Attack-defend
	    a. Aggressive: Confront-attack
	    b. Confront-withdraw (to avoid defeat)
	  2. Seek Win or Stalemate
	    a. Manipulate/Stalemate: Confront-depend
	    b. Limited Win/Stalemate: Reduce-depend
	  3. Seek to Accommodate: Reduce-withdraw
	  4. Avoid Conflict: Reduce-attack

25. Living in and content with each moment of one's life	25. Ambitious, always living in the future which holds promise of accomplishing some goal
26. Seek or promote situations where everyone wins; i.e., prefer Win/Win or no deal interpersonal interactions	26. Competitive; prefer Win/Loose = Zero sum game in interpersonal interactions
27. Egalitarian, communal where status is not important	27. Hierarchal; seek status within a hierarchy
28. Love others and want to be loved	28. Have Power over others and want to be respected

Definition of the Patriarchal Perspective.   From a purely rational point of view it is possible to discern patterns of thought in the progressive differentiation of control objectivity associated with patriarchy. I propose that the subordination of metaphorical, conceptual knowing to logical, conceptual knowing in the Jewish and Christian (Catholic) cultures that further differentiated or emerged after the death of Jesus, produced a pattern of thought which I call the *patriarchal perspective*. The *patriarchal perspective* is a socially constructed perspective in which CONTROL OBJECTIVITY represses or dominates the expression of PARTICIPATORY SUBJECTIVITY.

One may obtain an understanding this patriarchal perspective by looking at control objectivity associated with masculine persona vs. participatory subjectivity associated with feminine persona.

Emergence of Scientific Thinking

Partial Breaking from Catholic, Patriarchal Thinking

Turning from Ascent to Descent.   While Thomas Aquinas's philosophy (1225–1274) legitimized an extraverted interest in and knowledge of the world as a way of "glorifying God and ascending to Him," it pushed Western Christian culture toward a *new kind of disharmony* and despair. Aquinas's vision pushed Medieval intellectuals toward objective knowing and away from the empathic, subjective knowing that is the core of mystical contemplation. This is how Aquinas's ideas led to this turning away from ascent toward God to a descent toward the world viewed as independent of God (the One).

Aquinas retained the ideas of hierarchy of being created by Plato as the hierarchy of eternal forms and the distinction between natural and supernatural orders in reality. However, Aquinas anticipates one way of understanding an aspect of the modern theory of evolution. In evolution, a lower level pattern of organization is radically modified so as to be incorporated into a higher level pattern of organization. For example, the lower level pattern in bacterial cells including the molecular language (DNA → RNA → protein synthesis) was modified and incorporated into the higher level pattern in plant and animal cells, and likewise the central nervous system (CNS) pattern in apes was modified and incorporated into the CNS of hominids which in turn was modified and incorporated into the CNS of the first ego self-conscious humans. In like manner, according to Aquinas, "God's Grace" does not replace nature but rather transforms it so that it becomes incorporated into the supernatural order. With this "evolutionary understanding" of the transforming power of Grace, Aquinas distinguishes two legitimate kinds of knowing: secular knowing which understands the natural order and sacred knowing which understands the supernatural order. Sacred knowing builds upon and therefore depends on secular knowing, which a pagan such as Aristotle may possess to a far greater degree than most Christians. Thus, Aquinas radically separates secular and sacred knowing and then integrates them.

Aquinas' radical distinction helped initiate and then unify medieval culture. Now Christian intellectuals were encouraged to participate more fully in nature than in the first phase of Christianity, which embraced the Platonism of St. Augustine's philosophical theology. However, this participation was rational and objective rather than participatory and subjective. Thus, Medieval theologians had turned from participatory, subjective contemplation of God to a rational, objective knowing of an alienated world. This turning led to the diversity problem that laid the groundwork for initiating modern science.

The Diversity Problem.   Conflicts arising within objective knowledge result from the diversity problem which may be summarized as follows. One assumes a subject-object duality in which reality has an unchanging structure (form or order) that is to some extent potentially intelligible to a knowing human subject. The human subject has modes of knowing which provide knowledge models that represent the structure of reality. The model may be an absolutely true or an approximately true representation of some aspect of reality. For example, in Thomistic philosophy, knowledge of the nature of a being is gained by abstraction. By means of abstraction we know what it means to say that this particular table has a circular shape. We abstract the form in the table, its circularness, and convert it

▶ **Table 3.2**

Masculine Persona Control Objectivity	Feminine Persona Participatory Subjectivity
1. Control objectivity: objective knowing that gives one control over self and nature.	1. Participatory subjectivity: subjective knowing involving insights, feelings that may lead to engagement with another person or with nature.
2a. Possibility of approximately or absolutely true knowledge.	2a. No absolutely true knowledge, rather all knowing depends on context and is either valid or not valid based on agreed upon criteria.
2b. Mechanistic knowing in which any whole system is literally equal to the sum of its autonomous parts.	2b. Any system is a whole which must be understood in terms of a subjective intuition. Then, the parts are not autonomous but are interconnected to one another. Also a knowledge mutuality: the system is understood in terms of its parts and each part is understood in terms of an intuitive understanding of the whole.
3. Human relationships are maintained by abstract principles of morality that define justice and duty, e.g., in the relationship between a man and a woman, the man should dominate at all times and in marriage, the man should be ruler of the family.	3. Human relationships are sustained by feeling bonding, love, and mutuality, e.g., mutual understanding and agreement about the role and duties of each person in the relationship.
4. Structure and order are viewed as good; chaos is viewed as bad. If a system goes into partial chaos, every attempt should be made to reestablish the old order. Thus commitment to maintained order prevents creative change.	4. In some circumstances maintained structure and order is bad and chaos is good. Chaos provides the opportunity to transform to a new order, i.e., undergo a creative change. For example, subordination to England in the 1700s was viewed as bad and the Declaration of Independence which brought the chaos of the revolutionary war was good.
5. Human maturity is identified with developing rationality and learning rational knowledge. Alternatively, one learns how to solve problems and achieve utilitarian goals based on learned rational knowledge.	5. Human maturity is identified with developing subjective knowing that can be the basis of creativity. Alternatively, one via intro-spection, understands his/her feelings and express one's inner—"so-called true"—self.
6. One chooses rational, abstract spirituality, e.g., dogmatic religions that emphasize the rule of law.	6. One chooses subjective, feeling spirituality, e.g., religions that accommodate subjective, feeling insights.
7. Because of a patriarchal culture, humans view themselves as autonomous and radically different from all other life forms. Also humans feel justified in exploiting Nature; they fail to realize how interdependent humans and the rest of Nature are.	7. Humans view themselves as not autonomous but interdependent with all other life forms. As a result, humans participate mother Nature and revere her.
8. Human maturity is defined as one understanding and then obeying the dictates of masculine and feminine persona.	8. Human maturity is defined as conforming to personas in various circumstances but not identifying with any persona. This means that sometimes a mature person will choose not to conform to a persona.
9. Civilized males or masculinized females are viewed as superior to and rightfully dominate uncivilized males, feminine females, and children. The terms, civilized, masculine, and feminine are defined by one's culture.	9. There is an ethic of caring where all humans are viewed as members of a community.

into a concept in our minds. The pattern represented by this concept (called an intentional species in the mind of the knower) is identical to the structural pattern in the table. In modern science no two "round" tables are thought to be exactly alike. Empirical observation indicates that the two tables are similar and therefore put into the same category, "round table." Thus, we have two different models for understanding round tables. One is the philosophical model of Aquinas's science of nature, and the other is the empirical model of modern science. Even within modern science we have this diversity of rational models representing the world. Newtonian mechanics was thought to be an approximately true representation of the world involving a universal force of gravity between any two mass objects in the world. In Einstein's general theory of relativity there is no universal force of gravity between any two mass objects in space. Rather, mass objects appear to be attracted to one another as a result of their motion being determined by the properties of space. Any two models that claim to give an absolutely true or the best possible approximately true representation of the same aspect of reality must be equivalent to one another. If the models are not equivalent, they conflict with one another.

The diversity problem results from having to choose between two conflicting models each thought to be absolutely or approximately true based on a social consensus of appropriate criteria of truth. That is, each model is said to be "objectively true," meaning that the truth of the model is independent of any one person's subjective insights—what I call participatory subjectivity. However, "truth criteria" are expressions of a particular rational point of view whose "truth" is guaranteed by a more fundamental set of truth criteria whose "truth" depends on a still more fundamental rational model whose "truth" depends on . . . and so on to infinity. (Another more abstract way of saying this is that knowing and truth criteria for knowing, for example, ontology or science and epistemology, imply one another; neither can stand on its own so that it can prop up the other. Ontology by itself cannot give you a "true epistemology," and epistemology by itself cannot give you a "true ontology or science.") Aquinas, following Aristotle, assumed that the universe has an unchanging fundamental structure, the human mind is capable of understanding at least some aspects of that structure, and human understanding generates an absolutely (or even an approximately) true language representation of structure in nature.

But how do we know that these assumptions are true? What I saw as a 17-year-old college freshman and I believe what Thomas Aquinas also realized is that we don't know the assumptions are true. We may commit to them absolutely—as I did at one time and as I believe Thomas Aquinas did—by a leap of faith. Participatory subjectivity leads to such a leap of faith in one's personal understanding of some rational model. So long as the rational model "makes sense" and solves at least some of life's problems, then one will continue to make this leap of faith. Jacques Monod, a Nobel prize winner for his work in the 1960s in molecular biology, had the courage to realize and state that scientists make an analogous leap of faith in their commitment to mechanistic theories. Thus, Monod's realization is a special case of what Nietzsche boldly proclaimed that absolutely true knowledge results from a human act of will. Either true knowledge results only from a recognized or unrecognized leap of faith; or it results from a Will-to-power. If one is unable or unwilling to make a leap of faith, then there can be no True knowledge, and correspondingly, the universe has no logical, conceptual meaning; nor does one's life in the universe have this kind of meaning. If one demands to have meaning, then he/she must will to have meaning even though the person has no rational basis for asserting that there is meaning. Nietzsche, who believed himself to be the exemplar of his idea of the *superman,* did this Will-to-meaning, but then Nietzsche went mad.

The 17th Century Double Bind. By the 17th century, for example, during the lifetime of Descartes (1596–1662), the diversity problem had become a source of great despair. Intellectuals were convinced that the world is intelligible and humans could formulate absolutely true language representations of it. They also were convinced that an absolutely True knowledge of the world is necessary for a human to guide his/her moral and spiritual decisions. However, if two or more models are logically consistent and explain phenomena in the world but conflict with each other, how is one to choose one model over any other? No matter which model one chooses, he cannot have absolute certainty because all the conflicting models are equally valid, but therefore, none of them can be absolutely true. For example, if one model states that the earth is the center of the universe and the sun revolves around it and a second model states that the sun is the center of the universe and the earth revolves around it, both models cannot be absolutely true. The 17th century intellectuals were in a terrible double bind: either the world is not knowable or the world is rationally absurd—it leads to conflicting models about its structure. Either pole of the double bind leads to the despair-producing conclusion that somehow humans must make life and death decisions in a world that cannot be rationally understood. Of course the way out of this double bind is to transcend it as did Eastern thinkers such as Nagarjuna of Mahayana Buddhism and Shankara of the Vedanta tradition. However, this way out was not available either to the Western Medieval intellectuals[1] or to the initiators of the Enlightenment who, following Aristotle and Aquinas, had become totally committed to a rational and meta-rational (i.e., metaphysical) understanding of the world.

Partial Enlightenment: "The Scientific Revolution"

Nominalism: A Step Towards Transcending the Double Bind of the Diversity Problem.   Nominalism is a philosophical perspective judged to be heretical by the medieval Catholic Church. This perspective emerged in the following way. As described earlier, conceptual, mythical thinking emerged along with the emergence of any civilization, which dictated that a particular overarching mythical story is literally true; i.e., gods and goddesses really do exist and control the lives of humans. Classical Greek thought culminating in Presocratic philosophy transformed literally true, objective, conceptual mythical thinking into metaphorical, conceptual thinking. A metaphorical concept has a subjective and an objective aspect. The subjective aspect is that, as described in an earlier section, each individual formulates "subjective concepts." This is a remembered domain of experience that incorporates many subjective perceptions that are "seen" to be similar. The objective aspect is that objective, conceptual language emerged when a society of humans reached the consensus—the basis of objectivity—that a particular set of artificial symbols represents the subjective concept that each individual creates for oneself in a particular circumstance. For example, when several humans point to the same object and create for themselves a subjective perception of it, then they create an objective concept of it by agreeing to assign the symbol, *tree,* to this perception. All other similar perceptions will be referred to by the objective, language designation, *tree.*

Because of the subjective aspect of all metaphorical concepts, the Presocratic philosophies using this type of knowing were open to diverse interpretations. Sometimes the interpretations could contradict one another. This ambiguity led to the perspective of relativity of truth which, in turn, generated skepticism. As a result of Socratic dialogue, the chaos of Presocratic thought gave birth to logical, conceptual knowing. However, in rejuvenating philosophical thinking, Socrates-Plato-Aristotle and subsequently Catholic, medieval thinkers assumed that the subjective concepts resulted from a process of formal abstraction. That is, *forms,* such as circleness, that "eternally exist" in the real world that humans perceive are somehow abstracted from a concrete, material expression of the form. The abstracted form of circleness, for example, then becomes a subjective concept in the individual who perceives the concrete object or event. As described earlier, artificial symbolic language is a way of objectifying subjective formulations of concepts that truly represent forms really existing in nature. This theory of knowing by means of abstraction guaranteed that humans could have objective, absolutely true knowledge of nature. Socratic dialogue did this by eliminating metaphorical, conceptual knowing, thereby eliminating the ambiguity of subjective knowing.

Catholic Christianity required the rebirth of metaphorical, conceptual thinking as a way of talking about revealed "truths," for example, Jesus, the Christ, is truly human and also is truly God. However, down through the ages, most educated Catholics understood the Church's dogmas by means of the classical Greek notion of logical, conceptual knowing. These dogmas were extended to

---

[1]It is Thomas Aquinas's philosophy of nature following Aristotle that created the diversity problem. Aquinas's metaphysics = philosophy of Being is one way of expressing Shankara's formulation of the Vedanta tradition. Aquinas, of course, used very different terminology embedded in a very different culture and religious perspective, and Aquinas was not at all familiar with any of Eastern mysticism.

apply, not only to "revealed truths," but to all human perceptions of nature. Diverse and sometimes contradictory explanations of nature that were proposed in Europe between 1500 and 1700, generated the theoretical diversity problem. Nominalism was a first step to overcome this problem by rejecting the classical Greek and the Medieval theory of abstraction for knowing the world. This simultaneously undermined the notion that humans could obtain an absolutely true, logical, conceptual understanding of nature. According to the nominalistic perspective, Platonic forms or Aristotelian natures do not exist; or if they exist, we humans cannot know them. Humans "create" logical, conceptual representations of nature based on consensus, and that consensus can change as humans expand their perceptual experiences of nature. For example, the experiential fact for premodern humans was that the sun rose in the east and set in the west. The reasonable consensus based on these perceptions was that the sun circulates around the earth which was viewed as the center of the universe. The expanded perceptions of some European astronomers, e.g., Galileo, led to the interpretation that the earth as well as the other planets circulate around the sun which is the center of the universe.

Constructivism of Copernicus' Cosmology.   The Medieval Catholic Church held that Ptolemaic theory, created by Ptolemy (?100–168 A.D.), was absolutely true with respect to the idea that the earth is the center of the universe and the sun revolved around the earth. This fit very well with everyday observations in that the sun always arose in the east and set in the west. However, over the centuries accumulation of astronomical observations forced thinkers to revise the Ptolemaic theory to accommodate these new astronomical observations. As a result, the theory became hopelessly complex and unable to explain many new observations. Nicolas Copernicus (1473–1543), born in Torun, Poland wrote in about 1513 an account of cosmology that was much simpler and more elegant than the Ptolemaic theory. Copernicus modified his theory over a 30-year period and was reluctant to publish it, but relented as a result of the insistence of a German mathematics professor, George Rheticus; this heliocentric theory was published the year Copernicus died, 1543. This theory not only explained more of the new astronomical observations, it was much more useful for making navigation calculations. This was a time when European sailors were venturing out into the unknown seas to explore new continents, for example, Columbus landed in America in Cuba, 1492.

In Copernicus' cosmology the sun was the center of the universe and the earth revolved around the sun. Such a view contradicted one of the dogmas of the Catholic Church and Copernicus was a devout Catholic. His approach was that his theory was not really true; rather it distorted the true Ptolemaic theory in a way to make it useful. Thus, one can live with such a discrepancy by distinguishing between true knowledge and utilitarian knowledge. (Some Jesuits at St. Louis University in the 1950s tried to convince me of the same distinction with regard to true Thomistic philosophy of nature vs. merely valid natural sciences.) Other thinkers after Copernicus, for example, Galileo (1564–1642), became caught up in the spirit of Greek humanism that celebrates logical consistency which also was a hallmark of the Aristotelian-Thomistic synthesis. According to this view, practical sciences, that is, utilitarian sciences, are derived from theoretical sciences and, therefore, a true theoretical science will lead to a true practical science that achieves one's utilitarian goals. Also, if two theories contradict one another, one must be true and the other false. The Renaissance humanists rejected one Catholic dogma after another. But rejecting Catholic dogma did not solve their diversity problem. However, the distinction between true knowledge and valid knowledge provided a hint for a solution that was taken up and radically modified by the scientific revolution culminating in Newton's theory of motion published in the late 1600s. The nominalistic perspective coupled with Copernicus's constructivism led to a solution to the theoretical diversity problem. *Humans do not discover laws of nature. Rather, humans construct rational models to represent perceived regularities in nature.* Then one develops ways for testing the validity—not the truth—of these constructed representations. This is the core idea of what came to be known as "systematic experimentation."

Systematic Experimentation.   In the 16ᵗʰ and 17ᵗʰ centuries several thinkers culminating in Isaac Newton (1642–1727) created the mechanistic theory of motion. In so doing they participated in the creation of a modified version of objective knowing that enabled humans to obtain public consensus on the validity of the proposed solutions to certain types of problems. The ideas of operational definitions, empirical evidence, induction, analytic synthetic thinking involving rational intuition and deduction, and public consensus of objective truth were modified to fit in with a radically new approach to studying nature called systematic experimentation. The integration of all these ideas with systematic experimentation became known as the scientific method for producing objectively valid knowledge of the world.

I use the word, "producing," rather than "discovering" to indicate that this method involves two new features in knowing: Firstly, the knower interferes with nature, actually disrupts nature, in order to comprehend her. Secondly, both the knowing process and the knowledge that is produced are mechanical. The distinction between theoretical and practical knowledge so cherished by Aristotle is blurred. Theoretical knowledge no longer is wisdom defined as rational contemplative participation in nature. Such contemplation may be ecstatic, producing the highest form of happiness, and has value in itself. Contemplative knowledge is to the mind what health is to the body. Scientific knowledge, whether pure or applied, is like a set of directions explaining how things or events interact. The directions contain pieces of information.

There is nothing new about experimentation. All humans and even many other mammals do it. Experimentation is a kind of manual curiosity, irresistible in children, about one's environment: if I do this, what will happen? When the experimental observation is repeated many times under similar circumstances, it may lead to a useful generalization; for example, gas at a constant volume will always increase in pressure when the temperature is raised. Primitive science contains many generalizations of this sort obtained by experimentation. The facts and experimental generalizations in primitive science give one only a very limited control over nature. The key difference between primitive science and mechanistic science is that the one is grounded on mere experimentation whereas the other is grounded on systematic experimentation defined as: the association of "scientific mythology" with observations obtained by experimentation. Scientific mythology is a story that is taken to be literally true because the metaphorical concepts in the story can be related to operationally defined terms. This type of definition allows the story to be converted into a rational model that can be validated by empirical observations (see later description of the scientific method). In contrast, "non-scientific mythology" depends on metaphorical concepts which lead to diverse, subjective interpretations of the story. "Scientific mythology" may also involve metaphors (e.g., DNA is a master molecule), but one way or another the metaphors are transformed into concepts that lead to a uniform understanding of the story.

For example, the modern scientific thinker may tell the story that even though we cannot see gremlins, millions of them are in the air in this room. Some of these gremlins are evil in that if they somehow get into the blood stream of an animal such as ourselves, the animal will die in a day or two. These gremlins, so the story goes, are able to rapidly reproduce themselves when in a liquid nutrient broth. The scientist then associates this story with observations from two experiments. In the first he inserts a wire loop exposed to the air into a nutrient broth which is then sealed so that none of the gremlins can get out. After a day or so, the clear broth becomes cloudy, due to the many new gremlins, and then after several days becomes clear again as all the gremlins die and sink to the bottom of the test tube. In the second experiment, he injects some of the cloudy broth into the blood stream of a rat and observes the animal developing several symptoms of distress culminating in death. Of course the enormous practical value of this scientific myth is that surgeons sterilize their instruments and do whatever else is necessary to prevent evil gremlins from entering the blood stream of patients during an operation.

William Harvey (1578–1657) observed hearts dissected from dead humans and noted that the bulk of the heart is muscle and that the volume of the heart chamber is about four and a half ounces. Harvey reasoned that when the muscle contracts corresponding to a heart beat, blood would be forced out of the vessels leading away from the heart. Assuming that the heart beats an average of 75 times per min. and empties 4 and 1/4 oz per beat, he easily calculated that the heart empties 315 quarts per hour or 7,875 quarts per day. There is no way the body could produce 315 quarts of blood in one hour. Therefore, Harvey created the very useful and fundamental scientific myth that the heart causes blood to circulate in the blood vessels in the body. What is even more wonderful about scientific myths is that often individual stories fit together to form a grand myth with far reaching applications. The experimental observation of a burning candle going out some time after being covered by a bell jar is associated with the story that fire is the result of invisible oxygen chemically combining with some substance, in this case wax. A mouse enclosed by a bell jar will eventually die; this, of course, is associated with the story that breathing involves taking oxygen into the body where it is used to burn up some substance in the body (Claude Bernard referred to this as "life is death," i.e., the body stays alive by burning itself up). Other experimental observations and corresponding stories lead us to understand that hemoglobin in blood carries oxygen to all parts of the body along with the glucose that is dissolved in blood. Most cells take in glucose and oxygen, which combine to produce a kind of stepwise burning that releases energy. The released energy is used to carry out life processes in each cell that keeps it alive. Thus, the ancient myths involving the highly charged metaphors of fire, inspiration of breathing, and blood are related to interconnected scientific myths which use these terms to create a very powerful mechanistic story that relates to many kinds of animals.

Characteristics of the Scientific Method.   The scientific method is to all aspects of modern Western culture what water is to all life forms in the ocean. This thought process has profoundly influenced all our modes of thought and has transformed specialized areas of inquiry even in sociology, psychology, and some sub-disciplines in the humanities into science-like disciplines. To some degree, all these disciplines use the scientific method which is characterized by: 1) the use of operational terms, 2) empirical observation and induction, 3) the application of analytic-synthetic thinking, and 4) the use of scientific procedures for establishing public consensus of objective truth.

1. Operational Terms.   No matter how abstract and technical a scientific theory is, all its concepts must be derived from operational terms. Operational terms are defined by a process that will lead any two or more people who correctly perform the process to have similar sense perceptions. This characteristic of using operational terms allows only certain types of problems to be pursued. Thus, problems in such areas as religion, philosophy, and ethics are off limits to scientific discourse. For example, the abstract concept of beauty cannot be defined using operational terms.

2. Empirical Observation.   Science is empirical in that it deals with things or events that can be observed many times. Furthermore, the scientist may observe many instances of a type of thing or process and make the induction that all these instances have some characteristic in common. This empirical aspect of science is another limitation to the kinds of questions with which science deals. For example, one cannot observe instances of value; rather one observes things or events that he judges as good or bad, beautiful or ugly. Thus, science is without values other than the ones associated with the characteristics of the scientific method. As a result, scientists tend to avoid explicit value discussions in relation to scientific issues. It is much more difficult to precisely define values and reach a consensus about them than it is to resolve scientific problems.

3. Analytic-Synthetic Thinking.   The scientific approach to studying nature is to: (1) *analyze* a thing or process, i.e., imagine it to be composed of subunits; (2) describe the properties of the subunits; (3) describe how the subunits interact; and (4) make a synthesis, i.e., describe the whole thing or process in terms of the interactions among its subunits.

4. Scientific Public Consensus of Truth.   Public consensus of truth refers to the prescribed procedures for reaching a consensus about the truth of objective knowledge. In science there are two major criteria for scientific validity: (1) experimental prediction and (2) falsification. Three other criteria may also be used to judge the usefulness of a theory: (3) the theory accounts for and represents all of the thousands of relevant facts known so far; (4) the theory shows interrelationships among heretofore seemingly unrelated sets of information; and (5) the theory provides the mental disposition or receptivity to discover new facts or better still to see new unsuspected interrelationships among already known facts.

One type of systematic experimentation (described previously), which at the same time is a way of obtaining scientific public consensus of truth, is experimental prediction. This involves: (1) specifying a particular set of circumstances in terms of a scientific theory, (2) use the theory to predict what will happen under these circumstances, (3) set up the specified set of circumstances which is called an experiment, usually performed in a laboratory, (4) observe what happens in the experiment, (5) repeat the experiment several times, and (6) in some systematic manner conclude whether the experimental observations confirm or deny one's theoretical prediction. Experimental prediction is a type of hypothesis testing which may be exemplified by the process of rectifying the situation where a light bulb in a lamp no longer gives off light. The process may be divided into six steps:

1. *story (hypothesis):* the lightbulb is burned out (or some other hypothesis such as the fuse is blown).
2. *prediction based on the story:* If I replace the bulb with one that I know will give off light, then the problem will be solved.
3. *action (experiment):* I replace the lightbulb with one I know works.
4. *observe results of the action:* the lightbulb still does not give off light.
5. *compare observed results of action with the prediction:* the results of the action (experiment) contradict the hypothesis.
6. *conclusion:* the hypothesis is false.

Falsification is the process of observing an event that contradicts some aspect of a scientific theory. Lack of falsification is a basis for accepting the validity of a theory; however, it is important that the theory be formulated in such a way that it can be falsified. Aristotle's theory that change is the process of the potential of a being receiving a form (a potential being actualized) may be thought of as a "good" philosophical theory but it is a "bad" scientific theory. No observation can be made that could falsify this theory. Darwin's biological theory of evolution can be falsified. According to this theory multicellular plants and animals evolved from bacteria. If we ever discover fossils more ancient than fossil evidence of the first bacteria, about 3.5 billion years old, then Darwin's theory would be invalidated.

Ethic of Scientific Objectivity

Jacques Monod (1972) had the courage to follow the mechanistic perspective to its ultimate expression. Mechanistic philosophy is a will to control. It replaces Nature gods of primitive cultures, metaphysical realities of classical Greek culture, Pure Consciousness of Eastern cultures, and the transcendent God of Judeo-Christian religions with the mental self absolutely committed to scientific methodology, a commitment which Monod calls the "ethic of knowledge." The mechanistic interpretations of the past and its prescriptions for the future focus on a human's control of himself and of nature.

To his credit Monod indicated the ethical nature of this commitment to the scientific objectivist myth—the "ethic of knowledge."

> It [the principle of objectivity] cannot be objective: it is an ethical guideline, a rule for conduct. True knowledge is ignorant of values, but it cannot be grounded elsewhere than upon a value judgement, or rather upon an *axiomatic* value. It is obvious that the positing of the principle of objectivity as the condition of true knowledge *constitutes an ethical choice and not a judgment arrived at from knowledge, since according to the postulate's own terms, there cannot have been any 'true' knowledge prior to this arbitral choice.* In order to establish the norm for knowledge the objectivity principle defines a *value:* that value is objective knowledge itself. Thus, assenting to the principle of objectivity one announces one's adherence to the basic statement of an ethical system, one asserts *the ethic of knowledge* (Monod, 1972, p. 176).

Scientific Humanism.   (1) *Gross reductionism.* Science after the Newtonian synthesis became identified with *scientific humanism* that is grounded on the belief that all "legitimate" human problems can be solved by the scientific method. The criteria for deciding whether a problem is legitimate or not is either *gross reductionism* or *scientific positivism.* Gross reductionism is mechanistic analysis implying that any object of study is a whole system that can be totally understood in terms of the interactions among its autonomous parts. This is a defiant choice to completely suppress subjective, participatory consciousness based on the belief that in principle everything in nature is knowable by means of mechanistic analysis. Thus, though scientific mechanism limits objective knowing to a very narrow range of problems, theoretically it is applicable to all of nature.

Creative scientists, especially Einstein, would hasten to add that imagination and subjective insight are essential for producing new theories. However, subjective insights only present one with possible ways of understanding Nature; the insights themselves are neither true nor false. They only become valid knowledge when they are shown to be objectively valid knowledge by means of the scientific method. In fact, subjective participatory experience produces illusion which always must be replaced by "true knowledge" gotten by the method of science. This certainly was true for Newtonian mechanics and remains true today for those scientists that reduce all legitimate knowing to mechanistic analysis.

(2) Scientific Positivism.   This perspective may be defined-described by the following four statements: (1) all subjective insights must be doubted; (2) only those subjective insights that are converted into objective knowledge can possibly be valid or invalid, but objective knowledge never can be absolutely or approximately True; (3) only those forms of knowledge validated by means of the scientific method can be accepted as "legitimate" objective knowledge; (4) since scientifically tested objective knowledge involves specialized models, facts, and testing techniques, all of us must rely on the valid, specialized knowledge of experts.

Scientific Positivism as "Scientific Constructivism" Has Become a 20th Century Religion.   In the 13th century Thomas Aquinas, following Aristotle, proposed that any human could come to some absolutely true language representation of the eternal structure of the universe. This led to the generation of the diversity problem that put Renaissance thinkers and humanists of the Enlightenment in despair. The diversity problem results from having to choose between two conflicting models each thought to be absolutely or approximately true based on a social consensus of appropriate criteria of truth. The way Western thinkers beginning with people such as Copernicus, Kepler, Galileo, Descartes and culminating with Isaac Newton evolved to a resolution of the diversity problem is by creating the here-to-for unprecedented idea of systematic experimentation, which was described in relation to postmodern science.

A host of scientists who indirectly or directly guide the writing of high school and college science textbooks still talk about science as evolving toward ever-greater approximately true knowledge of the universe. However, a minority of creative and reflective scientists realize that all of modern science is not an accumulation of discoveries, but rather it is a network of mentally constructed, valid, language representations of the world. This approach, present at the very beginning of modern science exemplified by Newton's theory of universal gravitational forces, anticipates one of the three—according to Wilber—core assumptions of postmodernism.

> Reality is not in all ways pregiven, but in some significant ways is a construction, an interpretation (this view is often called *constructivism*); the belief that reality is simply given, and not also partly constructed, is referred to as "the myth of the given."[2]

The first modern scientists and many scientists today take this constructivism perspective to the extreme of control objectivity in two ways. First of all, while proclaiming that no one can have true knowledge of the universe, scientists proclaim that scientific constructivism is the only legitimate way of creating valid ways of understanding nature. This "ethic of objectivity" (i.e., see earlier quote of Monod) forbids all other ways of knowing, which, of course, undermines all religions, philosophies, theories of morality, and so on. Secondly, participatory subjectivity only is considered to be worthwhile when it leads to a validated, scientific theory. Once we have the validated theory we should ignore our subjective insights or pay attention only to those insights that allow us to apply the theory to concrete situations.

Scientific constructivism has evolved into what may be called a new world rationalized religion of the 20th century. As with any religion held to be universally true or exclusively valid, modern science as religion requires a leap of faith. In this case the believer must commit himself/herself

---

[2]Wilber, K. 2000. *Integral psychology*. Boston: Shambhala, p. 163.

absolutely to scientific constructivism. This scientific religion correspondingly requires absolute commitment to an evolving story of the universe that is increasingly valid and useful—for some this is interpreted to mean that science is approaching absolute truth ever more closely. Finally, the scientific mythology implies a set of attitudes and rules of conduct. The attitudes are specified by a scientific, positivistic interpretation of the patriarchal perspective described earlier in this chapter. The rules of conduct are specified by radical, utilitarian individualism associated with marketplace values and globalization. In short, the new world scientific religion admonishes each of its members to develop.

The Irony and Paradox of Mechanistic Science

The emergence of mechanistic science as a core idea of the Enlightenment is both ironic and paradoxical. Mechanistic science was the first, irrevocable break from the patriarchal perspective embedded in Judaeo-Christian and Islamic cultures. This break made and still is making a profound impact on the evolution of human self-consciousness and social structure. The founding fathers of science chose to side step traditional, dogmatic views and to create a subjective, metaphorical understanding of nature. The resulting subjective descriptions were analogous to the theories of the Presocratic thinkers of classical Greece, designated by Heidegger as "poetic philosophers." Like the Presocratic philosophies, the scientific descriptions were open to diverse interpretations. The radical innovation of the "fathers of science" was to reject turning these descriptions into literal, absolutely true philosophical theories. Instead they converted them into mathematical, physical theories. As a result, scientists could reach a consensus, based on a *scientific method*, of whether a particular theory is objectively valid or not. Thus, consensual, objective validity replaced dogmatic literal or analogical certitude.

The irony was and is that this scientific "enlightenment" early on, i.e., beginning with Newton, became associated with a philosophy known as positivism. Many scientists explicitly assumed this perspective to be absolutely true or, at least, more valid than any other perspective. A scientific community emerged and evolved to a tacit commitment that this perspective is a necessary defining characteristic of science. As described earlier, positivism includes four ideas. Firstly, only those subjective, metaphorical insights that lead to valid theories are legitimate. Secondly, once a validated mathematical physical (or empirical) theory has emerged, its subjective, metaphorical foundation can and should be discarded. Thirdly, scientific knowing is the only legitimate way of knowing or is the most legitimate type of knowing because it can be validated by the scientific method. All other kinds of knowing including theological and philosophical knowing generate the irresolvable paradox of the diversity problem. Fourthly, scientific theories are directly—the usual norm—or indirectly utilitarian. This idea combined with the third idea implies that the only legitimate types of theories of reality are those that give one objective control over some aspect of nature. The patriarchal perspective before the emergence of mechanistic science allowed for some participatory subjectivity and non-utilitarian knowing associated with philosophy, religious insights, and cultural-religious rituals and ceremonies. Mechanistic science, which emerged as a fundamental break from this perspective, ended up being the highest possible level of expression of the patriarchal perspective. Mechanistic science thought to be necessarily associated with positivism is totally objective and totally utilitarian.

The ironic aspect of mechanistic science is that it produced and is still producing a paradox that leads to a subjectivity double bind. Gross mechanistic science that emerged in the 17th and 18th centuries proclaimed a totally objective way of creating a valid understanding of nature. However, beginning in the middle of the 19th century, further evolution of science produced theories that at least tacitly included ever-greater degrees of subjective insights, e.g., thermodynamics involving the subjective idea of the irreversibility of time (see Chapter 1), biological theory of evolution involving the subjective idea of qualitative, hierarchal levels of organization (see Chapter 14), and systems theories involving the subjective metaphor of "self-organization" of matter (see Chapter 2). Nevertheless, scientists not only continue to use an objective scientific method but also focus only on objective interpretations, ignoring the subjective insights implied in their theories. Thus, we have the double bind: *on the one hand, scientists choose exclusive objectivity and ignore the subjectivity implicit in their theories as well as undermining the spiritual vision that is the basis for capitalistic democracy. On the other hand, scientists could acknowledge and further develop the subjective aspects of science, but this would undermine the perceived defining characteristic of modern science as well as undermine the power status of specialists valued for their expertise.*

Transcending the Subjectivity Double Bind of Science

I propose that a new theory of creativity, as briefly described at the end of Chapter 1, and as exemplified in Chapter 2, can begin to overcome this subjectivity double bind. The theory of creativity, then, must be expanded to include George Lakoff's theory of metaphorical, conceptual knowing. This forms the foundation for scientific constructivism expressed as systemic experimentation that includes the so-called scientific method of thinking. I expand the pattern of creative, scientific thinking to four styles of creative learning. This cyclical sequence of four styles may be applied to all situations in which humans seek personal meaning, explanations and solutions to existential or utilitarian problems.

Four Styles of Creative Learning

Background

I have taken the idea of a triune brain as the basis to propose that after the emergence of Mind, which is the basis for human ego self-consciousness, it also differentiates into three mind centers: Expressive Mind Center, (feeling perceptions and intuitions), Thinking Mind Center, and Doing Mind Center (action). I further propose in agreement with other thinkers that gene segments interacting with the "psycho-social environment" of the 5-year-old child leads to one of mind centers becoming the "preferred mind center." Further psychological development leads to the preferred mind center becoming either the dominant mind center or the repressed mind center. These ideas imply that there are 12 personality types wherein each type consists of (1) a *dominant mind center*, (2) a *support mind center* and (3) a *repressed mind center*. I have integrated my model of 12 personality types (based on my modification of the Enneagram as developed by Don Riso) with some neo-Jungian theories and with David Kolb's classic work on experiential learning to propose that there are four classes of mature individualism corresponding to four styles of experiential learning. Maturity refers to the idea that an individual has the beginnings of a well-developed dominant mind center and some development of the support mind center.

With the second law of thermodynamics integrated with the systems science theory of universal evolution as a base, I propose that creativity at all levels of organization in the universe including human creativity is an Order, Chaos, New Order process. Humans exemplify this process when an individual or institution goes into irreversible chaos, i.e., the chaos cannot be overcome by negative feedback mechanisms. The loss of structure to a system, i.e., the Chaos, always eventually shows possibilities that were not available when the intact structure was present. If the individual/institution mindfully endures the pain of "not knowing what to do," eventually they will see new possibilities and actuate some of them to produce a New Order that incorporates a modified Old Order. Based on the theories of David Bohm that were adopted by Peter Senge, I propose that when appropriate requirements are satisfied, dialogue becomes a creative process. Moreover, creative dialogue differentiates into a sequence of two types: (1) Vision dialogue and (2) Collaborative dialogue among people who share the same vision.

The ideas about personality types associated with experiential learning combined with the nexus of ideas associated with creative dialogue leads to the perspective that experiential learning is better described as creative learning. In this perspective there are four styles of creative learning corresponding to the four styles of experiential learning each of which corresponds to one of the four classes of

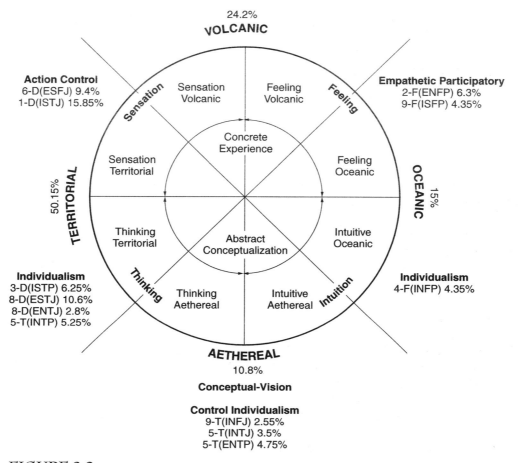

Existential-Vision
Control Individualism
3-T(ESFP) 5.6% 7-T (ESTP) 4.9% 1-D (ISFJ) 11.25% 6-F (ENFJ) 2.45%

24.2%
VOLCANIC

Action Control
6-D(ESFJ) 9.4%
1-D(ISTJ) 15.85%

Sensation Volcanic

Feeling Volcanic

Empathetic Participatory
2-F(ENFP) 6.3%
9-F(ISFP) 4.35%

Sensation

Feeling

Concrete Experience

Sensation Territorial

Feeling Oceanic

TERRITORIAL
50.15%

OCEANIC
15%

Thinking Territorial

Abstract Conceptualization

Intuitive Oceanic

Individualism
3-D(ISTP) 6.25%
8-D(ESTJ) 10.6%
8-D(ENTJ) 2.8%
5-T(INTP) 5.25%

Thinking

Thinking Aethereal

Intuitive Aethereal

Intuition

Individualism
4-F(INFP) 4.35%

AETHEREAL
10.8%

Conceptual-Vision

Control Individualism
9-T(INFJ) 2.55%
5-T(INTJ) 3.5%
5-T(ENTP) 4.75%

FIGURE 3.2

mature individualism, see Figure 3.2 The four styles of creative learning, in turn, may be described in terms of four types of creative dialogue, which are Vision dialogue and the sequence of 3 subtypes of dialogue that make up Collaborative dialogue. The "so called" scientific method is a special case of this sequence of four types of dialogue. In this dialogue perspective creative learning has at least two defining characteristics:

1. It involves an Order, Chaos, New Order process, i.e., Life, Death to one perspective, Endure the anguish of not knowing what to do in order to see new possibilities, and Rebirth to a New Life that incorporates the old life.

2. It involves Creative dialogue between two mind centers within an individual or between two or sometimes among three sets of personality types in an institution where the diverse personality types in each set represent the dominance of the same mind center.

I was able to correlate each of the 12 personality types in my model with the 16 personality types in the Myers-Briggs Type Indicator (MBTI). Many corporations, businesses and other institutions since 1980 have used the MBTI. Therefore, there is a great deal of data from which to infer the percent in the general population in the U.S. that represent each of the 16 personality types of the MBTI.

References used to correlate 12 personality types of Pribor's model and 16 personality types of the MBTI:

1. Don Richard Riso. 1987. *Personality types.* Boston: Houghton Mifflin Co.
2. Don Richard Riso with Fuss Hudson. 1996. *Personality types.* Boston: Houghton Mifflin Co.
3. Isabel Briggs Myers with Peter B. Myers, 1980. *Gifts differing.* Palo Alto, CA: Consulting Psychologists Press, Inc.
4. David Keirsey and Marilyn Bates. 1978. *Please understand me (character and temperament types).* Del Mar, CA: Prometheus Nemesis Book Co.

Reference for the percentages associated with each of the 16 personality types in the MBTI

1. R. Baron. 1998. *What Type Am I (Discover Who You Really Are).* New York: Penguin Books.

The Four Styles

I. **CREATIVE VISION DIALOGUE:** *Create Metaphorical, Conceptual Vision.*
   A. *Chaos Phase:* **Eros-chaos psychetypes expressing Empathetic, Participatory Individualism produce chaos by**
      1. Expressing what Kolb calls divergent thinking
      2. Allowing ideas from the Individual, Collective Unconscious to be acknowledged and looked at by ego consciousness
      3. The psychetypes are
         a. 2-F (ENFP)
         b. 4-F (INFP)
         c. 9-F (ISFP)
   B. *Order Phase:* Eros-order psychetypes expressing Conceptual-Vision, Control Individualism
      1. Formulate vague ideas into metaphorical concepts
      2. Organize the metaphorical concepts into a narrative or vision
      3. The psychetypes are
         a. 5-T (INTJ)
         b. 5-T (ENTP)
         c. 9-T (INFJ)
   C. *Dialogue Mutuality*
      1. Between the Expressive and Thinking Mind Centers
      2. A + B = (15%) + (10.8%) = 25.8% [see fig.3.2]

II. **CREATIVE, COLLABORATIVE, CONCEPTUAL DIALOGUE:** *Convert Metaphorical, Conceptual Vision into a Logical, Conceptual Vision*
   A. *Chaos Phase:* Eros-chaos psychetypes expressing Empathetic, Participatory Individualism
      1. Explore *diverse* implications of metaphorical concepts and the metaphorical vision
      2. The appropriate psychetypes are
         a. 4-F (INFP)
         b. 9-F (ISFP)
   B. *Order Phase:* Eros-order psychetypes expressing Conceptual-Vision, Control Individualism
      1. Select from the diverse metaphorical concepts those that will be represented by unambiguous, logical concepts
      2. Convert the metaphorical vision into a logical, conceptual vision
      3. The appropriate psychetypes are
         a. 5-T (INTJ)
         b. 5-T (ENTP)
         c. 9-T (INFJ)
   C. *Dialogue Mutuality*
      1. Between Expressive and Thinking Mind Centers
      2. A + B = (8.7%) + (10.8%) = 19.5% [see fig.3.2]

III. CREATIVE, COLLABORATIVE, STRATEGIC DIALOGUE (first phase of HYPOTHESIS TESTING): *Convert Logical, Conceptual, Vision into a Logical, Conceptual, Strategic Plan*

    A. *Chaos Phase:* Eros-order psychetypes expressing Conceptual, Vision, Control Individualism

        1. "See" *diverse* implications, possibilities, and problems associated with the logical, Conceptual Vision

        2. The appropriate psychetypes are

            a. 5-T (INTJ)

            b. 5-T (ENTP)

            c. 9-F (INFJ)

    B. *Order Phase:* Eros-order psychetypes expressing Action, Control Individualism

        1. Select those implications and possibilities that can be operationally defined in relation to the capabilities of the individual or institution in a particular context, i.e., the vision must be doable

        2. Select those solutions to problems that can be operationally defined in relation to the capabilities of the individual or institution in a particular context

        3. Create a hierarchy of implications, possibilities, and solutions to problems, e.g., what implications or possibilities should be explored first and what solutions should be tried first; what are the backup solutions

        4. Combine operational definitions and hierarchy into a strategic plan

        5. The appropriate psychetypes are

            a. 8-D (ESTJ)

            b. 8-D (ENTJ)

            c. 3-D (ISTP)

            d. 5-T (INTP)

    C. *Dialogue Mutuality*

        1. Between Thinking and Doing Mind Centers

        2. A + B = (10.8%) + (24.9%) = 35.7% [see fig.3.2]

IV. CREATIVE COLLABORATIVE TASK DIALOGUE: *Convert Logical, Conceptual, Strategic Plan* **into the** *Concrete Events of a Task that Accomplishes the Goals of the Strategic Plan* **as Judged by** *Operationally Defined Criteria*

    A. *Chaos Phase:* Eros-order psychetypes expressing Action, Control Individualism

        1. "See" diverse implications and possibilities and possible problems associated with the strategic plan

        2. The appropriate psychetypes are

            a. 6-D (ESFJ)

            b. 1-D (ISTJ)

            c. 3-D (ISTP)

            d. 8-D (ESTJ)

            e. 8-D (ENTJ)

            f. 5-T (INTP)

    B. *Order Phase:*

        1. Create: (1)individual skills, (2) functional teams with team skills and (3) dialogue teams that solve unforeseen, existential problems as they arise and use this integrated creativity to convert the logical, conceptual, strategic plan into the non-conceptual, existential set of events that achieve the goals of the strategic plan.

        2. *Hypothesis Testing:* determine via objective criteria

            a. Whether the existential events accomplish the goals of the strategic plan

            b. To what extent accomplishing the goals of the strategic plan is less than desired

            c. Whether the individual/institution needs a new

                (1) Strategic plan

                (2) Logical, Conceptual Vision

                (3) Metaphorical, Conceptual Vision

   3. The appropriate psychetypes of #1 of the Order Phase: psychetypes expressing Existential-Vision, Control Individualism
      a. 3-T (ESFP)
      b. 7-T (ESTP)
      c. 1-D (ISFJ)
      d. 6-F (ENFJ)
   4. The appropriate psychetypes for #2 of the Order Phase: psychetypes expressing Action, Control Individualism
      a. 3-D (ISTP)
      b. 8-D (ESTJ)
      c. 8-D (ENTJ)
      d. 5-T (INTP)
C. *Dialogue Mutuality*
   1. Existential, Non-conceptual aspect of the Thinking Mind Center or Feeling Mind Center AND the Conceptual Aspect of the Thinking Mind Center in Collaboration with the Doing Mind Center
   2. $A + B_{phase\ 1} = 50.15\% + 24.4\% = 74.55\%$ [see fig.3.2]
   3. $A + B_{phase\ 2} = 50.15\% + 24.9\% = 75.05\%$ [see fig.3.2]

Scientific Thinking Involves the Four Styles of Creative Learning

Exemplifying the first learning style, the fathers of modern science broke away from the order defined by the dogmatism of the medieval Catholic Church. The vision that emerged from the chaos of the theoretical, diversity problem (e.g., "What theory is absolutely True: the earth or the sun is the center of the universe?) is that humans could create valid—rather than absolutely true—descriptions of nature in terms of mathematics. Exemplifying the second learning style, thinkers like Galileo and Newton converted the scientific, Enlightenment vision into mathematical equations, which are logical, conceptual models. Then various Enlightenment thinkers created strategic plans, that is, they designed experiments, that could validate or invalidate these mathematical descriptions of processes in nature. Finally, some people carried out these experiments and determined to what extent, if at all, the mathematical descriptions are valid. This sequence of four styles of creative learning still describes the practice of scientific research today. Some scientists carry out all four learning styles. However, there also is specialization where some scientists specialize in the first and second learning style, and other scientists specialize in making minor extensions of a validated theoretical perspective and then engage in "hypothesis testing." This, of course, exemplifies the third and fourth creative learning styles. This cyclic pattern of scientific, creative thinking can be expanded to all human creativity including creating works of art, e.g., paintings, sculpture, music, and mystical insights and practices. The pattern remains the same but the product of action creativity is judged to be valid or invalid according to criteria specific to the type of creativity that is involved.

Tearout #1

Directions:

1. Describe the transition from *participatory embedded knowing* (=non-conceptual mythical knowing) to *conceptual, mythical knowing.* (Piaget calls this *concrete operational thinking.*)

2. Describe the difference between *conceptual, mythical knowing* (concrete operational thinking) and *metaphorical, conceptual thinking.*

3. Give an example of metaphorical, conceptual thinking other than the "key-lock" example described in the text.

4. Describe the meaning of nominalism.

Tearout #2

Directions:

1. Describe the meaning of systematic experimentation in relation to metaphorical, conceptual thinking.

2. Describe the meaning of systematic experimentation in relation to *scientific constructivism.*

3. Describe an example of the creative learning cycle that involves four styles of creative learning.

Homeostasis: One Defining Characteristic of Life

Stress in Terms of Homeostatic Machines

The idea of homeostasis was first introduced by Claude Bernard (1813–1878) with respect to properties of the circulating blood in mammals. For example, the temperature of blood in humans is maintained near 97.6° F even though extremes in the environment tend to elevate or lower this temperature. In the now classic book, *The Wisdom of the Body,* first published in 1932, Walter Cannon further developed ideas relating to physiological homeostasis in mammals, especially humans. This influential book began to make a dramatic impact on scientific thinking during World War II, and thereafter homeostasis progressively became one of the dominant themes of 20th century thought. Homeostasis may be defined more precisely in terms of the idea of stability.

Types of Stability

Stability.   Stability is the quality, state, or degree of being fixed, steadfast, not changing or not fluctuating. Science distinguishes two types of stability: one due to a state of equilibrium and the other due to a steady state condition.

Equilibrium: Type 1 Stability.   Equilibrium is the unique, special relationship between two things (or two systems or two points in space) where the potential for change of a particular quantitative property is zero. Very often potentials for change involve two things (or two points in a system) having different concentrations of a substance or different pressures or different temperatures. If there are no barriers to the interaction of these two points, then the substance will move from the point of higher concentration to the point of lower concentration, or fluid will flow from the point of higher pressure to the point of lower pressure, or thermal energy will move from the point of higher temperature to the point of lower temperature. These interactions will occur until the two points have the same concentration of substance, or same pressure, or same temperature. When the concentration of substance or the pressure or the temperature has become the same, there is no longer the potential for net change, and the two points are said to be in equilibrium with respect to these quantities. For example, one creates a disequilibrium by placing a warm bottle of Coke into a refrigerator; the temperature of the Coke is higher than the air and other contents of the refrigerator. Thermal energy

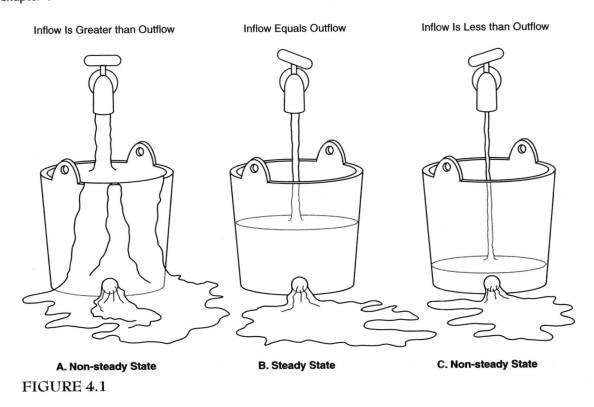

FIGURE 4.1

moves from the Coke to all other contents of the refrigerator. As a result, the Coke becomes colder and the refrigerator contents become warmer. These two opposing trends will collide, so to speak, at some intermediate point where now all contents of the refrigerator will be at the same temperature. When this happens, all contents of the refrigerator are in equilibrium with respect to temperature. At equilibrium there is no net transfer of thermal energy among the contents. The temperature everywhere is the same; it does not fluctuate, and it will not change until some outside influence disturbs the equilibrium.

Steady State: Type 2 Stability. A system is said to be in steady state with respect to a quantitative property when that property shows no net change with time in spite of potentials which tend to change the property. For example, in figure 4.1, the bucket of water is in steady state with respect to the amount of water in it in spite of the incoming water which raises this amount and the out flowing water which lowers it. The steady state results from the inflow equaling the outflow.

Equilibrium vs Steady State. These concepts are similar in that they refer to a quantitative property that does not change with time. The ideas differ in that equilibrium refers to a relation between two systems (or points) whereas steady state refers to the special condition of a single system. A second difference is that a quantitative property of a system in equilibrium with its environment remains constant because the potential between the system and its environment is zero. The same system in steady state remains constant over time *in spite of* a non-zero potential between it and the environment. Neither of these ideas implies one another. Two systems may be in equilibrium with one another but each is not in steady state, see figure 4.2a, or each system may be in steady state but not in equilibrium with one another, see figure 4.2b. Two systems may be in equilibrium with one another and each is in steady state. We call this special situation steady state equilibrium, see figure 4.2c.

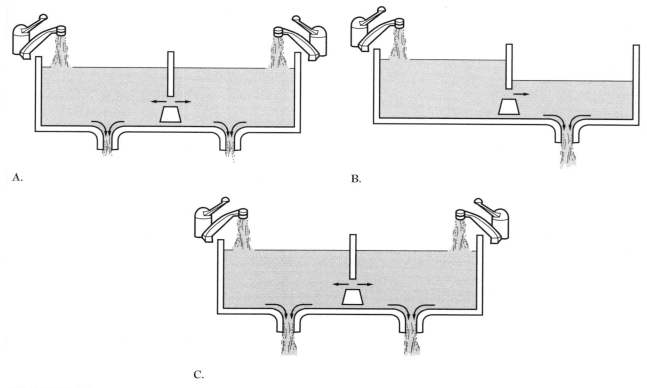

A.

B.

C.

FIGURE 4.2

Homeostasis: Auto-Regulated Steady State

Basic Definitions

Homeostatic System.   When a system is in steady state with respect to a quantitative property, interactions which increase the property are offset by interactions which decrease the property. For example, in figure 4.1, the water leaving it offsets the water flowing into the bucket. Any additional interactions that either increase or decrease the property will disturb this balance and cause the system to no longer be in steady state. With respect to figure 4.1, this means, for example, that opening the water faucet more or closing down the opening for water outflow would cause the water to rise, thus disturbing the steady state. For example, the human body is a homeostatic system with respect to temperature. If one's body temperature begins to rise as a result of being in a warm environment, the individual loses more heat energy by sweating thus bringing the temperature back toward the steady state value. If the body temperature begins to fall as a result of one being in a cold environment, the individual begins to shiver to produce more heat energy thus bringing the temperature back toward the steady state level. A homeostatic system is one that automatically makes adjustments to any interaction that disturbs steady state. The interaction that disturbs the steady state is called a stimulus which leads to an adjustment called a response, and the overall process is a linear sequence of interactions called a homeostatic stimulus-response process. The automatic adjustments are specified in terms of preset value, effectors, and negative feedback regulation.

Preset Value.   A homeostatic system may exist in one of many possible steady state values. For example, the temperature of one's house may be set at 65°, 66°, 67°, or some other higher or lower temperature. The particular steady state level, or desired steady state level, is called the *preset value.*

Effectors.   Effectors of a homeostatic system are those mechanisms that raise or lower the value of a regulated property. For example, a temperature homeostatic system may contain the following effectors: heating and cooling mechanisms that can raise and lower respectively the temperature of the system.

Negative Feedback Regulation.   The mechanism by which a homeostatic system autoregulates a quantitative property is called negative feedback regulation. This phenomenon is not a thing or specific event; rather it is a process which can be understood only in terms of the operation of a homeostatic system as a whole. We may obtain a preliminary understanding of this process by first describing its two component processes: negative regulation and feedback regulation.

Negative Regulation Has Been a Dominant Idea for Centuries.   Families, tribes, and societies sometimes regulate human behavior not by telling a person what he or she can do but rather by telling him what he must not do. These negative regulators are in the form of taboos, moral codes, and laws which seek to eliminate those behaviors that would disturb the peaceful coexistence of people living together. Very often when one is learning a new activity such as how to swing a tennis racket, he/she is told, "No, not that way; that's too much, and no, not that way; that's too little." Out of all these negations emerges a mean suited to the physical and mental characteristics of the person learning tennis. With respect to a homeostatic system, negative regulation negates any trend away from the preset value. If the trend is to go above the preset value, effectors are activated to decrease the property; if the trend is to go below the present value, effectors are activated to increase the property. For example, if the room temperature rises above the preset value, the air conditioner automatically turns on. This, of course, lowers the temperature. Conversely, if the room temperature falls below the preset value, the air conditioner automatically turns off, and the heating unit turns on resulting in raising the temperature. In terms of previous definitions we say that a stimulus to a homeostatic systems leads to a negative regulation as the response.

Feedback Regulation Refers to the Cyclical Interactions Involved with a Homeostatic System Regulating Itself.   A particular negative regulation always involves an "evaluation" of a trend away from a preset value and then an "appropriate command" to counteract this trend. As a result, the system makes an adjustment; i.e., negative regulation is an adjustment. This adjustment then is "fed back" into the system's evaluation-command processes thus producing a second adjustment to the first adjustment. For example, a homeostatic system may activate effectors to decrease its temperature as a first adjustment to the temperature being above the preset value. This adjustment, which may bring the temperature below the preset value, is fed back into the system resulting in the system making a second adjustment which raises the temperature. Hence feedback regulation is the perpetual cycle: adjustment #1 leads to an adjustment to adjustment #1, which leads to an adjustment to adjustment #2, which leads to an adjustment to adjustment #3, and so on. In general, any particular negative regulation of a homeostatic system is a response that simultaneously is a new stimulus to that system. Thus, negative regulation is the linear sequence of interactions beginning with the stimulus and ending with the homeostatic response and may be represented by a straight line. When the line is bent around so that the end point is superimposed on the beginning point, the straight line becomes a circle. In like manner, when the negative regulation response becomes a new stimulus, the linear homeostatic stimulus-response process becomes a homeostatic cycle, and the structure of the system that makes the negative regulation response a new stimulus is the feedback loop that converts a linear process into a cycle.

Negative Feedback Regulation Then is the Perpetual Cycle of One Negative Regulation Leading to a Second Negative Regulation Leading to a Third and So On Indefinitely.   As a result of this perpetual cycle, the property of a homeostatic system oscillates about a particular preset value. These oscillations represent the efficiency of autoregulation. A slow, wide oscillation represents low efficiency; a fast, narrow oscillation represents a high efficiency. We may obtain a better understanding of this regulating cycle by studying the operation of a homeostatic system.

Operation of a Homeostatic System

Component Processes.   Any homeostatic system has five "parts": (1) a property that can increase or decrease, (2) a sense receptor, (3) an evaluation unit, (4) a command unit, and (5) effector(s). Some properties of a system are not subject to autoregulation. For example, humans can regulate their weight but humans cannot regulate their height. A sense receptor represents the value of a property and transmits this information to the evaluation unit. For example, a mercury thermometer records various temperatures by varying the height of a column of mercury. The thermometer may be connected to wires that vary the amount of current sent to an evaluation unit according to the height of the mercury column. Such a "wired" thermometer would be a sense receptor. The evaluation unit compares each piece of information sent to it to a preset value. In terms of the previous example, a particular preset temperature would correspond to a particular amount of current. Temperatures above the preset value would cause the sense receptors to send an amount of current greater than this preset amount. Temperatures below the preset value would cause the sense receptor to send an amount of current less than the preset value. The command unit receives information from the evaluation unit and stimulates the effectors to make the appropriate adjustment. As already indicated, effectors bring about changes of the system with the net result that the property either increases or decreases. If the property is above the preset value, the effectors cause a net decrease of the property. If the property is below the preset value, the effectors cause a net increase of the property.

These five parts are interconnected to produce a sequential and circular set of communications. The property of the system always stimulates the sense receptor which stimulates the evaluation unit which then stimulates the command unit which causes the effectors to change the property. This completes one cycle which immediately leads to another cyclical set of interactions, see figure 4.3.

An Example: Autoregulation of Room Temperature.   Consider the operation of a temperature-controlled room with a heating unit and a cooling unit. Suppose we want the room to be at 68°F (20° C). The temperature of the room will be recorded and sent to the evaluation unit which will compare this temperature to the preset value. The evaluation will be sent to the command unit which will cause the effectors to make the appropriate adjustment. If the temperature is above 68° F, the command unit will cause the heating unit to turn off and the cooling unit to turn on. If the temperature is below 68° F, the command unit will cause the heating unit to turn on and the cooling unit to turn off. This negative

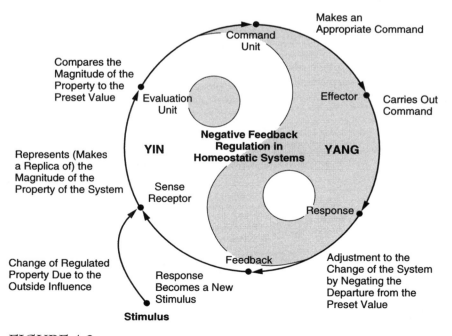

FIGURE 4.3

regulation will be fed back into this cycle of interactions; that is, the changed room temperature will stimulate the sense receptor thus starting a new cycle.

Positive Feedback

The human body is a biochemical machine that maintains a steady state temperature (97.6° F) by means of negative feedback regulation. Chemical reactions in cells throughout the body continually release thermal energy which raises the body temperature. The body has adjustment mechanisms, e.g., sweating, that negate this continual rise in temperature. The body's tendency to always increase its temperature results from a positive feedback process in the following way: (1) chemical reactions release thermal energy resulting in an increase in body temperature, and (2) increase in temperature "causes" chemical reactions to speed up, thus more rapidly releasing thermal energy and thus more rapidly increasing body temperature which leads to a still greater rate of chemical reactions and so forth. In this example, the positive feedback occurs within an organism but itself does not involve a stimulus-response process. Rather the chemical reaction process—like any process in nature—releases thermal energy which increases temperature. For purposes of description, we say the chemical reactions is an internal event of the system that leads to an output of thermal energy that simultaneously is an input to the system that results in an increase in the rate of the chemical reactions. The internal event also may be represented as a linear sequence: rate of chemical reaction → rate of release of thermal energy → rate of increase of temperature. The output of the system simultaneously is an input that enhances the internal events of the system and so is called a positive output because it increases rather than negates the internal event. Because the positive output which equals the increase in temperature simultaneously leads to an increase in the rate of the chemical reactions, we say the positive output is "fed back" to the first "happening" of the linear sequence of "happenings" representing the internal event. The positive output which is "fed back" to become a new input which always increases the internal event is thus called a positive feedback loop. The overall cycle in which this occurs is called a positive feedback process.

Negative feedback mechanisms of the body oppose this positive feedback cycle and thereby prevent it from continuously repeating itself. That is, the body maintains a steady state near its preset value in spite of the presence of a positive feedback loop that could bring the body temperature to escalate above this value. However, if for any one of many possible situations, the positive feedback loop overpowers the negative feedback loop, for example, an old person on a very hot summer day confined to a downtown city apartment with no air conditioning, the body temperature would start to escalate until outside intervention or the person dies. Outside intervention could take any number of different forms, but all successful strategies would bring the temperature down and put the person in a situation where the negative feedback loop could again contain the positive feedback loop.

There are many systems in which the component positive feedback loops bring about a desired result; for example, an immune response to an invading bacteria or virus, formation of a blood clot, stimulation of a nerve leading to a nerve impulse (an action potential that then is transmitted down the nerve fiber), the growth of a new company, the intensification of a romantic relationship. Eventually, however, the positive feedback enhancement process must be contained by negative feedback regulation.

Tearout #1

Directions:

1. Using figure 4.3 as a reference, fill in all labels.

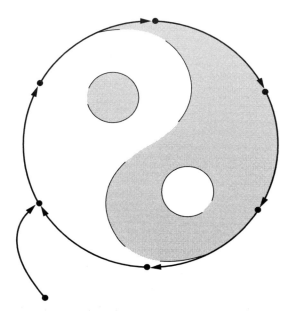

FIGURE 4.3

Tearout #2

Directions:

1. Using figure 4.3 as a reference, describe the negative feedback regulation of air pressure inside an airplane.

2. Describe the positive feedback mechanism for the growth of a business; note: the output of a business is money that can be fed back into developing the business.

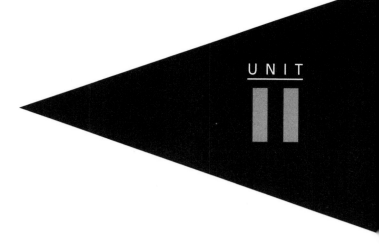
Biological Individuation as Control Of Stress

Cell as the Fundamental Unit of Life

Scientific Method in Terms of the Cell Theory

As described in chapters 1 & 3, the scientific method was first used to give a quantitative description of motion. The laws and concepts relating to this theory of motion require extensive mathematical background. Therefore, this first application of scientific method is not accessible to most people who do not have extensive mathematical training. In contrast, the cell theory is fairly easy to understand, and it exemplifies the most fundamental aspects of the scientific method.

The Cellular Perspective

To a first approximation, all living things can be classified either as a plant or an animal. This sets the stage for developing the cellular perspective which began to emerge in the 17th century. Based on external appearances alone, we may conclude that plants are very different from animals. Yet we now realize that the similarities among these organisms are more profound than their differences. How can a rose ever be considered similar to an elephant or a crab? Despite all differences in appearance, all animals and plants are alike in that they are made up of cells.

Over 150 years elapsed after the first cell was observed under a microscope before biologists were able to make this sweeping generalization. But what is a cell? The meaning of this term has changed during the emergence and expression of the cellular perspective. In 1665 Robert Hooke (1635–1703) described his observation of a section of cork as consisting of numerous cavities reminiscent of the structure of a honey-comb. Hooke imagined these cavities to be like prison cells and called them *cells*. About the same time, Marcello Malpighi (1628–1694) pointed out that many plants have this same "cellular structure." Accumulated observations over the next hundred years indicated that the framework for these cells (cavities) consisted of a fairly rigid, non-living material which surrounded living entities. The term, *cell*, was thereafter used to refer to these living entities, and the non-living material surrounding each cell was called the cell wall. Plant biologists realized that plant cells consist of: (1) an outer limiting cell membrane (adjacent to the cell wall), (2) a membrane enclosed central mass called a nucleus, and (3) a gel-like fluid between the nucleus and the cell membrane called cytoplasm.

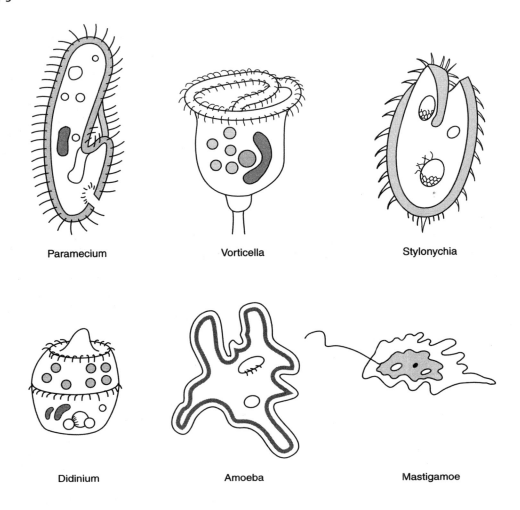

Paramecium   Vorticella   Stylonychia

Didinium   Amoeba   Mastigamoe

Single Celled Organisms

FIGURE 5.1a

Many types of tiny, simple organisms were observed to have these three characteristics, see figure 5.1a & b. Therefore, these life forms were called single-celled organisms. Like any living thing, they require nutrients, give off waste products, respond to stimuli, and can reproduce themselves. Some of these single-celled organisms have cell walls, others do not.

Based on dissection of many different types of plants, Matthias Schleiden concluded in 1838 that all plants consist of cells that somehow cooperate to carry out the life processes. One year later, Theodore Schwann concluded that all animals are composed of cells which cooperate to keep the organism alive. The two men exchanged views and postulated that all living things are composed of one or more cells.

In 1858, Louis Pasteur (1822–1895) demonstrated that "non-cellular" entities called bacteria (the gremlins), could live and reproduce themselves. Later, Pasteur demonstrated that some types of bacteria are the causes of infectious diseases. For example, a sore throat may be caused by bacteria (Streptococci) secreting a toxic substance (a toxin) which destroys the cells lining the throat. Bacteria consist of structural units which, like single-celled organisms, can survive and reproduce themselves independent of one another. However, a typical bacterium is about one-tenth the size of small animal cells (such as human red blood cells). Furthermore, bacteria do not have nuclei or any of the other inclusions found in most plant and animal cells.

Must we infer, then, that there are two types of living organisms, one cellular and the other bacterial? From various studies carried out in the second half of the 19th century and the first half of the 20th century, biologists concluded that the cytoplasm in plant and animal cells and the gel-

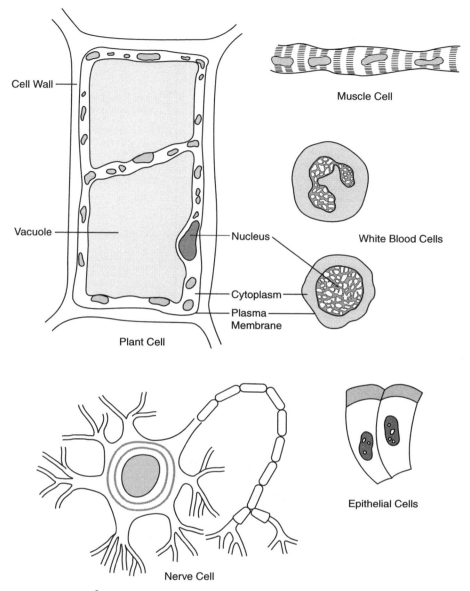

FIGURE 5.1b

like substance in bacteria are similar. This "life stuff" called protoplasm, has a similar composition and molecular organization in bacteria, plant cells, and animal cells. Thus, a cell should be redefined as a structural unit consisting of a membrane surrounding protoplasm.[1] Cells are broadly classified into two general types: prokaryotes (*pro* means before and karyote means nucleus) and eukaryotes (*eu* means true). Eukaryotes, the "true" cells, have a nucleus and other inclusions which are called cell organelles.

Armed with the above distinctions, we may summarize the most general version of the cell theory by means of four postulates. First postulate: all living things are composed of cells and cell derivatives. Second postulate: the function of all multicellular organisms may be understood in terms of the functions and interactions among cells. Third postulate: all living things begin life as a single cell derived from another living cell. Fourth postulate: the cell is the lowest level of organization that may be said to be alive. There are a few exceptions to one or more of these postulates; however, in general, the cell theory provides a useful framework for describing virtually all life forms.

---

[1]The current cell theory no longer uses the term, protoplasm, and gives greater significance to cell membranes.

Examples of Operational Terms

Cell. The four postulates that give a theoretical definition of a cell depend upon a more fundamental operational definition of the cell. To only define a cell as the fundamental unit of life is unscientific. This definition becomes acceptable and very useful when it is based on any variation of the following operational definitions: a cell is that "such and such" structure (e.g., enclosed vesicle with a nucleus) one sees when one prepares a portion of living matter in a particular way, sections it, stains it with some dye, and then observes the stained sections by means of a microscope.

Distance. Distance is a word signifying our experience of spatial reality. One would be hard pressed to define distance in a way that represents all subjective experiences of it. Each time we use the word it may signify something different to us, and of course, the meaning of the word will vary from one person to another. Lynne Walker, a poet friend of mine, wrote about this subjective aspect in this way:

> The following are three examples of subjective experiences of distance: (1) Someone you love is in California, neither of you has the resources to travel across the country. The distance between you is more than miles, it may as well be as far as the stars. Your loneliness and inability to cover the distance between loved ones makes you feel that the distance is great indeed. (2) Perhaps you just need to travel across the state to visit some dear friends you haven't seen in a long time. My subjective experience of this is that the distance seems increased on the way there because of expectations, of being anxious to see them. On the way back, the distance doesn't seem as long, as you are not as anxious to return. (3) Perhaps the distance is even shorter. Perhaps you see a friend across the street. You want to get their attention and yell, but they do not hear you because the distance between you is too far for the sound to travel. This distance is not far at all in terms of measurement, but is overwhelming because emotion is involved. In my mind I keep hearing the song by Carol King, "So Far Away". . . doesn't anybody stay in one place anymore. . . It would be so good to see your face at my door. . . You're so far away. . .

By defining distance in terms of the operation of measurement, we insure that everyone will have the same understanding of it. We pick some arbitrary length (a specific distance) as an agreed upon standard unit of length. We then compare this unit to any other distance, and in particular count how many of these units are contained in the "measured" length. Thus, the subjective relativity of distance experienced by Lynne becomes objectified:

> The distance to loved ones in California is 2,400 miles. The distance to friends across the state is 240 miles. The distance to the friend across the street is 20 yards.

But what is distance really? Mechanistic philosophy proclaims that distance torn from any particular context and defined by measurement is truly real. An expanded science recognizes that measurement imposes limitations on our experience of distance. In any attempt to control nature, the measurement definition of distance is absolutely necessary, but it also inhabits our participation in nature. To insist on an exclusively quantitative understanding of distance is to repress the direct experience of the incomprehensible reality in which it is emersed.

Time. We may discuss time in the same way. Time has meaning in relation to the particular context in which it is experienced, but for utilitarian purposes, time may be torn from any particular context and objectified by a process of measurement. Lynne Walker dramatically summarizes the contrast in this way:

> *Subjective:* The following are three examples of experiences of subject time: (1) It's Friday afternoon. It has been a grueling week at work. It is payday and there are big plans for the weekend. Just two more hours and I will be released from work. The clock seems to stand still. The minutes drag and two hours stretch to double that amount. (2) I had to get a shot in the arch of my foot. The doctor said it would hurt, but that it would only last a second. It was a very long second. (3) Many years ago when I was pregnant, people called me each day as my due date approached. "What are you doing?" they would ask. "I'm making a baby, and it seems to be taking a very long time. In fact, I think I'll always be pregnant." I wasn't always pregnant; I was in labor for sixteen hours. They were the longest hours of my life. I was convinced that I couldn't live through

another second of pain. Since the birth of my children, time has been passing faster and faster every year. I can't figure out where all the years have gone. It seems like yesterday that my sons were born and now they are men. Again, a song comes to mind, "Sunrise, sunset," . . . quickly flow the years. You're born, you grow old, you die. There isn't enough time.

Objective. There are 60 seconds in a minute, 60 minutes in an hour, 24 hours in a day, 7 days in a week, 365 days in a year. Time is constant when measured.

Empirical Observation

The first postulate of the cell theory is an example of induction from many empirical observations. This also is a limitation of what one can know "scientifically." Science cannot study the birth of Don Pribor or of Lynn Klapp Davies, but science can study the birth process in humans which is being repeated millions of times.

Analytic-Synthetic Thinking

A preliminary understanding of reflex activity in animal nervous systems is possible with relatively little technical background. This theory also nicely illustrates analytic-synthetic thinking in science.

Some Nervous System Definitions.   The nervous system consists of a network of intercommunicating cells called neurons. A neuron consists of (1) a cell body which contains the nucleus, (2) a single extension called an axon, and (3) one or several, usually short, cell extensions called dendrites, see figure 5.2. A neuron can generate electrical activity in response to a stimulus. This localized electrical activity may be transmitted to other parts of the neuron, which may respond by generating new electrical activity or by secreting specific hormone-like substances. The transmitted electrical activity is called a nerve impulse, and the cell generated electrical or chemical events are called neuron signals.

A neuron signal may stimulate a second neuron or an effector cell, which is a muscle cell which when stimulated contracts or a gland cell which when stimulated secretes a substance. Correspondingly, we say the first neuron "communicates" with the second neuron or with the effector cell to form a communication network. These communication networks may consist of many cells of which the component neurons are classified according to their function within the network as follows: (1) sensory neuron, (2) interneuron, and (3) motor neuron. A sensory neuron has an axon which terminates in one or several sense receptors which communicate with the external or internal environment of an animal; e.g., some pattern in the environment stimulates the sense receptor membrane to generate electrical activity that, in turn, is transmitted up the sensory axon. An interneuron is stimulated by a signal from one neuron and, in turn, stimulates another neuron to generate a neuron signal. A motor neuron communicates with one or more effector cells, see figure 5.3. The junction between two cells where neuron-neuron or neuron-effector communication occurs is called a synapse. Several interconnected neurons which as a group receive and transmit signals is called a neuron circuit. The simplest neuro-effector network of cells is called a reflex arc.

Analytic-Synthetic Model of Reflex Activity.   (1) Reflex activity is accomplished by a reflex arc which usually consists of the subunits: sensory neuron, interneuron, motor neuron, and effector cell. (2) Each of these components have specific structural and functional characteristics, and all of them have some characteristics common to all cells and specific characteristics necessary for generating and transmitting electrical activity. (3) The subunits are interconnected as shown in figure 5.4. How these cells interact could be described by the theory for transmission of electrical activity across a synapse. (4) The synthesis is as follows: A stimulus interacts with sense receptors which generate electrical activity. The resulting nerve impulse travels to the first synapse where it stimulates the interneuron to generate electrical activity. The interneuron impulse travels to the second synapse where it stimulates the motor neuron to generate electrical activity. The motor neuron impulse travels to the third synapse where it stimulates the effector to generate electrical activity which in turn stimulates the effector to carry out a reflex response, that is, muscle contraction or gland secretion. Thus, the whole reflex activity is understood in terms of interactions among the component parts of the system.

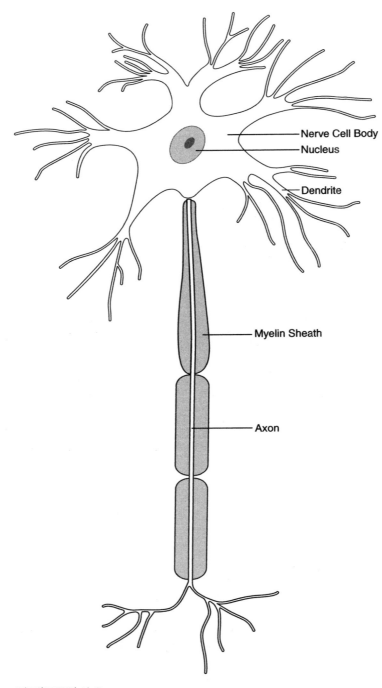

- Nerve Cell Body
- Nucleus
- Dendrite
- Myelin Sheath
- Axon

FIGURE 5.2

Experimental Prediction

Experimental prediction may be illustrated by testing the theory of osmotic behavior of hemolyzed red blood cells. Red blood cells, see figure 5.5, are red because of containing high concentrations of a protein pigment called hemoglobin which carries oxygen in the blood.

When these cells are put into water, they swell and are "hemolyzed," a process in which the hemoglobin molecules move out of the red blood cells. The resulting hemolyzed cells are called *ghosts* because they are difficult to see under a microscope after having lost most of the red pigment that made them visible in the first place. One hypothesis proposes that these ghosts have holes in the

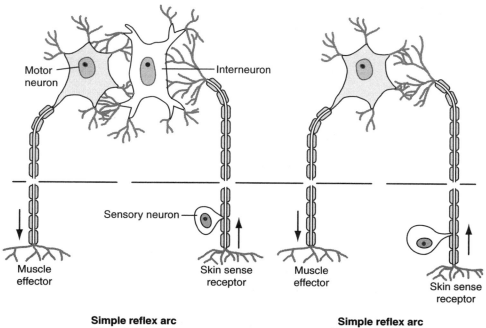

**Simple reflex arc
with an interneuron**

**Simple reflex arc
without an interneuron**

FIGURE 5.3

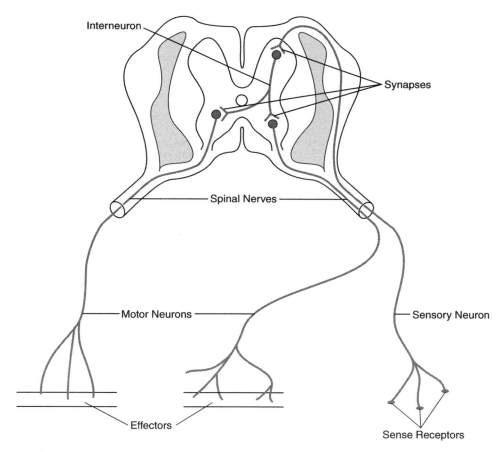

FIGURE 5.4

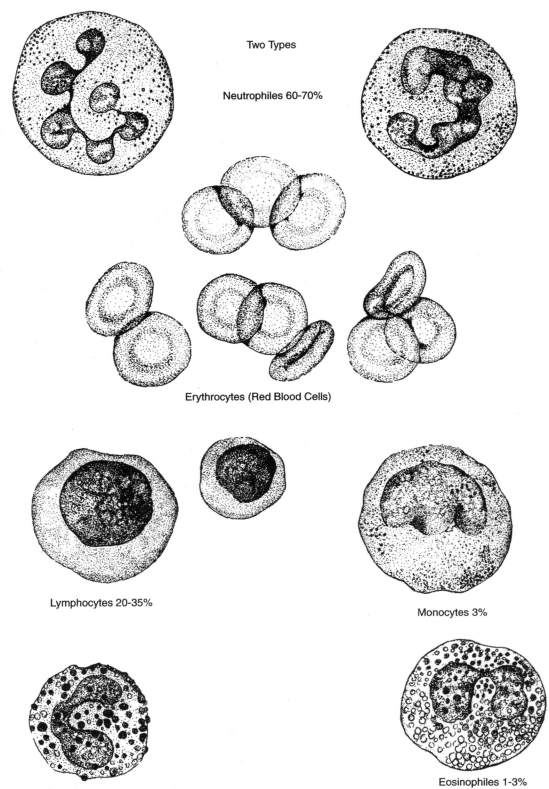

Two Types

Neutrophiles 60-70%

Erythrocytes (Red Blood Cells)

Lymphocytes 20-35%

Monocytes 3%

Basophiles 0.3%

Eosinophiles 1-3%

FIGURE 5.5

cell membrane through which hemoglobin moved out. Another hypothesis proposes that the cell membrane of ghosts become reconstituted implying that there are no holes in the membrane. Rather the membrane again becomes "semipermeable" meaning that water can freely move across the membrane but a somewhat larger molecule like sucrose (table sugar) cannot move across the membrane. We can test which of these hypotheses is correct, but first we need the concepts of diffusion, water solution, and osmosis.

Diffusion is the movement of a substance from an area of higher concentration to an area of lower concentration. Wind results from air moving from an area of higher to an area of lower concentration, that is, from higher pressure to lower pressure area. A sugar cube at the bottom of a cup of coffee eventually diffuses throughout the coffee so that the concentration of sugar is everywhere the same. Before this occurs we say there is a diffusion potential, that is, a potential for diffusion to occur. Diffusion potentials with respect to a cell are specified by a cell membrane that separates a water solution inside the cell that is very different from a water solution outside the cell.

A water solution consists of one or more substances "dissolved in water." "Dissolved in water" means that each molecule or atom of the dissolved substance is surrounded by and kept separate from other molecules or atoms of the same type by many water molecules. For example, a solution of sugar such as glucose consists of individual glucose molecules separated from one another by many water molecules, see figure 5.6. The dissolved substance is called a solute, and the fluid in which the solute is dissolved is called a solvent—water is the solvent of water solutions. The number of solute molecules per unit volume is called solute concentration and for water solutions, the number of water molecules per unit volume is the water concentration. As can be seen from figure 5.6, the greater the solute concentration, the less the water concentration, and conversely, the less the solute concentration, the greater the water concentration.

Osmosis is the diffusion of water across a membrane, such as a cell membrane, from an area of lower solute concentration (higher water concentration) to an area of higher solute concentration (lower water concentration). Osmosis usually refers to a situation where the membrane is semipermeable, that is, it allows water to move across but does not allow a solute to move across. We now can illustrate systematic experimentation.

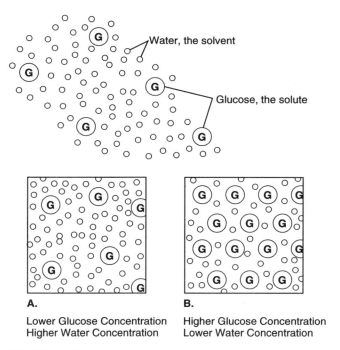

A.
Lower Glucose Concentration
Higher Water Concentration

B.
Higher Glucose Concentration
Lower Water Concentration

FIGURE 5.6

In systematic experimentation we put the red blood cell ghosts in a sucrose solution and observe what happens. Our observations can be related to the scientific ideas of cell theory, molecular theory, diffusion, water solution and theory of osmosis.

In particular, we may test in three steps the hypothesis that red blood cell ghosts have reconstituted cell membranes. First, we would predict that according to this hypothesis and the already known scientific myths, if we place ghosts in a sucrose solution, osmosis will occur where water inside the ghosts will move out resulting in the ghosts shrinking. Second, we would set up an experiment where we could put ghosts in a sucrose solution and then observe what happens, or better still, measure the average ghost volume change. Third, after repeated experimentation we should be able to conclude that the hypothesis is correct because the ghosts consistently decreased in volume; or the hypothesis is incorrect because the ghosts consistently did not change in volume.

Recapitulation

There are at least four characteristics of the scientific method: (1) operational terms, (2) empirical observations, (3) analytic-synthetic thinking, and (4) scientific public consensus of truth. The first three characteristics and some version of public consensus of truth characterize other types of objective knowing. The distinguishing feature of mechanistic knowing is systematic experimentation which specifies the first three characteristics in such a way that they lead to scientific public consensus of truth.

Mechanistic thinking involves using the above four characteristics to guide one in devising a specific procedure to solve some problem. Working out such a procedure may require only a little knowledge and some common sense, for example, finding out whether a light bulb is burned out or not; I leave it to the reader to devise an appropriate procedure for doing this and to see how the procedure exemplifies the four characteristics of the scientific method.

Mechanistic thinking of a "successful" scientist requires considerable specialized knowledge and training and various kinds of creativity. First of all, one must have enough knowledge, technical training, and creativity to choose a problem that has a high probability that it can be solved in the time period one has available. Many would-be scientists failed to obtain a graduate degree or failed in their science careers because they were unable to pursue a problem that was solvable with respect to their abilities and available resources. After choosing a solvable problem, a person must create one or more of the following: new technology, new experimental procedures, new mental skills for representing and analyzing empirical observations, and new scientific theories. The next chapter will indicate some of the great technical, mathematical, and physical-conceptual innovations that produced mechanistic solutions to the problem of describing motion.

Let me conclude this section with an important qualification. Scientific thinking often is said to involve the scientific method, but this is only partially true. On the one hand, all scientific solutions to problems will exhibit the four characteristics described earlier. On the other hand, each problem makes its own peculiar demands on the would-be scientist. In other words, there is no one scientific method; rather there are as many scientific methods as there are different problems about how things or events interact in nature.

The Chemical Makeup of Cells

The chemical composition of cells has been the subject of much biological investigation in the 20th century. Over 99 percent of the complex structures of all cells are made up of various combinations of just 10 elements listed in Chapter 8. Sodium, potassium, calcium, and chlorine are present as ions in or surrounding cells. The first six elements (CHNOPS) are present in thousands of different types of (organic) compounds, the simplest and most abundant of which is water. All the other organic compounds may be grouped into four types: (1) sugars (carbohydrates), (2) lipids, (3) proteins, and (4) nucleic acids and nucleotides. This chapter describes the biological significance of the general properties of water and of these four types of organic compounds.

Water

A large portion of a cell (usually about 40 to 60 percent) consists of water, and an active cell must be surrounded by water. This substance has many unique properties that make it essential for all life forms. First, it takes a relatively large amount of heat energy to raise the temperature of liquid water or to cause water to evaporate. One way to appreciate this fact is to contrast water with ethyl alcohol (C_2H_5OH), which as one can see from its formula contains more elements and is therefore larger than a water molecule. Because alcohol molecules are larger than water molecules, one would expect that it would take more energy to make alcohol evaporate than to make water evaporate. However, if one pours alcohol on his arm, almost immediately he feels cool at that place. This is so because alcohol evaporates right away and in the process removes heat energy from the arm. Water, in contrast, would remain on one's arm much longer because it takes much more heat energy to convert liquid water into a gas.

As a result of the preceding property, water tends to stabilize the temperature of its environment wherever it is found. This stabilization is very important for living things because life processes can go on only within a fairly narrow temperature range. In mammals the optimum body temperature is around 37° C (98.6° F). The large amount of water in mammals prevents the body temperature from easily decreasing or increasing. The body must lose a great deal of heat energy before the temperature goes down very much, and conversely, the body must gain a great deal of heat energy before the temperature will greatly increase.

A 2.8 to 5.6° C (5 to 10° F) rise in body temperature is much more dangerous to life than a temperature drop of the same magnitude. Chemical reactions in all the millions of cells of the body continually release heat energy which would raise body temperature considerably if there were not an efficient mechanism for removing excess heat energy. In hot climates mammals cool themselves by producing a salt solution know as sweat. Heat energy from the body is used to make sweat evaporate, thus cooling the body by drawing off excess heat. Since it takes a relatively large amount of heat energy to make water evaporate, sweating is a very effective cooling process. If the air already contains a lot of water vapor (i.e., if it is very humid), then evaporation of sweat is greatly retarded. In such cases, we remain overheated and become progressively more uncomfortable.

A second equally important property of water is that it is a good solvent for a large number of biologically important compounds. A solvent is a liquid in which another substance dissolves. When a salt such as sodium chloride is put into liquid water, the resulting solution is composed of water and positive sodium and negative chloride ions. Sodium and potassium ions are essential for the production and transmission of nerve impulses. In addition, these ions are the means by which active transport of essential substances occurs across cell membranes. Moreover, because water dissolves many different types of compounds, it provides a medium for chemical reactions to occur. The potential reacting molecules have a better chance of bumping into one another as a result of being dissolved in water. Furthermore, many chemical reactions occur much more rapidly at a particular acidity. An *acid* is a substance that separates in water solution into positive hydrogen ions and some negative ions. Thus, degree of acidity is determined by the amount of acid dissolved in the water. Some substances, such as hydrogen chloride gas, become acids when dissolved in water, that is, the hydrogen chloride molecules separate into positively charged hydrogen ions and negatively charged chloride ions, resulting in a highly acidic solution called hydrochloric acid. Hydrochloric acid secreted into the stomach facilitates many of the chemical reactions known as digestion.

Third, water is a very stable compound. As a result, even though diverse types of chemical reactions occur in it, the water molecules remain intact throughout the reaction.

Fourth, water is transparent to light. Plants use light energy to build organic molecules. Without light, plants would die. Because water is transparent, plants can survive several feet below the surface of water. Correspondingly, animals, which feed on these plants, also are able to survive.

Fifth, water has a high surface tension. This property relates to the fact that water molecules tend to bind to one another because of their unevenly distributed electrical charges, see figure 5.7. As a result, it is relatively difficult to remove a single water molecule from the surface of this liquid. All

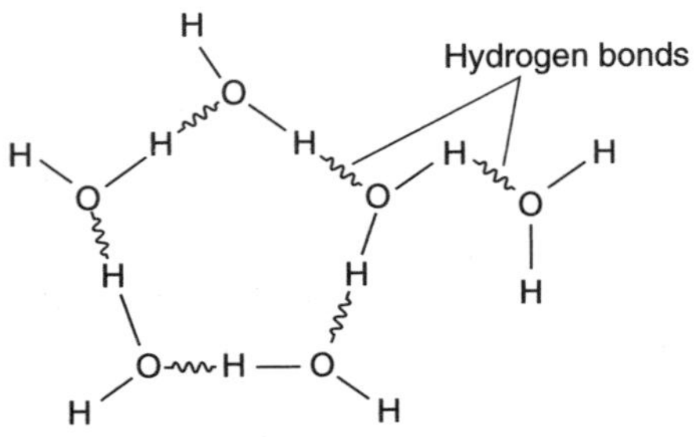

FIGURE 5.7

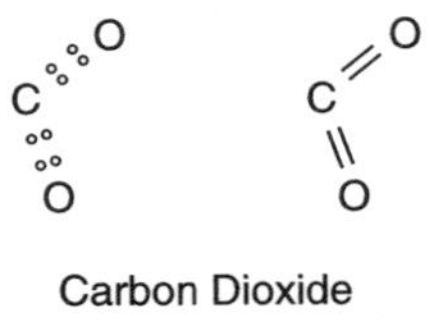

Carbon Dioxide

FIGURE 5.8

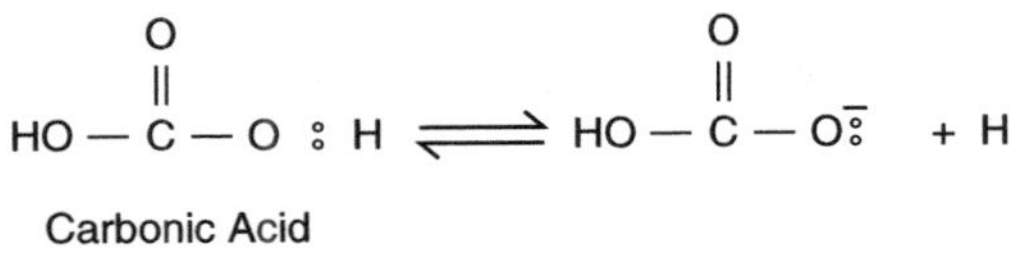

Carbonic Acid

FIGURE 5.9

the molecules tend to "cling" to one another, see figure 5.7. This property leads to water "wetting" the surface of containers. For example, we have to dry dishes we wash or wait a long time before the water evaporates. In order to dry a dish in a hurry, one should rinse it in alcohol; the alcohol evaporates almost immediately. This surface tension property of water is very important to land plants, particularly trees. As a result of it, water moves from the soil into roots and up long tubes to leaves that may be 15.2m (50 ft) above the ground.[2]

Sixth, liquid water at or below 4°C is denser than is the solid at or below 0°C. Consequently, in a body of water such as a lake, ice will form from the top down rather than from the bottom up. During winter this allows organisms to survive in deeper waters.

Organic Compounds

The other types of biologically important compounds are called "carbon compounds" because they all contain carbon in some form or another. Under ordinary conditions, carbon tends to share electrons with other atoms or with other elements via covalent bonds. It can share these electrons in a number of ways. For instance, in carbon dioxide each of the two pairs of carbon electrons teams up with an electron pair from an oxygen atom to produce a double bond (two pairs of shared electrons) with oxygen, see figure 5.8. In carbonic acid, the carbon is double-bonded to one oxygen and has two single, somewhat polar bonds with two other oxygen atoms, see figure 5.9. Carbon also can form triple bonds with itself and with other atoms. In addition, carbon also can form straight or branched chains of atoms and ring shapes containing from as few as three atoms to as many as thousands of atoms.

Carbon is a very chemically active element, and the bonds it forms are very strong. These properties make carbon unique among elements and result in a virtually limitless variety of carbon compounds. Carbon compounds involving at least one carbon atom on a nonpolar bond with some other atom are called organic compounds, organic because some of them are the major constituents of organisms. Compounds not classified as organic, even if they contain carbon, are called inorganic compounds.

Usually organic compounds consist of several connected carbon atoms that have attached to them various atoms or groups of atoms known as side groups. Both the nature and the position of these side groups determine the properties of the organic compound; so two such compounds containing the same number and type of atoms may have quite different properties. For example, the sugars, glucose, fructose, and galactose have the same chemical formula but have different properties corresponding to different arrangements of side groups on the carbon chain, see figure 5.10. Thus, in representing organic compounds one must indicate the relative position of the side groups. Certain conventions are used to do this. A single dash line represents a single nonpolar or polar covalent bond, and a double line represents a double covalent bond. Inspection of the list of side chains below shows that by convention dash lines are not always used in picturing some of these bonds.

---

[2]Note: This statement is based on the *cohesion-tension theory* of water movement in plants.

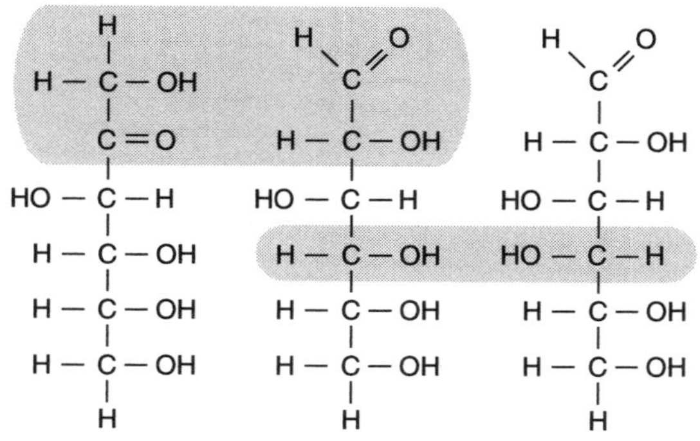

Fructose   Glucose   Galactose

FIGURE 5.10

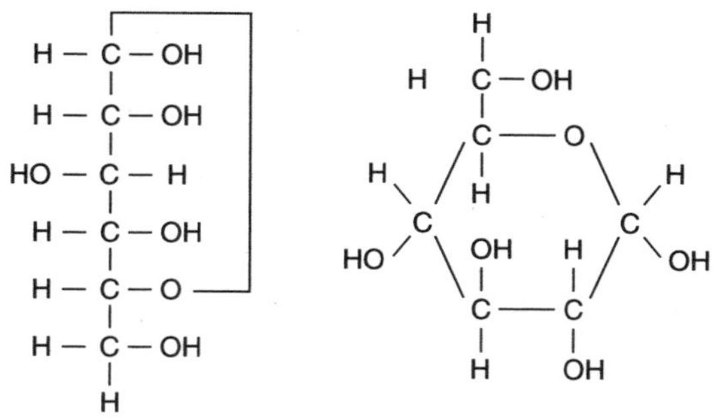

Different Representations of Glucose

FIGURE 5.11

-CH₃	methyl group
-OH	alcohol group. Ethyl alcohol which has two carbons with one alcohol group is the alcohol in beer, wine, and hard liquor

O
-C   carbonyl group
       1) O
       -C-H if the carbonyl group is at the end of a carbon chain, it is an aldehyde group
       2) O
       -C-C if the carbonyl group is surrounded by carbons on each side then it is a ketone group

O
-C-OH this is called a carboxyl group and sometimes is represented as COOH. The bond between the oxygen and hydrogen is more polar than in the alcohol group; so in solutions some of the carboxyl groups dissociate into a hydrogen ion and a COO⁻ ion.

-NH₂   amino group. The nitrogen in this group has a pair of unshared electrons, i.e.

$$H$$
$$N:$$
$$H$$

which tend to make it negative. As a result the nitrogen attracts positive hydrogen ions; so amino groups behave as bases (they bond to the positive hydrogen ions produced by acids in water solution).

Sugars

Sugars often are called carbohydrates because they are chains of carbon atoms that have been hydrated; that is, water molecules, which may be represented as H-OH, have been added to them; an H atom goes on one side of a carbon and an OH group goes on the other side. More precisely, sugars are defined as carbon chains with at least two hydrated carbons and another carbon double-bonded to an oxygen. Therefore, the smallest sugar molecule has three carbon atoms. Some carbohydrates, called compound sugars, can be broken down into simpler molecules which also are sugars. A simple sugar also called a monosaccharide (mono means "one"), is a substance that cannot be broken down into simpler molecules that have the defining characteristics of a sugar. Simple sugars may have as many as 10 carbon atoms and thus be fairly large molecules. Cells contain four or five carbon sugars, but the most abundant sugar is glucose which contains six carbons ($C_6H_{12}O_6$), see figure 5.11.

Not only is glucose the most abundant sugar in cells, it has properties that make it ideally suited as the primary source of energy for cells. First, it is quite soluble in water because of its 4 OH groups which are somewhat polar and because it is a relatively small molecule. Second, it is small enough to

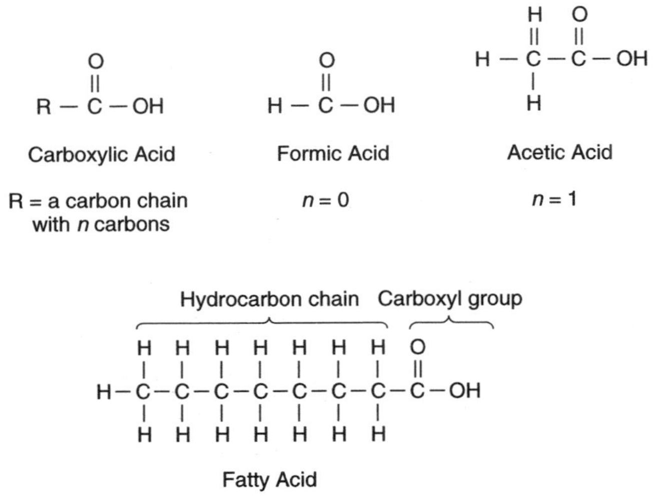

$$O$$
$$\|$$
$$R - C - OH$$

Carboxylic Acid

R = a carbon chain
with n carbons

$$O$$
$$\|$$
$$H - C - OH$$

Formic Acid

$n = 0$

$$H \quad O$$
$$\| \quad \|$$
$$H - C - C - OH$$
$$\|$$
$$H$$

Acetic Acid

$n = 1$

Hydrocarbon chain   Carboxyl group

$$\begin{array}{ccccccccc} & H & H & H & H & H & H & H & O \\ & | & | & | & | & | & | & | & \| \\ H- & C- & C- & C- & C- & C- & C- & C- & OH \\ & | & | & | & | & | & | & | & \\ & H & H & H & H & H & H & H & \end{array}$$

Fatty Acid

FIGURE 5.12   Fatty acids; derivative of carboxylic acids

diffuse across membranes into cells. Third, the pairs of electrons in the carbon-hydrogen bonds in glucose are at fairly high energy levels. When these electrons are transferred to lower energy levels in other molecules, a great deal of energy is released.

A compound consisting of two simple sugars is called a disaccharide. Polysaccharides are molecules consisting of three or more simple sugars; i.e., they are polymers of simple sugars. Disaccharides and polysaccharides may consist of any of the simple types of sugars, and the constituent monosaccharides need not be identical. Thus, whereas maltose consists of two glucose molecules, common table sugar, sucrose, consists of glucose and fructose. Polysaccharides provide the cell with a convenient way of storing energy. The cell gets energy to sustain life by continually burning glucose. Excess glucose molecules would get in the way of these reactions. The cell "solves" this problem by uniting several hundred glucose molecules to form very large polysaccharides. In plants, such macromolecules (*macro* means "large") are called starch molecules. In animals, the glucose units are arranged somewhat differently than in plants, and so the resulting macromolecules are a kind of starch called glycogen, which are produced and stored in granules in liver and muscle cells.

Lipids

Lipids are a very diverse group of fat-like compounds such as fats, phospholipids, and steroids that share one property: they are insoluble in water. One important type of compound, called a hydrocarbon, consists of a chain of carbon atoms bonded to hydrogen atoms. Since the electrons in the carbon-hydrogen bonds are in high energy levels, these compounds contain a large amount of chemical potential energy. Hydrocarbons present in petroleum and coal provide a major energy source for industry and personal needs. However, these very nonpolar substances (because of the predominance of C-H bonds) are not useful to cells because they are insoluble in water. As a result, they cannot participate in reactions that would make their high chemical potential energy available to cells. In fatty acids this problem is partially overcome because the hydrocarbon chain is bonded to a carboxyl group (COOH), see figure 5.12. The carboxyl group makes one end of the fatty acid molecule soluble in water. It accomplishes this because the carboxyl end of the fatty acid tends to separate in a water environment into a hydrogen ion and a negatively charged fatty acid ion. Because of this, one end of fatty acid molecules can enter reactions that will release some chemical potential energy. Such reactions (oxidations) convert the original molecule into a smaller molecule also having a carboxyl acid at one end, as shown in figure 5.13. As a result, this smaller molecule also can release a portion of its chemical energy to the cell. In effect, the ends of fatty acids are chopped off thereby releasing energy and producing new smaller fatty acids. This process goes on until the whole molecule is used

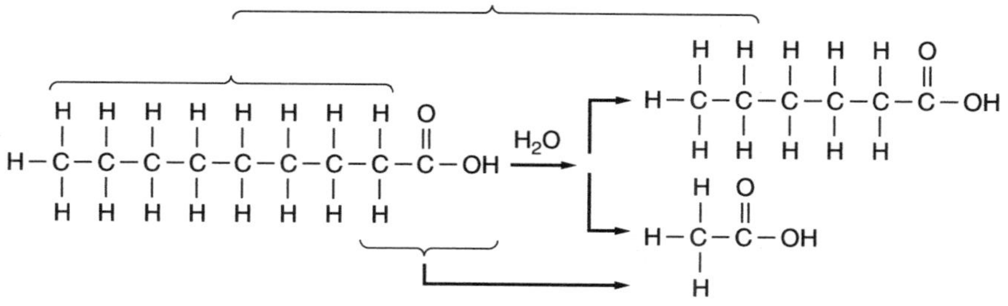

FIGURE 5.13   Biological oxidation of fatty acid

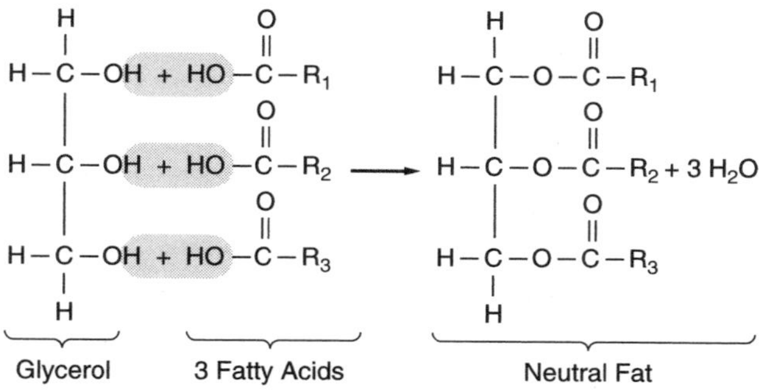

Glycerol   3 Fatty Acids   Neutral Fat

FIGURE 5.14   Formation of neutral fat from glycerol and fatty acids. The hydrocarbon chains in the three fatty acids need not have the same number of carbons.

up. In this way the original fatty acid can release a great deal of its chemical potential energy to the cell. However, it releases this energy in a roundabout way because only one end of a molecule ever is soluble in water.

By virtue of having more carbon-hydrogen bonds, fatty acids have more chemical potential energy than glucose molecules. However, the chemical energy in glucose is much more available to cells than the energy in fatty acids. Glucose is completely soluble in water and can readily diffuse into and out of cells. Moreover, glucose dissolved in blood plasma is easily carried to all the cells in the body. On the other hand, the high amount but low availability of chemical potential energy in fatty acids makes them ideal for storing chemical energy in the form of fat. The acid ends of three fatty acid molecules react with glycerol (an alcohol) to produce a large, water-insoluble molecule of neutral fat, see figure 5.14. Being insoluble in water, the neutral fat will tend to enter reactions that would break it apart. Suppose, as a result of a low supply of glucose, the cell needs the chemical energy in neutral fat. The cell activates a special mechanism that breaks neutral fat into glycerol and three fatty acid molecules. The fatty acids are oxidized in the roundabout way described above to release energy. The quickly available energy in glucose and the not so available energy in fat have some important dietary implications. We tend to accumulate fat instead of glucose when we eat more than we need over a long period of time. The neutral fat we accumulate is difficult to remove. Many reducing diets impose a very limited intake of both sugar but especially of fat so that the body is forced to dip into its fat energy storage "depot."

The nonpolarity of fatty acids (because of the many carbon-hydrogen bonds) makes them ideal structural components of the cell membrane. The fluid content of cells, namely the cytoplasm, is mostly water, and cells usually are surrounded by an aqueous medium. Therefore, it is important that cell membranes consist of nonpolar substances so that water and water-soluble substances do not easily move into or out of the cell. As a result of this structure, the cell can more easily regulate what passes into or out of it.

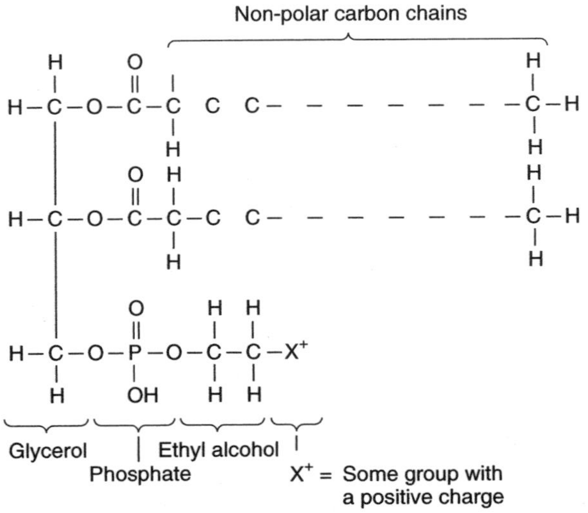

Non-polar carbon chains

FIGURE 5.15a

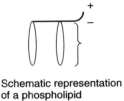

Schematic representation
of a phospholipid

FIGURE 5.15b

On the other hand, many biologically important chemical reactions occur at the inner and outer surfaces of membranes. For example, some hormones influence cell activity by entering into interactions at the cell membrane's outer surface. Also, active transport of substances across the cell membrane depends on interactions at the outer and inner membrane surfaces. Therefore, these two interfaces of the cell membrane should be soluble in water so that these water-dependent interactions can occur. By virtue of having two long, nonpolar hydrocarbon chains connected to a short polar group, phospholipids are ideal building blocks for membranes, see figure 5.15. These molecules are similar to neutral fat molecules composed of one glycerol molecule reacted with three fatty acid molecules, except that the third fatty acid molecule is replaced by a phosphate group bonded to some polar group. The two ends of the phospholipid molecule are therefore very different; the fatty acid-glycerol end is nonpolar and does not dissolve in water, while the phosphate-polar group tends to be attracted to water. This polarity causes these phospholipid molecules to aggregate into a double-layered membrane. The long, nonpolar carbon chain ends form a semifluid core, and the polar ends of the molecules dissolve in the water at the membrane's inner and outer surfaces. This arrangement helps to stabilize the membrane and provides an interface for the membrane to interact with dissolved substances inside and outside the cell.

Some types of proteins known as enzymes regulate all the chemical reactions that occur in cells. From what was stated previously, we would expect to find proteins at the inner and outer surfaces of membranes. Proteins are large molecules that contain nonpolar, polar, and charged regions. Therefore, it is reasonable to assume that the nonpolar regions are dissolved in the lipid core of the cell membrane. The polar and charged regions are dissolved in the water at each interface of a membrane. Thus, proteins appear like icebergs floating in a sea of lipid, as shown in figure 5.16. This theory of the interaction of phospholipids and proteins to form membranes is known as the fluid mosaic model. The model explains many properties of membranes, such as stability, permeability to fat-soluble substances, low permeability to some water-soluble substances, and enzyme activity at the membrane interfaces. The fact that membranes are very permeable to water and somewhat permeable to ions such as sodium, potassium, and chloride suggests that the membrane has pores which allow water and small ions to pass through it.

The lipids described so far are biologically important primarily because of their long chains of carbon atoms containing many carbon-hydrogen bonds. There is another type of lipid which is a modified version of a core of carbon atom rings known as steroid ring structure. These types of compounds are called steroids, see figure 5.17. The steroid, cholesterol, is thought to collect in plaques on the inner surface of small arteries and to contribute to atherosclerosis. Cholesterol also is found in many cell membranes. It helps to keep the lipid core of the membrane fluid, thus allowing greater lateral mobility of the proteins embedded in it. Some important hormones such as androgen and estrogen (male and female sex hormones) and cortisol secreted by the adrenal cortex are steroids.

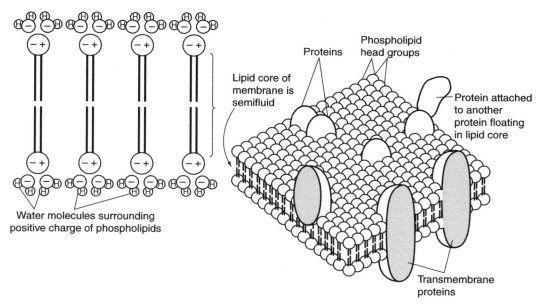

FIGURE 5.16

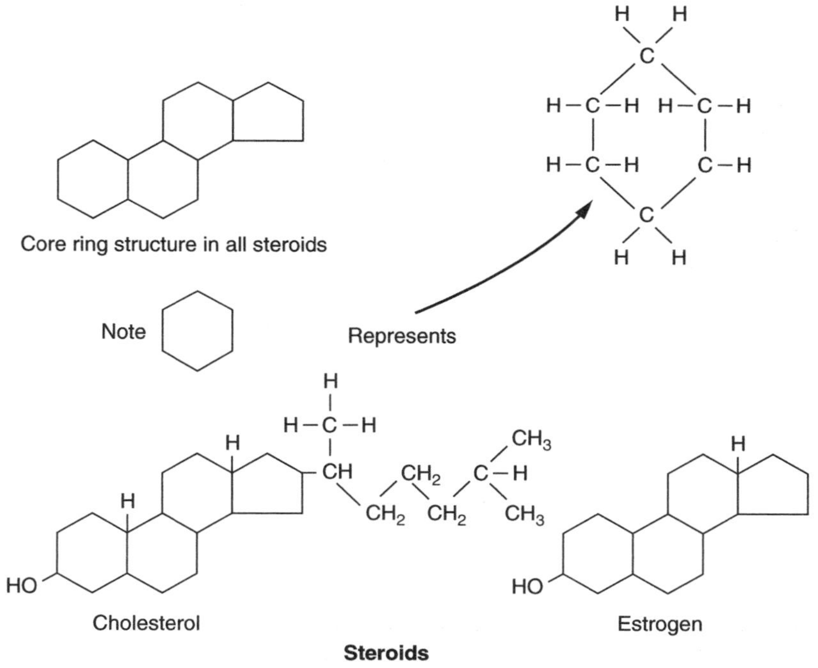

Core ring structure in all steroids

Note ⬡ Represents

Cholesterol

Estrogen

Steroids

FIGURE 5.17

Proteins

The organic molecules discussed so far have been linear molecules with fairly simple structures. Steroids are only somewhat more complex with two dimensional cyclic structures. However, a living thing is characterized by a complex internal organization that coordinates many diverse processes. The cell is the level at which nonliving chemical and energy processes are organized and coordinated to produce a relatively autonomous living thing. Therefore, one would expect to find a relatively complex type of substance to bring about this internal order and coordination. Protein is such a type of substance. The class of proteins know as enzymes act as catalysts, facilitating virtually all the

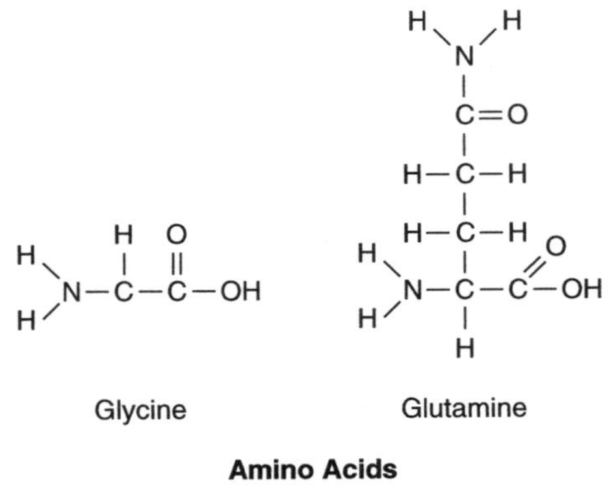

Amino Acids

FIGURE 5.18

chemical reactions that go on in cells. Other proteins known as collagens are chemically inactive but form a 3-dimensional structural framework for many tissues throughout the body. Elastic fibers such as those found in large arteries and lung alveoli are composed of another type of structural protein. The mucus secreted by glands lining the intestinal tract and the respiratory tubes consists of muco-proteins. Albumin, which is so important for tissue fluid balance, is a protein found in the blood plasma. Other blood proteins include antibodies which form complexes with foreign proteins, and thrombin & fibrinogen which are involved in forming blood clots. The multistranded protein, hemo-globin that is in red blood cells, carries oxygen to cells throughout the body. There are other transport proteins such as myoglobin which carries oxygen in muscle and lipoproteins which transport fatty acids in the blood. The contractile fibers in muscle cells and in cilia and flagella consist of contractile proteins. Finally, there are some important protein hormones such as insulin, ACTH (AdrenoCorti-coTrophic Hormone), and the growth hormone.

How can one type of substance carry out such an array of processes? The answer lies in the structural diversity that proteins exhibit. Proteins are long, complex chains composed of many subunits called amino acids. As the name implies, these compounds contain an amino group ($-NH_2$) and an acid group, COOH.

The amino group attracts a positive hydrogen ion to it and therefore tends to become positively charged. The carboxyl group tends to separate into a negatively charged ($-COO^-$) group and a posi-tive hydrogen ion just like the carboxyl group in fatty acids. As a result, an amino acid in solution can be positively charged, negatively charged or electrically neutral (due to the positive charge at one end balancing the negative charge at the other. The charge of the amino acid, which affects some of its properties, depends on the acidity of the solution.

Figure 5.18 shows two examples of different amino acids. Glycine is the simplest possible amino acid; glutamine is a somewhat more complex type. There are 20 types of amino acids found in pro-teins, but not all proteins have all 20 of these acids. The chemical bond between two amino acids is called a peptide bond. When three or more acids are bonded, the resulting molecule is called a polypeptide chain (poly means many). All proteins are polypeptide chains of more than 100 (some-times thousands of) amino acid units.

Since there are 20 different kinds of amino acids, the number of different combinations of these molecules is virtually infinite. As a result, each type of protein has a unique sequence of amino acids which gives it a very specific set of properties. The amino acids tend to interact with one another causing the polypeptide chain to become coiled. The coiled polypeptide chain may fold back on itself to give the protein a three-dimensional spatial configuration, as shown in figure 5.19. This three-dimensional structure of protein is a kind of "internal organization" that leads to special properties. This structure may specify that the protein will bind to only one or a few types of substances. For example, an enzyme regulates a specific type of reaction because it will only bind to a particular type of molecule (called the substrate) that could enter that reaction. The three-dimensional structure may

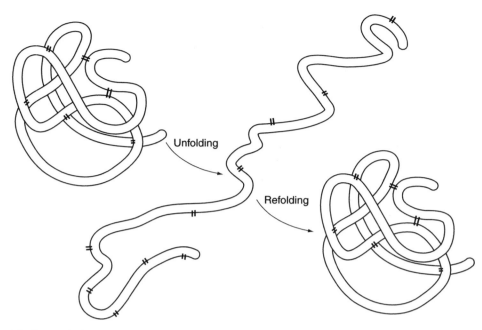

FIGURE 5.19   Three-dimensional structure of protein which can be altered by changes in its environment, such as acidity. This alteration may be reversible.

make the protein soluble or insoluble in water, depending on the distribution of polar and nonpolar regions in the macromolecule. Membrane proteins have a three-dimensional structure with one end being fat soluble and therefore embedded in the lipid core of the membrane and the other end being soluble in the water at the interface of the membrane.

The acidity of the water solution, the presence of various ions, and other physical or chemical factors can alter the three-dimensional structure of a protein. For example, pepsinogen, an enzyme secreted by the stomach, is inactive at low acidity but is converted to the active enzyme at the high acidity resulting from the additional secretion by the stomach of hydrochloric acid. The high acidity alters the protein's structure. This capacity of proteins to change their three-dimensional structure under altered circumstances may be thought of as similar to information processing. The particular sequence of amino acids in a polypeptide chain may be thought of as representing information. Environmental factors such as acidity, etc. cause the amino acid sequence to take on a particular three-dimensional structure. This is similar to a stimulus-response process in which a cell responds to some aspect of its environment. Thus, proteins are "life-like" in having a three-dimensional structure that changes in response to certain changes in the protein's immediate surroundings. However, since the early part of the 1950s, we have known that proteins cannot duplicate themselves as living cells can, but another kind of macromolecule can.

Nucleic Acids and Nucleotides

Nucleotides are made up of three types of molecules: (1) a group of carbon and nitrogen atoms that forms one or two rings, (2) 5-carbon sugar that is called ribose sugar, and (3) phosphate groups. In a nucleotide, the nitrogen-ring compound is connected to a ribose sugar which in turn is connected to a phosphate group. Some nucleotides capture energy from energy producing reactions and make this energy available for doing work in the cell. Figure 5.20 shows a diagram of the major types of nucleotides involved with capturing energy for immediate use by the cell. The simplest one, adenosine monophosphate (AMP), has relatively little usable chemical energy in it. The word, adenosine refers to the nitrogen ring compound attached to a ribose sugar. The word, monophosphate, refers to adenosine being bonded to only one phosphate group. The AMP molecule greatly increases its

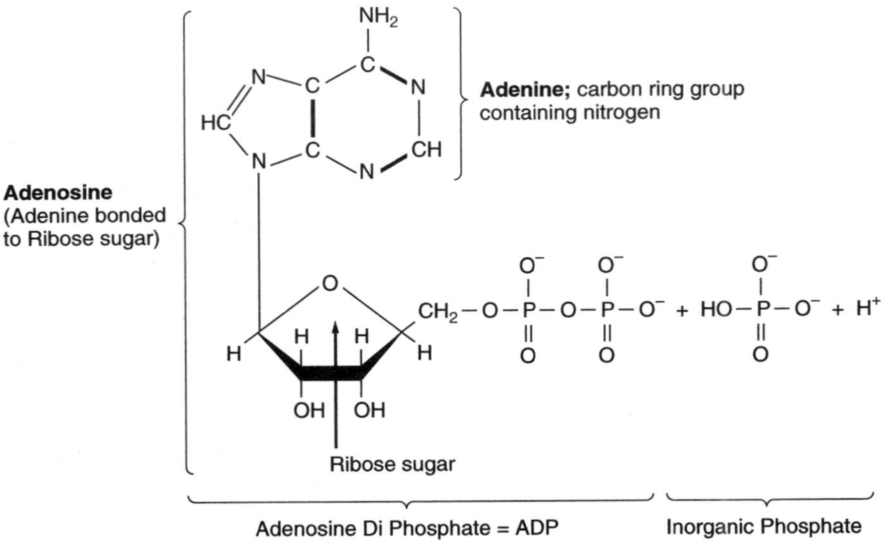

FIGURE 5.20a

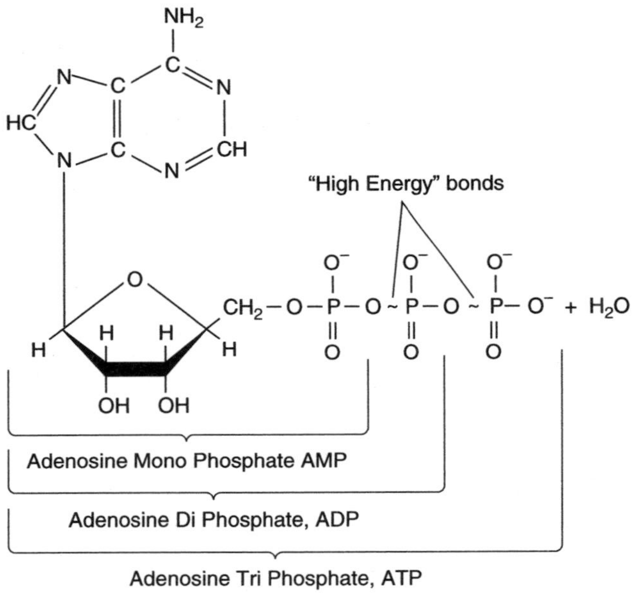

FIGURE 5.20b

chemical potential energy when a second and third phosphate group is added. These latter two compounds are called adenosine diphosphate (ADP) and adenosine triphosphate (ATP) respectively. Note that the phosphate groups are connected to one another in sequence. ATP molecules provide the immediate energy source for virtually all cellular processes. Many of the energy producing chemical reactions in the cell are coordinated to produce ATP.

Under the right conditions a type of compound known as nucleic acid, which consists of two sequences of nucleotide subunits can duplicate itself. There are four types of nucleotides in any particular nucleic acid. Since nucleic acid molecules consist of many hundreds of subunits, there are many different possible sequences of these four types. Just as was true of amino acids in proteins, the sequence of nucleotides represents information. In this case, however, the most common nucleic acids in the nuclei of cells are deoxyribonucleic acid (DNA) and ribonucleic acid (RNA). These nucleic acids store information and transfer it from parents to their offspring and direct the synthesis of proteins. The nucleic acids in the sex cells from each parent come together to make up the nucleus of the fertilized egg formed by the union of these two cells. If the fertilized egg (the zygote) successfully develops into a normal adult, then half of the DNA information (genetic information) present in all the cells of each parent has been transferred to this offspring. The information present in these nucleic acids directs the synthesis of all proteins. Since enzymes are proteins and regulate all chemical reactions in the cell, nucleic acids indirectly control all cellular processes that result from these chemical reactions.

Some people argue that nucleic acids are more fundamental units of life than cells. It is true that nucleic acids store and process information and duplicate themselves, but in nature these processes only occur in cells. The cell provides the energy, the suitable conditions, the chemical components, and the regulatory enzymes for this molecular information processing and duplication to occur. Scientists have caused molecular duplication to occur in a test tube; however, they had to provide the same things that cells ordinarily provide. A nucleic acid does not have the relative autonomy that a cell has. Moreover, the nucleic acid molecule is not a system that exhibits negative feedback control which is found in all cells. The cell controls when nucleic acids duplicate themselves and what information from all that is stored in the nucleus will be expressed in the life functions. Thus, even though nucleic acids have life-like properties, the cell is the lowest level of organization that can be said to be alive.

Cell Structure

Eukaryotes, such as plant and animal cells, are much more complex than prokaryotes, such as bacteria; however, both are called cells because they share universal cell structures and both are homeostatic chemical machines which contain DNA that duplicates itself and RNA that directs protein synthesis. After contrasting prokaryotes with eukaryotes, this chapter describes the universal cell structures and then describes the major structural features of the eukaryotic cell. A few years ago eukaryotic cell structure was one of the hottest topics in biology. The last section of this chapter describes the current change in perspective about cell structure which may affect all other disciplines in biology.

Prokaryotes vs Eukaryotes

Universal Cell Structures

Prokaryotes and eukaryotes share certain structures which therefore may be called universal cell structures.

1. All cells have an outer limiting membrane called the plasma membrane.
2. Most prokaryotes and most plant cells have a cell wall.
3. Most prokaryotes have internal membranes which, however, do not form into the organelles as found in most eukaryotes.
4. Both prokaryotes and eukaryotes store genetic information in DNA. In bacteria, the DNA is a single molecule which is many times longer than the cell. This means the bacteria DNA molecule is folded many times, and it is possible to visualize a portion of the cell, called the nuclear region, where DNA is concentrated. The nuclear region in eukaryotes, of course, is the membrane bound nucleus.
5. Both bacteria and eukaryotes have RNA formed into particles called ribosomes, which are part of the protein synthesizing machinery of the cell.
6. Both prokaryotes and eukaryotes have a fluid matrix called protoplasm. The fluid matrix was called cytoplasm in eukaryotes but now is referred to as cytosol.

Major Differences between Prokaryotes and Eukaryotes

The major differences between prokaryotes and eukaryotes are the structures they do or do not contain. These two types of cells also differ in the size or form of the universal cell structures which they have in common. The following is a list of descriptions of the major differences between prokaryotes and eukaryotes. The new terms present in this list will be defined and described in a later section of the chapter dealing with eukaryotic cell structure.

1. The cell wall of prokaryotes is quite different in structure from that of any eukaryote.
2. The internal membranes in prokaryotes do not divide the cell into several discrete compartments; in eukaryotes, the internal membranes divide plant cells into eight compartments and divide animal cells into seven compartments—animal cells lack plastids.
3. Bacterial DNA is a single molecule; the genetic information of eukaryotes is stored in several DNA molecules which in turn are bound to a specific type of protein, the histones.
4. The ribosomes of bacteria are somewhat smaller than the ribosomes in the cytoplasm of eukaryotes.
5. The nuclear membrane and nucleolus found in eukaryotes are not present in prokaryotes.
6. Microtubules and perhaps even the semblance of a cytoskeleton is absent in prokaryotes but present in eukaryotes.
7. In those cells that move via one or more flagella, the bacterial flagella are much smaller and less complex in structure than eukaryotic flagella.
8. The largest bacteria overlap with the smallest eukaryotes, but in general, prokaryotes are ten times smaller than eukaryotes.

Description of 1st & 6th Universal Cell Structures

The six universal cell structures were listed in the preceding section. The structures corresponding to numbers 4 and 5 of this list are discussed in this chapter and in Chapter 6. Discussion of the chemical makeup of cell walls involves technical details that need not concern us. Though internal membranes may be quite complex in photosynthetic bacteria, in general, internal membranes are not a distinctive feature of prokaryotes. Therefore, these membranes will be discussed in the section on eukaryotes. Accordingly, this section only will describe 1 and 6 cell structures on the list.

The Plasma Membrane

According to the most commonly accepted theory of evolution of life from nonlife, the formation of a plasma membrane was the key event that converted a nonliving system of proteins and nucleic acids into living cells. The plasma membrane is not merely a passive barrier separating the inner world of the cell from its environment. Rather the cell membrane is a dynamic sub-cellular system that plays a key role in a cell functioning as a homeostatic machine, cell division, cell evolution, cell ecology, and cell communication.

As is true of so many other concepts in science, our understanding of the plasma membrane has drastically changed corresponding to changes in perspective from mechanistic to process and now to an ecological perspective. Initially (1930s), the plasma membrane was postulated to be a passive permeability barrier with "pores" in it. Lipids making up the core of the membrane keep water and water soluble substances inside the cell separate from water solutions bathing the cell. Proteins were thought to reduce the surface tension between the lipid core and the water at the inside and outside membrane surfaces; that is, proteins were bound to each surface of the lipid core and were partially dissolved in the water at each membrane surface, see figure 5.21. The existence of "pores" were postulated to explain the rapid passage of small ions such as potassium and chloride (and sodium ions during electrical activity, i.e., action potential of nerves and muscles) across the membrane. In the

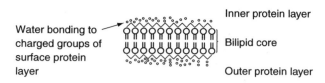

Water bonding to charged groups of surface protein layer

Inner protein layer

Bilipid core

Outer protein layer

FIGURE 5.21   Old (Danielli) model of cell membranes.

1950s and 60s cell biologists studied enzymatic activity and active transport mechanisms associated with the plasma membrane. These studies were integrated with the concept of homeostasis that became prevalent in biological thinking at this time. As a result, the plasma membrane was thought of as a homeostatic mechanistic system.

However, the mechanistic model of membrane structure could not account for various cell processes such as change in shape, cell motility, cell division, cell-to-cell contact interactions, and membrane surface changes when a normal cell becomes a cancer cell. After a few years of proposing many alternative cell membrane models, biologists quickly reached a consensus on the fluid mosaic model proposed by S. J. Singer and G. L. Nicolson in 1972. Some of the elementary chemical aspects of this model were described earlier in this chapter. This model also is a phenomenological description of the plasma membrane which now is thought of as a dynamic fluid system. Surface proteins, called integral proteins, are like iceberg; they can move laterally; i.e., in the plane of the cell membrane. Some membrane proteins change their configuration and interaction with the lipids; this in turn changes the membrane's properties such as (1) membrane causes the cell to change shape; (2) nonmobile cells become mobile because of membrane fluidity and ability to glide over surfaces; (3) membrane once not permeable to sodium ions now becomes very permeable to these ions. In general, all the cell properties that were inconsistent with the mechanistic model of membrane structure are now at least conceivable with the fluid mosaic model.

As an example of the dynamic aspect of the plasma membrane system, consider the membrane transport of macromolecules and particles. Many cells, for example, insulin producing cells in the pancreas, synthesize macromolecules and "package" them in membrane bound vesicles. These vesicles move to the inner surface of the plasma membrane, and in a process called exocytosis, the vesicles fuse with the plasma membrane and open to the extracellular space, thereby releasing the macromolecules to the outside, see figure 5.22a. Many cells can ingest macromolecules and particles by a somewhat similar mechanism. As the substances to be ingested are drawn to specific sites on the cell, the membrane becomes indented. This leads to the substances progressively being enclosed by a small portion of the membrane which eventually pinches off to form an intracellular vesicle containing the ingested material. This process is called endocytosis and consists of two types: pinocytosis which involves ingestion of fluid and/or macromolecules via small vesicles, and phagocytosis which involves ingestion of large particles such as viruses or bacteria, see figures 5.22b and 5.23.

In the later part of the 1970s and 80s molecular biologists focused their attention on the plasma membrane. One of the key ideas developed during this time was that the plasma membrane contains regulator molecules other than enzymes. These regulator molecules called receptor sites are complementary to various specific substances that may reach the external surface of the cell. A lock-key interaction is analogous to this molecular complementarity: a specific receptor site is like a lock and the specific substance complementary to it is like a key that can fit into the lock. There are hormone receptor sites; receptor sites for ingestion of specific substances that the cell can use, such as, cholesterol which the cell incorporates into its manufactured cell membranes; receptor sites for "recognizing" other cells similar to it, and receptor sites for "recognizing" foreign substances such as bacteria or "excommunicated cells" such as cancer cells. In other words, the plasma membrane regulates extracellular communication analogous to DNA regulating intracellular communication.

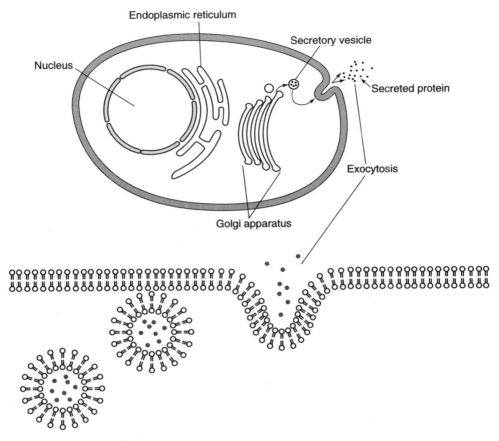

FIGURE 5.22a

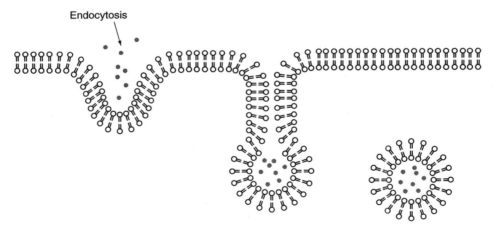

FIGURE 5.22b

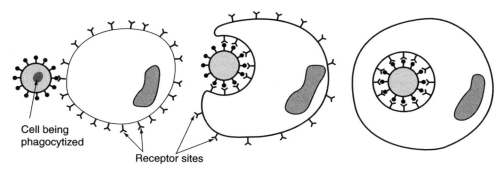

Cell being
phagocytized

Receptor sites

FIGURE 5.23

Thus, a communication perspective of the plasma membrane has emerged. Not only do we view the plasma membrane as a self-regulating mechanical system and as a set of membrane processes, we appreciate that it is like a brain in having sense receptors, short term memory, and decision centers. This, of course, is the communication life theme. The last section of this chapter will briefly summarize the emerging but as yet controversial ecological aspect of the plasma membrane. In this view, the plasma membrane is not a subsystem distinct from other cell compartments. Rather, it is the outer portion of a complex three-dimensional network of proteins, called the cytoskeleton, that fills the whole cell.

Protoplasm

Though prokaryotes and eukaryotes differ greatly in internal organization, they are remarkably similar in chemical makeup, chemical mechanisms for obtaining and using energy to carry out cellular processes, DNA duplication, and protein synthesis. Many chemical reactions occur in association with membranes, but others occur in the protoplasm. In fact, before the discovery of extensive internal membrane systems (in the late 1940s), all cells were thought of as bags of dissolved substrates and enzymes in which occurred thousands of different chemical reactions. However, this internal fluid, protoplasm, was no ordinary solution. Somehow it coordinated these diverse sequences of chemical reactions with observable cell processes such as motility, shape change, cell division, and so on. A key to understanding how protoplasm could coordinate these cell processes was the realization that protoplasm is a colloid. Ordinary water solutions are homogeneous, and their overall properties are dominated by the greater concentration of the solvent, water, that prevents dissolved ions or molecules, the solutes, from interacting with one another. However, when the solutes are macromolecules, they occupy the same or even greater space than the water, and the solute ions or molecules begin to interact with one another. Such "solutions" are called colloids. Solutes of intermediate size give rise to mixtures that have some properties of ordinary solutions and some properties of colloids, but true colloids are very different from ordinary solutions.

The chemistry of colloids is a sub-discipline unto itself, but for our purposes, we need only describe one important phenomenon, sol-gel transformations. Under some conditions, the macromolecules aggregate to form small particles called micelles, which remain in suspension separated from one another by water. A change of factors such as solute concentration, ion concentration, acidity, temperature, etc., causes the micelles to begin interacting with one another producing an intermediate stage called gel-solution. Further changes of the solution properties leads to micelles or individual macromolecules forming a three-dimensional network that may appear as a fibrous framework, called a gel, see figure 5.24. The transition of a colloid from a sol state to the gel state is a sol-gel transformation, see figure 5.24.

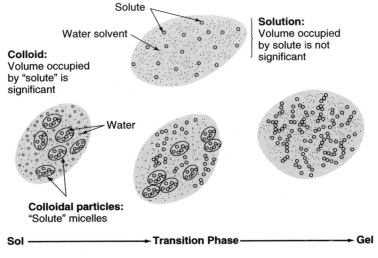

FIGURE 5.24

The idea of sol-gel transformations plus the concept of protein being able to change shape suggests a perspective for postulating how protoplasm coordinates diverse chemical reactions into cell functions. The cell is seen as a dynamic system in which some portions of the cytoplasm are in the sol state, others in the intermediate state, and still others in the gel state. Which portions of the cytoplasm are in one of these three states depends on the environmental conditions and the corresponding functions the cell is carrying out. In other words, the cytoplasm is constantly changing. Furthermore, specific, repeated sol-gel transformations were postulated to explain cell functions such as separation of chromosomes during cell division, cell motility, shape changes, and the flow in streams of cytoplasm that can be seen in some cells.

As biologists became preoccupied with plasma and internal membrane systems, the colloid properties of protoplasm were neglected. However, the recent emergence of the idea of a cytoskeleton and the dynamic fluidity of membranes may be integrated with colloid chemistry to produce a revolutionary new view of the cell. This will be briefly described at the end of this chapter.

Idealized Eukaryotic Cell

Changes in perspective have caused biologists to change the way they view cell structure. Early in the 20th century, mechanistic biologists viewed the eukaryotic cell as composed of parts: nucleus, cytoplasm, granules, mitochondria, and chloroplasts which could be seen with the light microscope. These cell parts were "really there" independent of biologists looking at them. Then in the late 1940s biologists used the electron microscope to discover heretofore unknown internal membrane systems and new structural features of nuclei, mitochondria, and chloroplasts. However, these new structures were no longer simply there for one to observe. Scientists had to carry out elaborate procedures for viewing tissues with various types of electron microscopes. Biologists took many thousands of pictures of hundreds of different types of cells. To the untrained eye, electron microscope pictures usually don't make any sense whatsoever. One has to be specially trained to take a two-dimensional picture showing a pattern of lines and dots and imagine how this flat image represents a particular three-dimensional structure in a cell. Furthermore, the picture one sees depends on the way the tissue is prepared and the type of microscope one uses.

The internal membrane are "really there" in the sense that if one takes certain types of cells, prepares them in a particular way, and views them with a particular type of electron microscope, she consistently will obtain a certain type of picture. Biologists have reached a consensus about what many of these pictures represent, but there still is room for radically different interpretations. Some biologists, such as H. Hillman and P. Sartory[3], believe these structures "seen" with the electron microscope do not represent structures really present in the living cell. Rather the picture of structures represent changes that researchers produced in cells in order to look at them with an electron microscope. Even if one does not adopt this extreme view, one qualification generally is conceded by cell biologists: our picture of a cell is no longer truly objective; rather internal cell structures are "something there" plus experimental treatment of a population of cells plus subjective interpretations of what one sees. Incidentally, this situation is analogous to how the quantum mechanics physicists view atoms and molecules.

Cell structure used to be thought of as separate though related to cell function. Biology students studied cell structure in one course using one textbook and cell physiology in another course using a completely different textbook. This all began to change in the mid 1960s. Cell structure and cell functions are distinct but are like two sides of a coin; they imply one another. Instead of talking about cells in mechanistic terms, biologists began discussing cells from a phenomenological perspective. In single courses and textbooks, biologists began describing cell phenomena in terms of structural aspects and mechanistic processes. In other words, the emphasis has shifted from cell structures that interact in specific ways to cell phenomena that imply or are consistent with certain structural patterns and process patterns. The phrase, idealized eukaryotic cells, represents the biologists' phenomenological picture of eukaryotes.

Internal membranes divide the eukaryotic cell into seven compartments in animal cells and eight compartments in plant cells. A compartment plus the internal membrane surrounding it is called a cell organelle. There are seven types of organelles: (1) nucleus, (2) endoplasmic reticulum (ER), (3) Golgi apparatus, (4) mitochondria, (5) plastids (found only in plants), (6) lysosomes, and (7) peroxisomes (microbodies). The seventh compartment in animals or the eighth compartment in plants is the cytosol which is surrounded by the plasma membrane and which bathes the organelles. The old term, cytoplasm, now means cytosol plus organelles other than the nucleus plus cytoplasmic inclusions plus protein filaments that form networks in the cytosol. The filament networks are like the "bone and muscle" of the eukaryotic cell and therefore are called the cell's cytoskeleton. Figures 5.25 and 5.26 represent current composite pictures of animal and plant cells respectively.

Cell Organelles

Nucleus. The cell nucleus is a spherical body more or less in the center of the cell. It consists of nucleoplasm separated from the cytoplasm by a nuclear envelope which is a double membrane each consisting of a bilipid layer. The outer nuclear membrane is continuous with the endoplasmic reticulum and is often studded with ribosomes, see figure 5.27. At several points the inner and outer nuclear membranes are connected thus forming *nuclear pores*, see figure 5.28. Each pore is surrounded by a disc-like structure, the nuclear pore complex, consisting of eight large proteins arranged in an octagonal pattern. Exchange of materials between nucleoplasm and cytoplasm occurs through nuclear pores and is regulated by nuclear pore complexes.

Under the light microscope the nucleus appears as a dense mass of material, but special treatment and closer examination reveal fine threadlike structures called chromatin in the nuclear material. During one of the phases of cell division (M phase), the chromatin threads become thicker, separate from one another, and are readily seen with only the light microscope. When this happens, they are referred to as chromosomes, see figure 5.29. In other words, a chromosome is one form of a chromatin thread. A chromatin thread consists of a single, enormously long DNA molecule tightly bound to an equal mass of histones, a special type of structural protein. A chromatin thread always shows a high degree of folding so that the thread which, when extended, is many times longer than the diameter of the

---

[3]H. Hillman and P. Sartory, *The Living Cell.* (Chichester, England: Packard Pub. Limited, 1980)

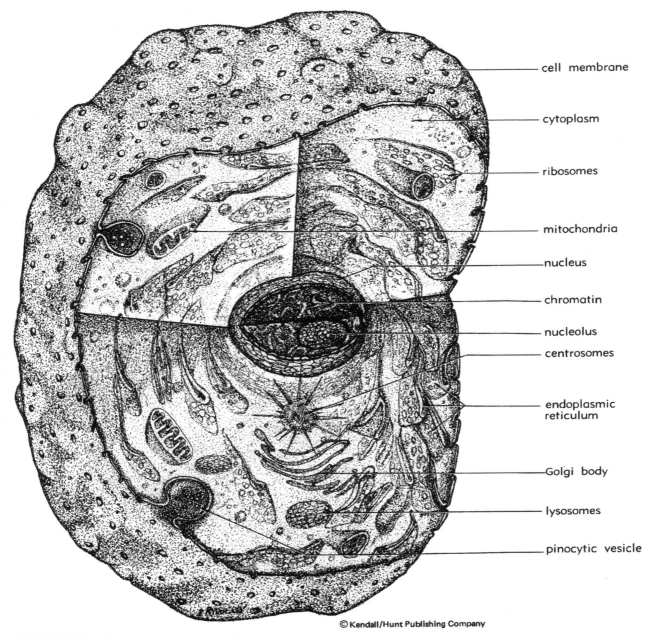

cell membrane

cytoplasm

ribosomes

mitochondria

nucleus

chromatin

nucleolus

centrosomes

endoplasmic reticulum

Golgi body

lysosomes

pinocytic vesicle

© Kendall/Hunt Publishing Company

FIGURE 5.25   Idealized animal cell.

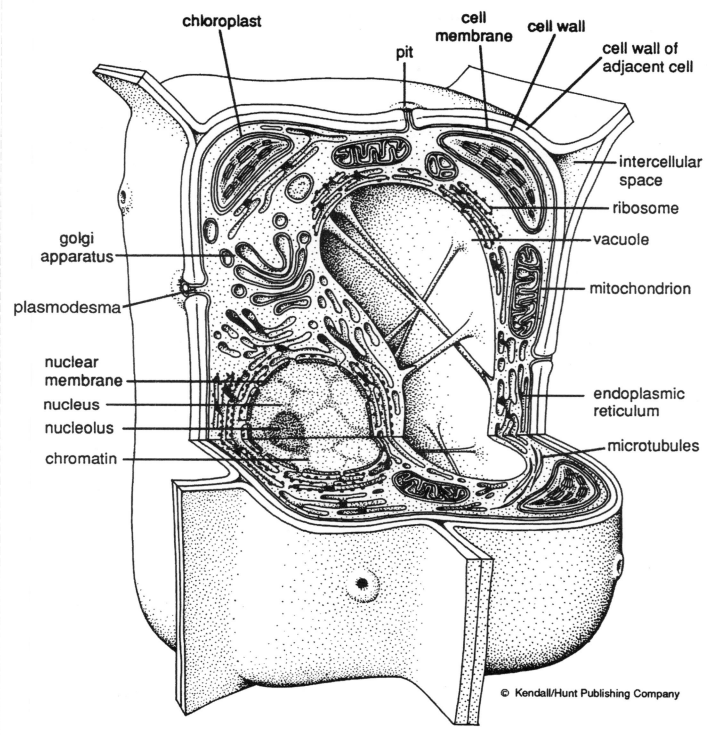

chloroplast

cell
membrane   cell wall

pit   cell wall of
adjacent cell

intercellular
space

ribosome

vacuole

golgi
apparatus

mitochondrion

plasmodesma

nuclear
membrane

nucleus

nucleolus

endoplasmic
reticulum

chromatin

microtubules

© Kendall/Hunt Publishing Company

FIGURE 5.26   Idealized plant cell.

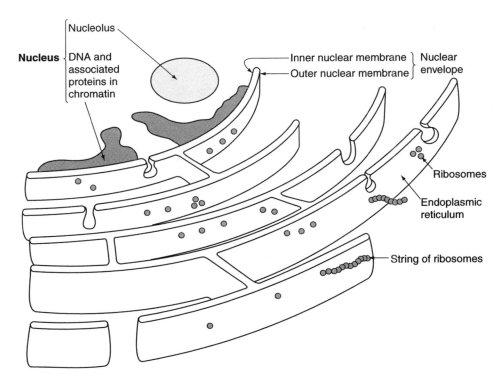

FIGURE 5.27 Outer nuclear membrane continuous with endoplasmic reticulum.

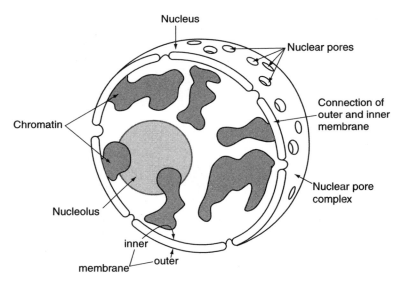

FIGURE 5.28 Inner and outer membrane connected nuclear pore complex.

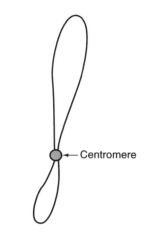

FIGURE 5.29   Chromosome.

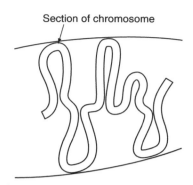

FIGURE 5.30

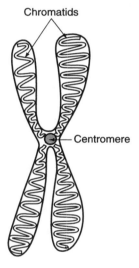

FIGURE 5.31   Single chromosome consisting of two chromomatids each containing a DNA with associated proteins.

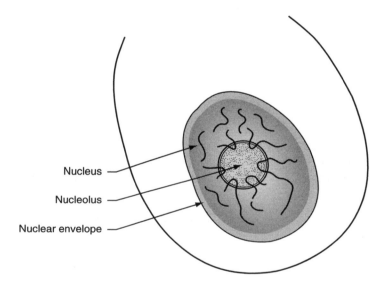

FIGURE 5.32   Nucleolus consisting of loops of different chromosomes.

nucleus, can be stored in a small volume, see figure 5.30. The chromatin takes on many forms depending on which portions of the DNA are transferring information to another molecule (mRNA). The chromosome form of a chromatin thread consists of two identical DNA molecules each associated with histones. These two threads, called *chromatids,* are joined at a relatively rigid region called the centromere, see figures 5.29 and 5.31.

The nucleolus is a spherical or ellipsoid intranuclear organelle seen in cells that are not dividing, see figure 5.32. This intranuclear organelle is a good example of how the study of cell structure has had to change from the mechanistic substance (thingness) perspective to the process perspective. What was once believed to be an independent structure is now known to be a permanent or a recurring event in the life of a cell. That is, the nucleolus consists of one or more loops of DNA, each loop from a different chromatin thread. These loops are organized into a particular pattern and are associated with nuclear proteins, but the size and shape of the resulting nucleolus changes depending on the activity state of the cell. The nucleolus is the site of synthesis of a type of RNA (rRNA) and of the assembly of all the cell's ribosomes. During cell division when chromosomes are separating from one another, the nucleolus disappears and of course no ribosomes are made.

All of the cell's RNA molecules and ribosomes are made in the nucleus and actively transported to the cytosol through the nuclear pores. All the proteins involved in nuclear processes are synthesized in the cytosol and are actively transported through the nuclear pores into the nucleoplasm.

Endoplasmic Reticulum (ER), see figures 5.25 and 5.26.   The endoplasmic reticulum is a tortuous network of membrane tubes that is continuous with the outer nuclear membrane. The portion of this tubular system closer to the nucleus has ribosomes embedded in its membranes and is called rough endoplasmic reticulum; the remaining portion which does not have ribosomes is called smooth endoplasmic reticulum. The ER produces proteins and lipid components for most of the cell's organelles. Newly formed proteins pass into the interior of these tubes and then to other parts of the cell or to the outside if the protein is a substance which the cell secretes.

Golgi Apparatus.   The Golgi apparatus, usually located near the nucleus, consists of stacks of disc-shaped membrane sacs, resembling stacks of plates. The Golgi apparatus separates out different newly formed proteins from one another and regulates the distribution of these proteins to various other sites within the cell.

Lysosomes.   Lysosomes are a diverse group of organelles whose common feature is a high content of enzymes that regulate the breakdown of various organic substances. The Golgi apparatus produces membrane sacs called primary lysosomes, which only contain digestive enzymes initially produced by the ER. These sacs fuse with a variety of membrane bounded substrates to produce secondary lysosomes, in which digestion occurs. Secondary lysosomes break down worn out organelles as well as substances or bacteria taken into the cell by endocytosis. Some lysosomes function analogous to our digestive system. For example, lipo-protein molecules are taken into the cell by endocytosis and broken down by secondary lysosomes. Some of the end products are used for synthesis of cholesterol. For those cells which the organism "decides" are no longer needed, mass production of lysosomes are the means by which cells destroy themselves. For example, during human development, arms and legs start out as limb buds with solid masses of cells at their tips. Selective cell destruction carves out the fingers and toes.

Peroxisomes (Microbodies).   Peroxisomes are sacs of diverse sizes which pinch off from smooth ER and contain enzymes that regulate chemical reactions that utilize oxygen. Some of these reactions generate hydrogen peroxide, hence the name peroxisomes. Other enzymes present regulate reactions which both use and destroy the hydrogen peroxide.

Mitochondria.   Virtually all eukaryotes have mitochondria, which suggests that these organelles are essential to cell survival. Mitochondria are elongated bodies consisting of an outer smooth membrane and a separate inner membrane that is folded on itself to form *cristae*. Mitochondria are the power houses of the cell. Reactions take place in the mitochondria which release energy that converts ADP plus phosphate into ATP. The ADP-ATP system is the major energy coupler in cells.

Plastids.   One of the distinguishing characteristics of plant cells is that they contain *plastids*. These are similar to mitochondria in that they have an outer and an inner membrane, but their function is quite different. There are three types of plastids which are named according to whether they have colored pigments or not. Leucoplasts are colorless and are sites where sugars, fats, and proteins are stored. Chromoplasts synthesize and contain pigments which give many fruits, vegetables, flowers, and leaves their bright colors. Chloroplasts contain the green pigment chlorophyll which captures sunlight and transforms it into chemical potential energy. This is the first step of photosynthesis, the process in which light energy is used to synthesize organic substances from carbon dioxide, water, and other inorganic substances.

Cytoplasmic Inclusions

Various substances aggregate within the cytosol to form particles which can be seen with the electron microscope. These so-called cytoplasmic inclusions are: glycogen particles and lipid droplets. In fat cells, which are among the largest cells in the human body, the fat droplets coalesce into a single droplet which occupies most of the interior of the cell. Some ribonucleic acid particles lie free in the cytosol and therefore also are cytoplasmic inclusions. Proteins sometimes aggregate to form crystals which have been reported to occur in all compartments of the cell including the nucleus.

Living Colloid

The cytosol may be thought of as a "living colloid" due to its continuous interactions with the other compartments in the cell. A nonliving colloid may be described in terms of the size of its suspended macromolecules, types of gel formations that occur in it, and the factors that bring about sol-gel transformations. In contrast, the cytosol is more analogous to the web of organism-environment interactions described by ecologists. In other words, the "living aspect" of the cytosol colloid means that the cytosol is a set of interactions that are expressed as cell structures of varying degrees or permanence. The three major types of protein involved with expression of these cell structures are: actin, tubulin, and fibrous subunits, see figure 5.33. Actin is a globular protein that can bind to other actin proteins to form a beaded chain. An actin filament results from the interaction of two beaded chains of actin molecules that become arranged as a helix. Tubulin is a globular protein that can form beaded chains called protofilaments which further interact laterally to form hollow cylinders called microtubules. A single microtubule plus a partial microtubule form a doublet microtubule; addition of a second partial microtubule produces a triplet microtubule. Fibrous subunits are thread-like molecules, usually insoluble in water, which may interact laterally to form rope-like structures called intermediate filaments. These filaments are called intermediate because their diameters are intermediate between the larger microtubules and the smaller actin filaments.

Relatively permanent structures include intermediate filaments, small muscle fibers made up of actin filaments and myosin molecules found in all three types of muscle, and microtubules that make up cilia, flagella, centrioles, and basal bodies. A flagellum is like a much larger version of a cilium. Both carry out side-to-side movements which either move the cell through a fluid medium or move fluid past the surface of a cell. Both cilia and flagella are anchored to basal bodies immediately beneath the cell surface. The cilia and flagella consist of doublet microtubules in a 9 + 2 arrangement, and the basal body consists of triplet microtubules, see figure 5.33.

Transient Cytoplasmic Structures.   Suspended actin and tubulin proteins may rapidly form actin filaments and microtubules and then just as rapidly disassemble to their molecular components. These transformations are analogous to sol $\rightarrow$ gel $\rightarrow$ sol transformations discussed earlier. However, transient formation of filaments and microtubules are much more specific, diverse, and controllable by regulator proteins than are sol-gel transformations. Therefore, the assembling and disassembling of these structures may more readily explain many of the cell's dynamic properties. Actin filaments may aggregate in various patterns thereby structuring the cytosol in different ways. For example, cross linkages among separate actin filaments can result in a sol-gel transformation in the cytosol. Regulator molecules in the cytosol or on cell membranes may determine the structure of cytosol in parts of the cell. The diverse and continually changing structuring of cytosol is influenced by cellular communication with its environment including other cells. The occurrence of transient cytoplasmic structures implies diverse structuring of cytosol which further implies that cytosol is a "living colloidal system" defined by thousands of cell compartmental interactions and extracellular communications.

Cytoskeleton.   The meaning of the term cytoskeleton depends on one's view of the structure of cytoplasm. The conservative view is that cytoskeleton represents local interconnections among one, two, or all three types of filaments present in the cytosol. An extremely conservative view is that the interactions among these three types of filaments are too transient to produce a network of fundamental structural significance to the cell. Cell biologists in this camp think of the three types of filaments as relatively independent populations of molecules. Accordingly, these scientists focus attention only on the kinds of interconnections among a particular type of filament.

The opposite extreme view is that the cytoskeleton is a three-dimensional network of fibers that pervades the cell and interacts with suspended macromolecules, ribosomes, filaments that are not part of this network, and proteins embedded in membranes of cell organelles. In some cells this "universal network" consists of actin filaments. In other cells, there is controversial evidence for the existence of a network of extremely fine filaments presumably made up of proteins. This network, called the microtrabecular lattice, is thought to regulate all cell functions. Thus, not only is it the single "skeleton" of a cell, it is like an information containing nucleic acid with interconnected information

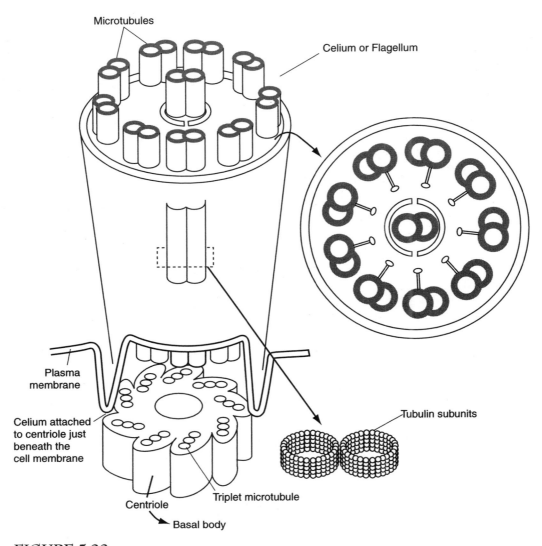

FIGURE 5.33

sequences spread out in all directions within the cell. A phenomenological version of this revolutionary perspective is that the cell is an interconnected set of intracellular and extracellular communications which are expressed in two coequal, complementary ways: (1) cell structure which is the various transient and permanent interconnections among cell compartments and (2) various cell functions.

Centrioles. Centrioles are cylinders very similar in structure to basal bodies, as shown in figure 5.33. Almost all animal cells have a structure located near the nucleus called a centrosome which consists of a pair of centrioles. These structures help organize the cytoplasm during cell division and separate into each daughter cell after cell division. The single centriole in each daughter cell duplicates itself in preparation for a possible new cell division. In some cells this pair of centrioles alternates between basal bodies for a pair of flagella and a centrosome which helps organize cell division. In other cells many centrioles arise by duplication of already existing centrioles and become basal bodies of cilia. However, there are cases where centrioles spontaneously arise from tubulin in the cytosol.

Plant vs Animal Cells

There are some cells and organisms which have some "plant-like" properties and some "animal-like" properties. However, in general, plant and animal cells differ in the following ways: (1) plant cells have cell walls whereas animal cells do not; (2) plastids which usually are present in plant cells are not present in animal cells; (3) in contrast to animal cells, most mature plant cells have a large central vacuole which occupies a major part of the total cell volume; (4) plant cells are usually larger than animal cells.

Vacuole.   A vacuole is a space in the cytoplasm containing water and dissolved substances and surrounded by a single membrane. Though found in some single celled organisms which are like animal cells, the presence of vacuoles is a characteristic feature of most mature plant cells. A young plant cell develops several small vacuoles each surrounded by a membrane called a tonoplast. As the cell grows to maturity, the several small vacuoles fuse to form a large single vacuole which may occupy 80 to 90 % of the total cell volume. Sometimes this large vacuole may break up into two or many smaller vacuoles. There are four major functions a vacuole may carry out in a plant cell. (1) It provides a storage place for materials not immediately required by the plant cell. (2) It is a place into which cellular wastes or poisonous substances can be released. (3) It is a water reserve in the cell. (4) It maintains the cell structure and rigidity when filled to capacity. It does this by exerting fluid pressure on the cell wall thus preventing the cell from collapsing or distorting its shape.

Cell Wall.   Prokaryotes and plant cells are surrounded by a cell wall composed of a "non-living" substance in contrast to protoplasm which is the "life stuff" of a cell. The cell wall in plant cells is made up of fibrils composed of cellulose embedded in a structureless gel-like substance. The fibrils, analogous to the intermediate filaments found inside cells, are nonelastic and provide rigidity to the cell wall. Particularly in those cells where the wall becomes fairly thick, the rigid wall supports the cell and prevents it from bursting due to high internal pressures resulting from the accumulation of water in the vacuole(s). The cell wall also helps prevent the invasion of organisms such as bacteria which could harm the cell.

Revolution in Cell Biology

The current revolution in cell biology may be stated in each of the three perspectives of science. In the mechanistic substance perspective the cell is considered to be a single three-dimensional cytoskeleton with transient or permanent connections with most other cell components including the plasma membrane. The cytoskeleton contains information that helps regulate all cell functions. In the process or phenomenological perspective, the cell is considered to be an interconnected set of intracellular and extracellular communications that are expressed in two coequal, complementary ways: (1) cell structure which is the various transient and permanent interconnections among cell components and (2) various cell functions.

In the ecological perspective (communications perspective) the whole cell is like the nucleolus. Recall the nucleolus is the structural expression of the cellular *event* of loops of DNA each from different chromatin fibers producing ribosomes. Structural changes correspond to changes in extent and type of ribosomal production. When this process ceases during cell division, the nucleolus disappears. The cell is an event which is determined by a set of communications at a particular time and context. Cell structure is the spatial representation of a particular *cell event* which is partially determined by the techniques for "seeing" the *cell event* and by the imagination of scientists visualizing the *cell event.* Cell functions are complementary to cell substructures and are processes partially determined by techniques and imagination of scientists studying a *cell event.* A scientist may hold this view and at the same time, for the purpose of designing experiments, he may narrow his view to the process-phenomenological or the mechanistic-substance perspective of a cell.

Tearout #1

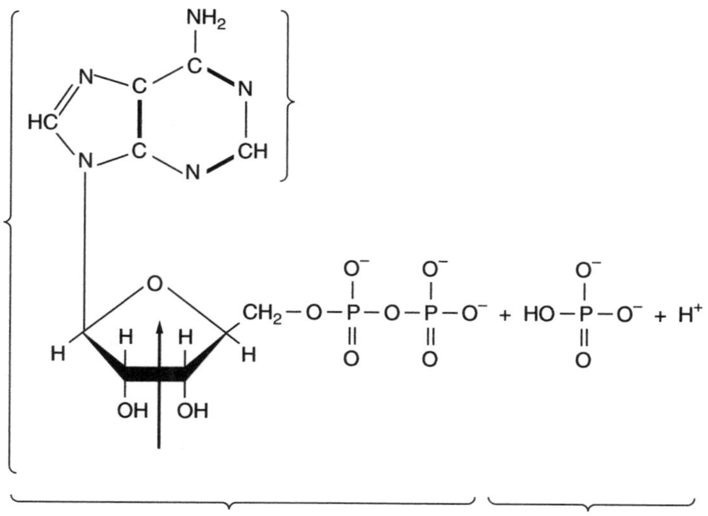

FIGURE 5.20a

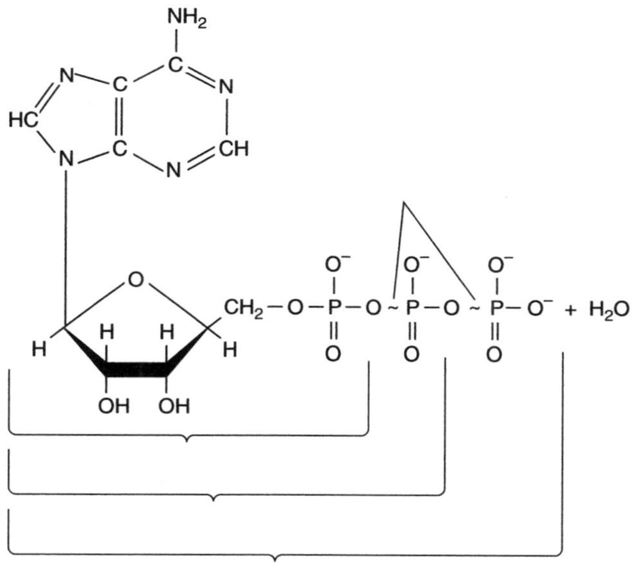

FIGURE 5.20b

Directions:

1. Using figure 5.20 as a reference, fill in all labels.

Tearout #2

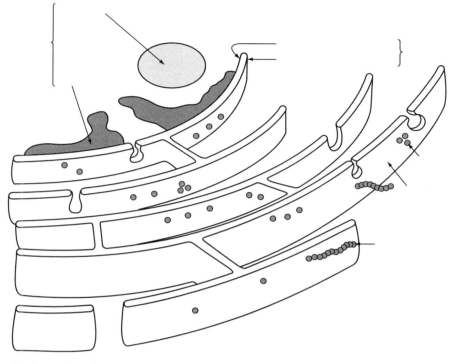

FIGURE 5.27

Directions:

1. Using figure 5.27 as a reference, fill in all labels.

Cell as a Chemical Machine
That Reproduces Itself

Simultaneous with the acceptance of the cell theory there developed an attitude that life may be understood in terms of physical and chemical processes. This attitude represents the mechanical view of life. Chemistry was viewed as the application of physics to chemical interactions, and biology was viewed as the application of physics and chemistry to life processes.

Concepts Necessary for Describing Chemical Reactions

Atomic Symbols and Formulas

Of the 103 known elements only 10 make up over 99% of the complex structures of all cells; the symbols and names of these are:

1. **C**   carbon
2. **H**   hydrogen
3. **N**   nitrogen
4. **O**   oxygen
5. **P**   phosphorus
6. **S**   sulfur
7. **Na** sodium
8. **K**   potassium
9. **Ca** calcium
10. **Cl** chlorine

Most elements occur in nature as collections of individual atoms, but some elements, called polyatomic elements, occur as collections of molecules wherein each molecule consists of two or more atoms of the same element. Six of the above 10 elements are polyatomic: H_2, N_2, O_2, Cl_2, P_4, S_8.

Formulas represent the resultant of a chemical reaction (or a series of chemical reactions) among two or more different elements. Often the products of a chemical reaction exist as autonomous molecules and the formula gives how many atoms of each element make up each molecule, e.g. water = H_2O is a molecule consisting of 2 hydrogen atoms and 1 oxygen atom; glucose = $C_6H_{12}O_6$ is a molecule consisting of 6 carbon atoms, 12 hydrogen atoms, and 6 oxygen atoms.

Atomic and Molecular Mass

One of the fundamental characteristics of an atom of any element is its mass which sometimes because of traditional usage is also called atomic weight. The mass of any atom is the ratio of its mass to the mass of some other atom designated as a reference. As of 1961, both physicists and chemists have agreed to define the mass of carbon that is found in ordinary elemental carbon as 12.00000 amu (atomic mass units). The mass of a hydrogen atom turns out to be 1.008 amu and that of oxygen is 16.00 amu. The mass of any molecule is equal to the sum of the masses of its constituent atoms; for example the mass of water, H_2O, is 2 × 1.008 (mass of hydrogen) + 16.00 (mass of oxygen) = 18.016. The calculation of the mass of a single atom depends on the concept of a mole.

The Mole Concept

For most scientific theories the most convenient units of measurement are: meter (m) for length, gram (g) for mass, and liter (L) for volume. For purposes of measurement it is convenient to introduce the concept of gram atomic mass. Again carbon will be our reference point. Just as *one* carbon atom has a mass of 12.0000 amu, then some large number, N, of carbon atoms will have a net mass of 12.0000 g. We define the gram atomic mass of carbon equal to 12.0000g as being the net mass expressed in grams of N carbon atoms. Then the gram atomic mass of any other element will be the net mass expressed in grams of N atoms of that element.

The number, N, is called Avogadro's number, and is used to define a new reference unit, a mole, that is useful for other types of measurements. A g-mole (gram-mole) is the amount of substance that contains the same number of subunits as there are atoms in 12.0000g of carbon; i.e., a g-mole is N subunits. For the remainder of this chapter we will use the term, *mole*, with the understanding that it refers to gram-mole (there is a more general definition of mole which need not concern us). Thus, we may refer to a mole of atoms or a mole of molecules or a mole of charge given off or received during electrolysis. We also now introduce the idea of gram molecular mass as the mass expressed in grams of N molecules of a substance. Thus, the gram molecular mass of N molecules of water is 18.016g; conversely, if we have 18.016g of water, we know we have N molecules of water.

Chemical Equations

Chemical equations are compact symbolic representations of three pieces of information: (1) the reacting atoms or molecules, (2) the product atoms or molecules of a chemical reaction, and (3) the number of each atom or molecule involved in the chemical reaction. For example, we represent the chemical interaction between hydrogen and oxygen to produce water as:

$$2H_2 + O_2 \rightarrow 2H_2O$$

This equation tells us that 2 molecules of hydrogen chemically reacts with 1 molecule of oxygen to produce 2 molecules of water.

Chemical Reactions

Chemical reactions are interactions among atoms which destroy one or more sets of relations among atoms and creates one or more new sets of relations. Chemical reactions always consist of some relative transfer of electrons among atoms resulting in breaking and creating atomic associations. Correspondingly, we will discuss compatibility between atoms, electronegativity which influences transfer of electrons, and chemical bonds which maintain atomic associations.

Atomic Compatibility

The theory of quantum mechanics provides rules for when and to what extent any two atoms may enter into a chemical reaction. For our purposes we may summarize these rules as consisting of two general features: (1) atoms are in a donor-receiver relationship, i.e., one atom donates electrons and the other receives them or (2) they both donate and receive electrons from one another. These relationships define their compatibility. An atom which only can donate electrons is chemically compatible only with atoms which can receive electrons. Likewise atoms which only can share electrons only are compatible with other atoms that can share electrons.

Electronegativity

Electronegativity is the potential of an atom in a molecule to attract electrons to itself. This property varies depending on the molecule in which the atom is a component. The electronegativity of metals such as sodium and potassium relative to nonmetals is low and therefore metals tend to loose electrons. Ionization energy characterizes isolated atoms in the gaseous state and is defined as the energy transfer in the work necessary to completely remove an electron from an atom. Electronegativity of nonmetals such as chlorine and oxygen relative to metals is high and therefore nonmetals tend to gain electrons. Electron affinity characterizes isolated atoms or ions in the gaseous state and is defined as the energy change when an electron moves into one of the orbits of an atom. Electron affinity is a measure of how strongly an isolated atom attracts electrons to itself. For high electron affinities (expressed as negative values) the electron acquires energy in moving into an orbit of another atom. For low electron affinities (expressed as positive values) energy must be added to the electron for it to be able to enter an orbit of another atom. For example, a neutral oxygen atom gaining an electron is the result of a high electron affinity, but then it has a low affinity for gaining a second electron—the first electron gained repels the addition of a second one. In fact, the overall process of adding 2 electrons to oxygen requires a net addition of energy.

Chemical Bonds

Ionic Bonds.   Ionic bonds are chemical bonds holding two atoms together as a result of an electron transfer from one atom to the other. The resulting molecules are ionic compounds; the one atom having lost an electron becomes a cation which becomes bonded to the second atom that became an anion as a result of gaining an electron. This bonding force between the ions is described by Coulomb's law:

$$F = \frac{-k(Z_1e)(Z_2e)}{r^2}$$

where Z_1 and Z_2 are the number of charges on the two atoms and e is the magnitude of the charge and r is the minimum distance between the ions. If r > min., the ions move closer together due to the force of attraction between unlike charges. If r < min., the ions move away from one another due to the force of repulsion between negative electron clouds around each atom. The energy, V, of an ionic bond is the energy in the work done in the ions coming to be r distance apart.

$$V = \frac{k(Z_1e)(Z_2e)}{r}$$

Covalent Bonds.   Covalent bonds are the result of two atoms sharing one or more pairs of electrons rather than an electron being transferred from one atom to the other. In fact, each participating atom partially donates and partially receives an electron; i.e., there is a partial transfer of electrons. As a result there is a continuum between overtly ionic bonds at one extreme and overtly covalent bonds at the other. In the middle of the continuum the distinction between ionic and covalent bonds is hazy.

In forming a covalent bond the electrons of participating atoms are influenced by three forces: (1) an electron is attracted by the positive nucleus of the other atom; (2) an electron is attracted to the positive nucleus of its home atom; and (3) an electron is attracted to a second electron with opposite spin in the other atom. The result of these forces is that each atom holds on to its electron but the two electrons move toward one another. This, in turn, causes the two atoms to move toward and remain bonded to one another. The distance between the nuclei in a covalent bond is r = min. When r > min., the attractive forces listed previously cause the atomic nuclei to move toward one another. When r < min., the repulsion of the positively charged nuclei cause the atomic nuclei to move apart.

There are three types of covalent bonds. (1) In nonpolar covalent bonds the pair(s) of electrons in the bond are "more or less" equally shared, that is, equidistant between the two bonded atoms. (2) In *polar covalent bonds* there is unequal sharing of the pair(s) of electrons, i.e., the electron pair(s) is closer to one atom than the other. The atom with the electron pair closer to it is more negative relative to the other which is more positive. Hence, overall the bond is polar though it also is electrically neutral since the positive and negative charges balance each other. (3) In a *hydrogen bond* the hydrogen nucleus, a proton, is covalently bonded to an atom in one molecule and it shares a pair of electrons with an atom in another molecule. There are two requirements for hydrogen bonding: (1) a hydrogen atom initially must be covalently bonded to a very electronegative small atom such as nitrogen, oxygen, or fluorine; (2) the second atom participating in the hydrogen bond also must be very electronegative and have an unshared pair of electrons available to attract the positive hydrogen nucleus, as for example in water, see figure 6.1. Hydrogen bonds are weaker than the other types of covalent bonds.

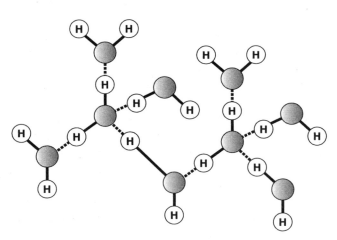

FIGURE 6.1   Hydrogen bonding among water molecules.

Oxidation-Reduction Reactions

Oxidation is the loss of one or more electrons in a chemical reaction in contrast to reduction which is the gain of electrons in a chemical reaction. Metaphorically speaking we might say that the atom "reduces" tension by gaining electrons to fill its outermost shell and thus becomes "happy." The atom that loses an electron(s) is oxidized. The atom that is oxidized is a reducing agent in that it supplies electrons that "reduce" the tension in some other atom. The reducing agent also is called a reductant. The atom that gains an electron(s) is reduced. By receiving an electron(s) from some other atom it is an oxidizing agent which also is called an oxidant.

Thermodynamics of Chemical Reactions

Thermodynamic Potential at Constant Temperature

The internal energy, **U,** of any system may be thought of as consisting of thermal energy, **E_T,** and non-thermal energy, **A.** Thus:

$$U = E_T + A$$

The potential of a system to do work as a result of its thermal energy is determined by its entropy, S, and its temperature, T. Work only will be accomplished if there is a change in T or S; i.e., $E_T = ST$. For a system at constant temperature, ST is the internal energy of a system that cannot be transferred away as work or heat so long as the temperature is constant.

For example, as shown in figure 6.2, consider liquid water at 100°C and 1 atom of pressure (liquid water is at its boiling point) which is transformed into water vapor. The thermal energy of an isolated system changes liquid into a gas with the consequent increase of entropy from S_1 to S_2. The amount of thermal energy of the system does not change, but the degree of organization does. S_1 represents greater organization due to the attractive forces among water molecules (hydrogen bond formation) when they are in the liquid state. S_2 represents less organization due to water molecules being independent of one another (therefore more random) when they are in the gaseous state. The transformation of water from liquid to a gas under these circumstances is spontaneous (entropy increases). The transformation of water from gas to a liquid would be nonspontaneous (entropy decreases). Work would have to be done on the gas to obtain a liquid; e.g., an increase in pressure and a decrease in volume, that is, ΔPV where ΔPV involves a force acting over a distance and thus is equal to work.

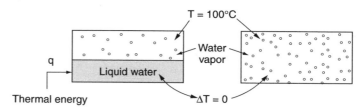

FIGURE 6.2 Liquid water becoming water vapor at constant temperature equal to 100°C. All thermal energy added to liquid water is used to convert it to water vapor.

S is a capacity factor of the thermal energy of a system at constant temperature. The greater S is the greater the mass of the system and conversely. The greater the mass (or entropy) the greater the capacity of the system to hold thermal energy at a particular temperature.

At a constant temperature, A is the internal energy of a system that can be transferred into work. For example, a coiled spring has a certain potential = A, to expend; the actual expansion would be work accomplished. The coil aspect of the spring is independent of the temperature; e.g., the degree of coiledness is independent of temperature. Therefore, a change in coiledness does not imply a change in temperature and in a reversible process all the energy of the coiledness potential could be converted into work.

In general, the internal energy, U, of a system can do work as a result of its thermal energy, $E_T = TS$ and as a result of its nonthermal energy, A. Thus:

$$U = A + TS$$

Enthalapy. The total energy of a system, called enthalapy, is defined as $H = U + PV$. P = pressure is the potential for material to expand and V = volume is a capacity factor; e.g., for an isolated system, the greater V is (and therefore the greater the mass of substance) at a particular pressure, the greater the potential for doing work of expansion. Thus, PV = potential for work = potential energy.

A change of H of a system held at constant pressure (which is the usual case for chemical reactions in the laboratory and in living systems) is equal to the amount of thermal energy that is absorbed or given off to the environment; i.e., $\Delta H = q =$ heat. Note: Δ means change.

1. $H = U + PV$
2. $\Delta H = \Delta U + \Delta(PV)$
3. $\Delta H = \Delta U + P\Delta V + V\Delta P$
4. $\Delta U = \Delta H - P\Delta V - V\Delta P$
5. $q = \Delta U + w = \Delta U + P\Delta V$ (1st law)
6. $q = \Delta H - P\Delta V - V\Delta P + P\Delta V$
7. $q = \Delta H - V\Delta P$
8. for constant pressure ($\Delta P = 0$) $\Delta H = q$

Free Energy. For many chemical reactions studied in chemistry and in biology, the temperature is held constant. Therefore, we define the concept of *free energy,* **G,** as the total potential for doing work by a system at constant temperature as

$$G = A + PV.$$

Thus, **G** is the potential of a system for doing work that is independent of its thermal energy. In particular, **G** may be associated with a particular set of restrictions on the motion of molecules; e.g., a particular *structure* such as bonding among molecules which gives the system the potential to do work independent of thermal energy. If the structure changes then **G** also changes. **G** by convention always is the *final state, f,* minus the *initial state, i;* so when $\mathbf{G} = \mathbf{G}_f - \mathbf{G}_i < 0$, **G** has decreased and the lost free energy shows up as work. None of this decrease shows up as heat because temperature is held constant.

Relation of Free Energy and Enthalapy.

1. $H = U + PV$
2. $U = A + TS$
3. $H = A + TS + PV = A + PV + TS$
4. $H = G + TS$ (since $G = A + PV$)
5. $\Delta H = \Delta G + \Delta TS$
6. at constant temperature, $\Delta H = \Delta G + T\Delta S$
7. thus: $\Delta G = \Delta H - T\Delta S$

Thermodynamic Potentials for Chemical Reactions

Heat of Reaction.   Let us consider the change in the total energy of a system in which atoms interact to form some compound. Since $H = U + PV$, $\Delta H = \Delta U + \Delta PV$ where ΔPV is the net work done as a result of the reaction which shows up as the expansion or contraction of reactants and products and ΔU is the net work done in the formation of chemical bonds; e.g., ΔU = the chemical energy of the molecules formed. Thus, overall, ΔH = the chemical energy in the molecules formed + work of expansion. When this chemical reaction is done at constant pressure, then as shown above, $\Delta H = \Delta U = q$ which we may define as heat of reaction.

$U = A + TS$; so at constant pressure, $\Delta H = \Delta U$ = nonthermal energy that is used to form a chemical bond + TS = thermal energy. If we consider atoms in the gaseous state, we then do not have to consider ΔS due to change of state (solid $\rightleftharpoons$ liquid or liquid $\rightleftharpoons$ gas). If we also specify the temperature for the heat of reaction, then ΔH = amount of energy given off when a chemical bond is formed = bond energy. Of course in order to break a chemical bond, the energy that was lost in producing the bond must be added to break it. Thus, the energy of the system after the bond is broken will be greater than before it was broken; i.e., $\Delta H = H_f - H_i$ = positive value. During chemical reactions under these conditions, some bonds will be formed while others will be broken; that is, energy will be given off during bond formation while energy will be used up during breaking of bonds. If the net ΔH = negative value, then the energy given off in the formation of new bonds is greater than the energy used up in breaking the old bonds. We say the chemical reaction is exothermic; i.e., energy is given off to the environment and the reaction is spontaneous. If ΔH = positive value, then the reaction is endothermic; i.e., energy is taken up from the environment to make the reaction occur. If the environment cannot supply this energy, then the reaction will never occur and thus it is non-spontaneous.

We often consider chemical reactions at constant temperature and pressure among atoms and molecules not in the gaseous state, and ΔH will be either positive (endothermic) or negative (exothermic). This often also will indicate whether the reaction is spontaneous or not, but not always. It is possible for a particular chemical reaction to be spontaneous but the **ΔH** is positive. Such an endothermic reaction is due to the products of the reaction changing state so that, for example, energy from the environment is absorbed and used to convert the product molecules from a liquid to a gaseous state. The net energy given off in the chemical reaction is less than the energy needed to bring about the change of state. In general, if the products of a spontaneous chemical reaction have more molecules in the gaseous state than the reactants, the reaction is endothermic; otherwise it is exothermic.

Free Energy Changes Determine Direction of a Reaction.   As described in a previous section:

$$\Delta G = \Delta H - T\Delta S \text{ (at a constant temperature)}$$

One can derive from the above equation the following relationships:

1. for an ideal gas where P_2 = final pressure and P_1 = initial pressure:
   $\Delta G = nRT \ln\{P_2/P_1\}$; (ln = natural log), n = number of moles
2. for ideal solutions where C_2 = final concentration and C_1 = initial concentration:
   $\Delta G = nRT \ln\{C_2/C_1\}$

If ΔG = negative value, the reaction will occur until $\Delta G = 0$; i.e., the reaction is spontaneous. If ΔG = positive value, the reaction will not occur unless energy is added so that the net change in G is negative; i.e., for $\Delta G > 0$ the reaction is non-spontaneous.

Cell as a Chemical Machine

Review Concept of a Machine

A machine takes in energy and focuses it, so to speak, on accomplishing a specific task. In a simple machine, the key component that does this focusing is an energy coupler which connects an energy source with a specific task that requires energy. We say this component "couples" the energy source to the task. A hammer is an example of an energy coupler; the muscles in a person's arm which raises and lowers the hammer is the energy source. Driving a nail into a block of wood is an energy requiring task. The hammer couples the up and down movement of the hand with the task of driving a nail into a block of wood.

As stated in Chapter 1, machines exhibit three characteristics: (1) Energy flows from an energy source into the machine, through it, and out again as work and heat. (2) Incoming energy causes an energy coupler to go from a low to a high energy level; then in going from a high to a low energy level, the coupler accomplishes some task. (3) A large task is accomplished by the sum of many small tasks each accomplished by one cycle of an energy coupler.

Cell Machine

All cells are machines because they satisfy the three defining characteristics of a machine described in a previous section. Cells take in energy, usually in the form of glucose and convert it into smaller molecules, usually carbon dioxide and water. This conversion is similar to burning paper or burning gas in car engines. In all cases, chemical energy contained in whatever is burned up is released. The unique thing about cells is that this burning is carefully controlled so that packets of energy are sequentially released rather than the energy being released all at once. The packets of energy are temporarily stored in the cell until they are used to accomplish specific cell functions. As always, during energy transfer or conversion of energy from one form to another, some energy is diverted from accomplishing work to be "lost" as heat. Thus, as in any machine, energy from the outside flows through the cell and out again to the environment.

There are special chemical systems in a cell which operate as chemical energy couplers. These systems *receive* packets of energy from cellular burning of molecules such as glucose. These systems also *transmit* packets of energy to some cellular task that requires energy, as shown in fig. 5.20. The major chemical energy coupler in cells is the ADP/P_i (P_i represents a phosphate molecule) system; see figure 5.20 for the structure of ADP and ATP. When the ADP/P_i system receives a packet of energy, phosphate is "energized" to combine with ADP to form ATP. This reaction: ADP + phosphate + received packet of energy → ATP, is the transformation of the ADP/P_i system from a low to a high energy level. The ATP is analogous to the hammer held above a nail (the hammer has potential energy due to its position and to the chemical potential in the arm muscles; it has its highest kinetic energy just before it hits the nail). In hitting the nail the hammer goes from a high to a low energy level. Likewise when ATP breaks down into ADP + phosphate, the ADP/P_i system transmits energy to some cellular task and moves from a high to a low energy level; see figure 6.3. The ADP/P_i system may be thought of as a population of ADP and phosphate molecules (P_i) that may alternate between

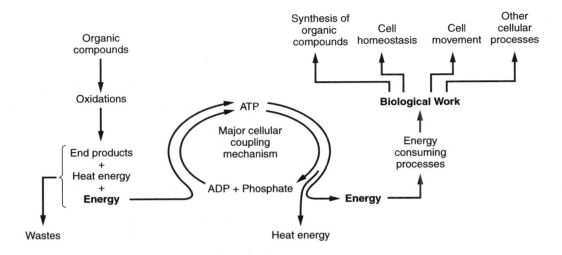

FIGURE 6.3

two patterns. In one pattern which is the lower energy state corresponding to a lower degree of organization (i.e., higher entropy representing greater randomness in the population), the ADP and P_i molecules are independent of one another. In the other pattern which is the higher energy state corresponding to a higher degree of organization (i.e., lower entropy representing less randomness in the population), each ADP molecule is chemically bonded to a phosphate molecule to produce a single molecule called ATP. The ADP and P_i molecules are not independent of one another. The chemical bonds between ADP and P_i are the new structure (i.e., pattern) in the population corresponding to a greater organization and less randomness (lower entropy) in the ADP/P_i system, see figure 6.3.

There are several thousand ADP/P_i systems in every cell. These systems continually receive packets of energy and transmit some of the received energy to other systems that accomplish particular cellular tasks. For example, a muscle cell consists of fibers which can shorten in a sequence of stages. Each shortening is brought about by a packet of energy. The whole series of shortening steps of all the fibers in the muscle cell is the task of muscle contraction. Several packets of energy are produced by the cell converting glucose into carbon dioxide and water (36 packets of energy per glucose molecule). Each of these packets of energy converts an ADP/P_i system from a low to a high energy level; that is, ATP is formed. When each of the ADP/P_i systems at a high energy level breaks down, it transmits a somewhat smaller packet of energy to a muscle fiber which then undergoes one stage of shortening. The release of packets of energy from several thousand ATP molecules causes the muscle contraction. Note: the energy packet released when ATP breaks down is somewhat smaller than the packet of energy needed to convert the ADP/P_i system into ATP. The reason for this is that some of the energy in the original packet is diverted to heat and the remaining energy does the work of converting ADP + P_i into ATP. Thus, the amount of energy diverted to q is not available when ATP breaks down into ADP + P_i.

The cell viewed as a machine with the **ADP/P$_i$** system as its major energy coupler exhibits the three characteristics of any machine as follows.

1. Energy flow through the cell
   a. Chemical energy (C.E.) in organic molecules (O.M.) e.g. glucose, flows into the cell.
   b. Organic molecules are oxidized (burned up) producing chemical energy (C.E.) and thermal energy (T.E.).
   c. Energy couplers focus C.E. to accomplish biological work.
   d. Some or all of the incoming energy in organic molecules leaves the cell as heat (q) + work (w).

2. Cycle of the major cell energy coupler: ADP/P_i system
   a. Chemical energy from oxidized organic molecules bring about the release of energy which converts a population of ADP and P_i molecules into a population of ATP molecules; i.e., the ADP/P_i system is converted from a low energy level to a higher energy level.
   b. ATP binding to sites within the cell may initiate accomplishing a small task.
   c. ATP ($ATP + H_2O$) guided by enzymes to break down into $ADP + P_i + C.E. + T.E.$ ($ADP + P_i + C.E. + T.E.$).
   d. The C.E. from the breakdown of ATP completes the small task.
   e. Some of the energy from oxidized molecules cause $ADP + P_i$ to form ATP thus starting a new cycle.
3. A large biological task is accomplished by the sum of many small tasks each accomplished by one cycle of the energy coupler ADP/P_i.

Cellular Oxidation of Glucose

Overview of Cellular Oxidation

The oxidation of glucose is the major energy producing process in all cells. A series of chemical reactions including four oxidations convert a single 6-carbon glucose molecule into two 3-carbon molecules called pyruvic acid. If no oxygen is present or if the cell is unable to use oxygen, pyruvic acid is converted to some other form. Mammalian cells convert pyruvic acid into lactic acid. Some bacteria convert pyruvic acid into acetone; whereas other bacteria and yeast convert pyruvic acid into carbon dioxide and ethyl alcohol. This latter process is what is usually associated with fermentation. The wine and beer industries make profitable use of this biological process. Thus, this whole sequence of reactions ending in lactic acid or acetone or alcohol is called anaerobic glycolysis or fermentation (anaerobic means "no air"). If the cell can use oxygen for receiving electrons from molecules being oxidized, e.g., glucose, pyruvic acid is converted into three carbon dioxide molecules, and all the hydrogen atoms in pyruvic acid are transferred to oxygen atoms thereby producing water (H_2O). In other words, a glucose molecule is broken apart, and some of the released hydrogen atoms combine with oxygen to produce water. The carbon atoms in glucose become carbon dioxide molecules. We can summarize the net result of this series of reactions as:

GLUCOSE + OXYGEN $\rightarrow$ CARBON DIOXIDE + WATER + ENERGY

The first phase of this sequence of reactions, the breakdown of glucose into two pyruvic acid molecules, is called aerobic glycolysis.

There are some interesting practical applications of these biochemical facts. During vigorous exercise, our muscles use up oxygen faster than the blood can supply it. These muscle cells then revert to anaerobic glycolysis. Muscle cells cannot utilize lactic acid, so this substance begins to accumulate. The increased acidity due to lactic acid inhibits muscle cell function and we get a cramp. The liver cells are able to convert lactic acid back to glucose. Thus, the faster the blood carries the lactic acid to the liver, the faster we get rid of the cramp. We can help this process along by soaking the sore muscles in hot water and/or by rubbing them or applying liniment. All these techniques increase blood circulation to the sore area.

We also now can appreciate what it means to be in above average physical condition. The lungs are more efficient in getting oxygen to the blood; the cardio-vascular system is more efficient in circulating blood; the muscle cells are more efficient in extracting glucose and oxygen from the blood; the muscles store a greater amount of glucose and by having a greater number of mitochondria, muscle cells oxidize glucose more rapidly.

Anaerobic glycolysis is associated with energy couplers to cause ADP and phosphate to form two ATP molecules. Cellular respiration of glucose to carbon dioxide and water is associated with energy couplers to cause ADP and phosphate molecules to form 36 ATP molecules. Obviously cells that can

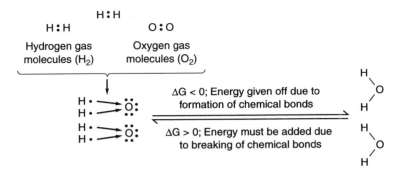

FIGURE 6.4

use oxygen are much more efficient chemical machines than cells that cannot use oxygen. The rate at which cells can carry out life functions is limited by the rate at which cells can bring ADP plus phosphate back to the higher energy level in ATP.

Biochemical Details

Energy Producing Processes in Cells.   All cells are like cars in that cells regulate the continuous burning of organic molecules. This intracellular (intra means "within") energy producing process feeds energy into several different series of energy couplers which terminate in the various cellular functions. Burning is a type of oxidation-reduction reaction.

The forming of water from the reaction of hydrogen with oxygen and the breakdown of water into hydrogen and oxygen are oxidation-reduction reactions. In forming water an electron in a higher energy level moves from each of two hydrogen atoms to unoccupied lower energy levels in an oxygen atom. This oxidation-reduction reaction is an energy producing process. The energy produced is equal to the difference between the energy level in hydrogen and the energy levels in oxygen. In contrast, the breakdown of water is an energy consuming oxidation-reduction reaction. Energy must be added to move electrons from the lower energy levels associated with oxygen in the water molecule to the higher energy level in each of the two hydrogen atoms, see figure 6.4.

Hereafter all oxidation-reduction reactions will be referred to simply as "oxidations" with the understanding, of course, that whenever there is an oxidation, there must also be a reduction.

Organic Molecule Oxidations.   The most common type of oxidation of organic molecules involves the transfer of a negative hydrogen ion containing two electrons from one atom to another. Consider, for example, the combustion of gasoline. Gasoline consists of long chains of carbon atoms with many carbon-hydrogen bonds. Burning gasoline involves the breaking of carbon-hydrogen bonds in which a pair of electrons are shared between carbon and hydrogen. A pair of electrons in a high energy level in the gasoline molecule moves along with the hydrogen nucleus (a proton) to a lower energy level in an oxygen atom to form $H : O^-$. The neutral oxygen atom has gained two electrons and one proton; so it has a net gain of one negative charge. A second hydrogen nucleus is attracted to and attaches to this negative OH group to produce a neutral H_2O molecule. In effect the oxidation of gasoline involves the transfer of negative hydrogen ions from carbons in gasoline to oxygen atoms, see figure 6.5.

Cellular Energy Producing Oxidations.   Cells produce energy by transferring pairs of electrons in carbon-hydrogen bonds to lower energy levels in some other atoms or molecules which are called electron acceptors. Electron acceptors which receive pairs of electrons by combining with negative hydrogen ions are called hydrogen acceptors. The first step of the oxidation of sugars, lipids or proteins involves the transfer of negative hydrogen ions from these organic molecules to nicotin (the vitamin, niacin) hydrogen acceptors. There are two types of nicotin hydrogen acceptors: NAD (*n*icotinamide *a*denine *d*inucleotide) and NADP. The NADP is an NAD molecule with an added phosphate (P) group. Cells with mitochondria oxidize small molecule derivatives from sugars, lipids, or proteins to produce several NAD-H and NADP-H molecules; hereafter we will refer to these mol-

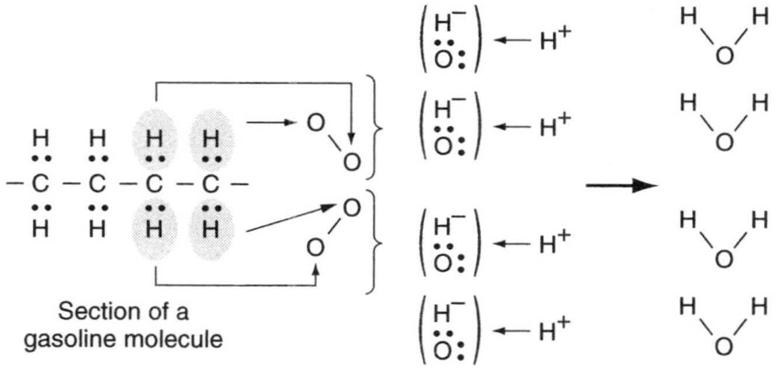

FIGURE 6.5

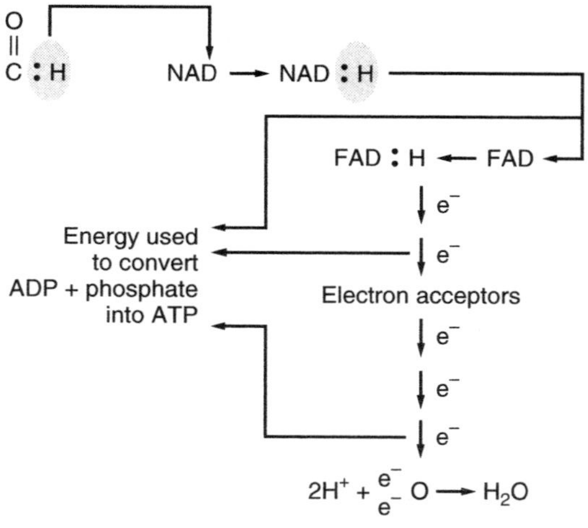

FIGURE 6.6

ecules simply as NADH and NADPH.[1] These molecules transfer the hydrogen with two electrons to a lower energy level in another hydrogen acceptor (FAD). This, in turn, transfers single electrons to electron acceptors (cytochromes). These electrons enter a sequence of electron transfers always from a higher energy level in a one-electron acceptor to a lower energy level in another. At last the electrons move to the lowest energy level in the final electron acceptor which is oxygen. Each transfer of negative hydrogen ions and each transfer of electrons between electron acceptors releases some energy, as shown in figure 6.6.

Cellular Production of ATP. The release of energy from glucose usually occurs in three main phases: (1) glycolysis, (2) the citric acid cycle (Krebs cycle), and (3) the electron transport chain. These phases can vary according to the type of cell or organism and the conditions under which oxidations occur. For example, glycolysis occurs in all living cells, but the citric acid cycle can occur only where oxygen is available.

Glycolysis. Glycolysis is the sequence of chemical reactions that adds phosphate groups to glucose, breaks glucose in half, and then oxidizes each half twice. The energy released in each oxidation is used to transfer phosphate to ADP to produce ATP, see figure 6.7.

Two ATP molecules are used to add a phosphate to each end of a glucose molecule. This process is called phosphorylation. Then this enlarged glucose molecule is broken in half so as to eventually produce two identical 3-carbon molecules, called phosphoglyceraldehydes (PGAL), each containing a phosphate group. Each of these molecules is oxidized by the removal and transfer of a negative hydrogen ion to the hydrogen acceptor NAD to produce NADH. Some of the energy released in this oxidation is used to add another phosphate derived from inorganic phosphate within the cell to the PGAL carbon atom that lost the hydrogen. The resulting molecule (called 1,3 diphosphoglycerate) is a high energy molecule with two phosphate groups, but the second, newly formed phosphate bond is unstable. This high energy molecule therefore spontaneously breaks down, transferring its unstable phosphate to ADP to produce ATP, as shown in figures 6.8 and 6.9. The remaining 3-carbon molecule is chemically rearranged (with the help of an enzyme) and oxidized again to produce a second high energy molecule containing an unstable phosphate bond. This molecule spontaneously breaks down, transferring its phosphate group to ADP to produce ATP. Having lost its last phosphate, this molecule becomes the 3-carbon compound, pyruvic acid, see figures 6.8 and 6.9. Glycolysis involving one glucose molecule produces four ATP molecules from four ADP and four phosphate molecules. During this process two ATP molecules are used to add phosphate to the glucose molecule, and two

---

[1]Note: The NAD (and NADP) has a positive charge. The negative hydrogen ion with its two electrons neutralizes the positive charge and causes one carbon in the molecule to have a pair of unshared electrons giving the carbon an overall negative charge. Positive hydrogen ions (i.e., a proton) are attracted to these sites. Therefore, NADH becomes $NADH_2$.

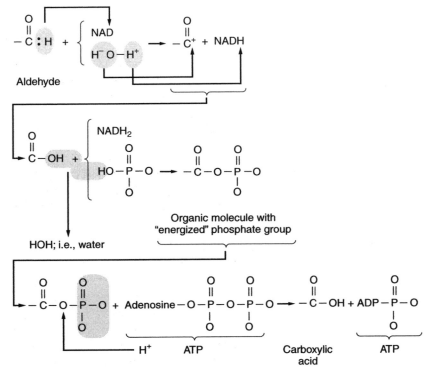

Biological oxidation coupled to production of ATP from ADP + phosphate

FIGURE 6.7

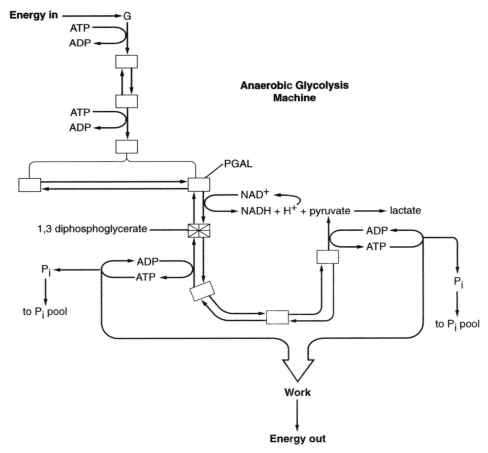

FIGURE 6.8

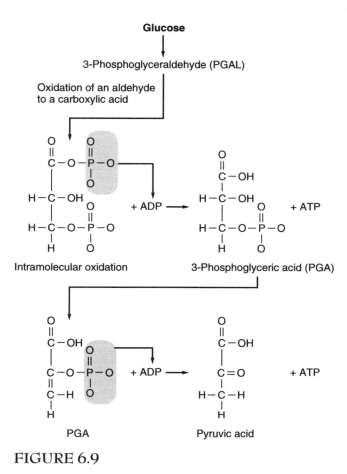

FIGURE 6.9

hydrogens are released to two NADs to produce two NADHs. Therefore, glycolysis involving one glucose molecule produces a net gain of two ATP and two NADH molecules.

Citric Acid Cycle.   What happens to the pyruvic acid formed as a result of glycolysis in aerobic animals such as humans? Actually, pyruvic acid enters a series of reactions, called the citric acid cycle or Krebs cycle. As a result of the Krebs cycle most of the energy that pyruvate still holds is extracted from it.

In order to get pyruvic acid into the citric acid cycle, a chemical facilitator is needed. Ordinarily, such facilitators (or regulators) are enzymes which act as catalysts. In this case, however, a coenzyme, coenzyme A (which is synthesized from vitamin B) actually enters the reaction in contrast to enzymes that do not do this. In the presence of oxygen the 3-carbon pyruvic acid molecules are oxidized resulting in the transfer of hydrogens to NAD to form NADH. In addition, during this oxidation a terminal carbon is removed in the form of carbon dioxide. The remaining 2-carbon compound, called an acetyl compound, combines with coenzyme A to form an unstable high energy molecule, acetyl coenzyme A (acetyl CoA), as shown in figure 6.10. The 2-carbon acetyl group (in acetyl CoA) comes into contact with a preexisting 4-carbon compound, oxaloacetic acid, with which it reacts chemically to form a 6-carbon compound, citric acid, from which the cycle derives its name. This reaction readily occurs because of the high energy and relative instability of acetyl CoA. Now a series of seven reactions, four of which are oxidations, converts the citric acid into oxaloacetic acid. A carbon in the form of carbon dioxide is removed in two of the four oxidations. Two of the oxidations involve the transfer of two hydrogens each to NAD to produce NADH molecules. One of the oxidations transfers a hydrogen to flavine adenine dinucleotide (FAD) to produce FADH. In addition, one of the oxidations results in a phosphorylation that produces an ATP molecule. Thus, the net outcome of the citric acid cycle is three NADHs, one FADH, and one ATP molecule per acetyl group that enters the cycle, see figure 6.10.

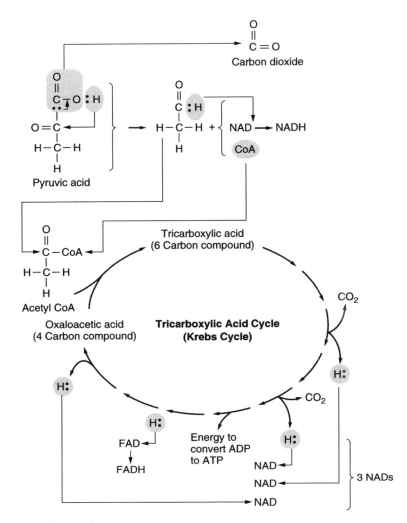

FIGURE 6.10

Most of the production of ATP occurs in the next phase of respiration, the electron transport chain or respiratory chain. This chain takes the electrons accepted by NAD and FAD in both glycolysis and the citric acid cycle and passes them down a series of electron acceptors so that their energy is gradually released to form ATP molecules, with the remaining electrons being passed to oxygen via water formation.

The Electron Transport Chain.   As a molecule of glucose passes through glycolysis and the citric acid cycle, some of its hydrogens are removed. As we know, these hydrogens are "picked up" by the hydrogen acceptors NAD and FAD, producing NADH and FADH. In fact, for the first phases of respiration (glycolysis and the citric acid cycle), eight NADH and four FADH molecules are produced in addition to the four ATP molecules. Now, in the electron transport chain these hydrogens are passed down a series of 10 hydrogen and electron acceptor molecules, resulting in the production of ATP and water. Since NADH is a higher energy molecule than FAD, it oxidizes and in the process passes its hydrogen to FAD, creating one molecule each of ATP and FADH for each NADH. When all the hydrogens have been joined with FAD to produce FADH, the electrons from these molecules are passed to a series of electron accepting pigments called cytochromes. As the electrons are passed down this series of cytochromes, energy is sequentially released at certain stages to ADP which picks up inorganic phosphate from the cell to produce ATP. For every hydrogen that passes through the electron transport chain, three ATP molecules are produced. The hydrogen nuclei (protons) that were released from FADH when the electrons were passed to the cytochromes seek out oxygen and along with the leftover electrons result in the formation of water, see figure 6.11.

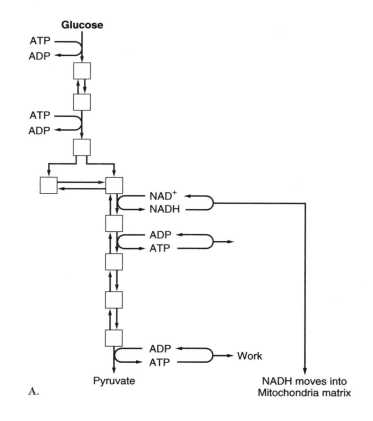

A.

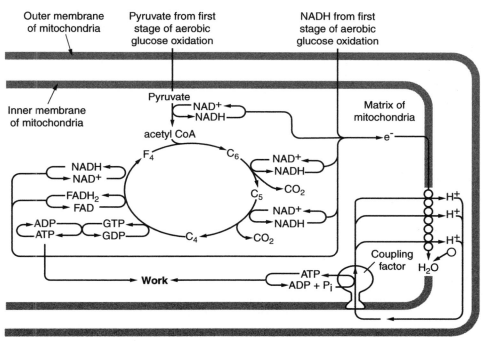

B.

FIGURE 6.11   First stage of aerobic glucose oxidation.

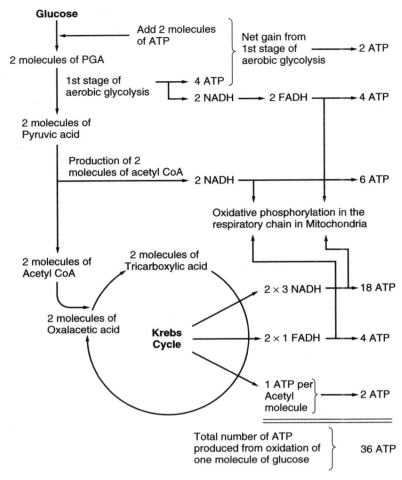

FIGURE 6.12

Net Result of Breakdown of One Glucose Molecule. Anaerobic glycolysis produces a net gain of two ATP and two NADH molecules. In time, these NADH molecules transfer hydrogen to FAD without producing any ATP, thus in effect producing two FADH molecules. Each of the two pyruvic acid molecules from glycolysis produces an acetyl CoA molecule and an NADH molecule. Each acetyl CoA molecule enters the citric acid cycle and produces one ATP, three NADH, and one FADH molecule. In other words, glucose, by means of the citric acid cycle, ultimately produces two ATP, six NADH, and two FADH molecules. The net result of the first two phases of the cellular respiration of glucose is four ATP (two from glycolysis and two from the citric acid cycle), four FADH (two from glycolysis and two from the citric acid cycle), and eight NADH (two from the production of two acetyl CoA molecules and six from the citric acid cycle). All oxidations beginning with NADH in the electron transport chain produce three ATP molecules, and all oxidations beginning with FADH produce two ATP molecules. The net ATP output of oxidizing one glucose molecule to CO_2 and water is: 2 (from glycolysis) + 2 (from the citric acid cycle) + {(8 × 3) = 24 (from NADH entering the respiratory chain)} + {(4 × 2) = 8 (from FADH entering the respiratory chain)} = 2 + 2 + 24 + 8 = 36 ATP molecules, as shown in figure 6.12.

Production of ATP From Molecules Other Than Glucose. As indicated in Chapter 9, the chemical energy in glucose is more available to cells throughout the body than energy in other organic molecules. This is why cellular respiration usually is described in terms of oxidation of glucose to carbon dioxide and water. However, under various circumstances, other organic molecules may enter the same oxidative processes that produce ATP for the cell. For example, liver cells detoxify ethyl alcohol by converting it into acetyl CoA, which enters the citric acid cycle and eventually produces 15 ATP molecules. When there is a shortage of glucose, cells will draw upon the fat deposits as energy sources. A terminal 2 carbon unit is successively broken off from a fatty acid and converted into acetyl CoA. Again, each acetyl CoA leads to the production of 15 ATP molecules. Thus, the number of acetyl CoA molecules that come from fatty acid decomposition multiplied by 15 equals the number of ATP molecules the cell can extract from the fatty acid. The cell even can use structural proteins as energy sources. Proteins are broken down into amino acids. The cell has enzymes to guide the conversion of many of the 20 different amino acids into acetyl CoA. As before, each acetyl CoA leads to the production of 15 ATP molecules.

Cell as a Homeostatic Machine

Transfer of Substances in and out of Cells

A cell, like any other thing in the universe, is exposed to potentials that tend to bring it into equilibrium with its environment. One essential feature of any living organism is that it maintains a radical separation from its environment; that is, life is the continuous struggle to prevent equilibrium with the environment. How a cell opposes any particular potential exemplifies how a cell operates as a homeostatic machine. Because cells contain and are surrounded by water solutions of many different substances, cells are continuously exposed to many diffusion potentials.

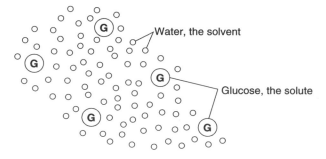

Transfer Due to Diffusion Potentials.   Diffusion potentials with respect to a cell are specified by a cell membrane that separates a water solution inside a cell that is very different from a water solution outside the cell. A water solution consists of one or more substances dissolved in water. Dissolved in water means that each molecule or atom of the dissolved substance is surrounded by and kept separate from other molecules or atoms of the same type by many water molecules. For example,

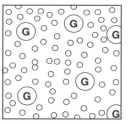
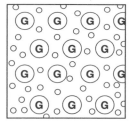

A. Lower glucose concentration higher water concentration   **B.** Higher glucose concentration lower water concentration

FIGURE 6.13

a solution of sugar such as glucose consists of individual glucose molecules separated from one another by many water molecules, see figure 6.13. The dissolved substance is called a solute and the fluid in which the solute is dissolved is called a *solvent;* water is the solvent of water solutions. The number of solute molecules per unit volume is called solute concentration and for water solutions, the number of water molecules per unit volume is the water concentration. As can be seen from figure 6.13, the greater the solute concentration, the less the water concentration, and conversely, the less the solute concentration, the greater the water concentration.

Salts are compounds formed by the chemical reaction of metals like sodium, potassium, and calcium with nonmetals like chlorine. In these chemical reactions the neutral metal atoms lose one or more electrons to become a positively charged atom, and the neutral nonmetal atoms gain one or more negatively charged electrons to become a negatively charged atom. Like charges repel one another, whereas unlike charges attract one another. In salts the positively charged metal ions are attracted to the negatively charged nonmetal ions. For example, sodium atoms lose electrons to chlorine atoms to form table salt, sodium chloride, see figure 6.14. The positive sodium ions are attracted to negative chloride ions which in turn are attracted to other positive sodium ions. These sodium ions, in turn, are attracted to other negative chloride ions. The positive and negative ions form a three-dimensional lattice network called a salt crystal, see figure 6.14. Water dissolves salts by separating

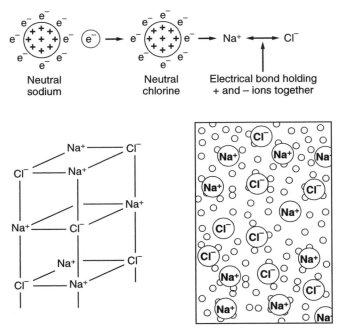

FIGURE 6.14

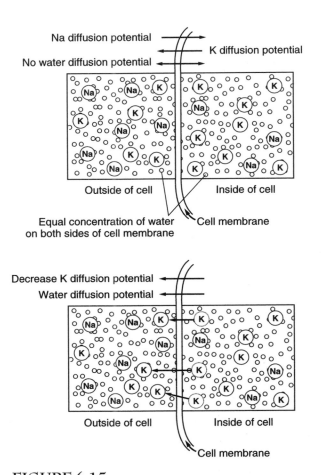

FIGURE 6.15

positive ions from negative ions and keeping each ion separate from any other by means of many intervening water molecules, see figure 6.14. Thus salt solutions always consist of at least two solutes: positive ion solutes and negative ion solutes.

Any membrane, and in particular a cell membrane, that allows a substance to pass into the membrane bounded system, e.g., into the inside of the cell, is said to be permeable to that substance. A membrane that is permeable to some substances but is not permeable to some other substances is said to be semi-permeable. Cell membranes are permeable to water, glucose, sodium ions, potassium ions, chloride ions but are not permeable to large molecules such as albumin (a protein that can be dissolved in water).

A diffusion potential is the difference in concentration of a substance between two points. In general, cells have: (1) a higher concentration of potassium ions on the inside than on the outside of the cell membrane; (2) a higher concentration of sodium ions on the outside than on the inside of the cell membrane; (3) an equal concentration of water on each side of the cell membrane. Unless acted on by some energy requiring process, substances diffuse from an area of higher concentration to an area of lower concentration. Thus, cells are exposed to a sodium ion diffusion potential which causes sodium ions to move from the outside into the inside of the cell and to a potassium ion diffusion potential which causes potassium ions to move from the inside to the outside of the cell, see figure 6.15. Usually cells make adjustments to eliminate a water diffusion potential; i.e., cells change until the water diffusion potential becomes zero, see figure 6.15.

Osmosis is the net movement of water across a membrane in response to a water diffusion potential. In other words, osmosis is the diffusion of water across a membrane from an area of higher water concentration to an area of lower water concentration. From the ideas already presented and from looking at figure 6.13, we readily see that the greater the solute concentration, the less the water concentration. Therefore, osmosis also may be defined as the diffusion of water across a membrane from an area of lower solute concentration (i.e., higher water concentration) to an area of higher solute concentration (i.e., lower water concentration). Thus, if for any reason, the cell should increase

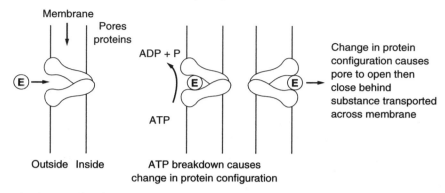

FIGURE 6.16

the concentration of solutes such as sodium ions inside the cell, a water diffusion potential would exist, and water would move into the cell toward the greater solute concentration. If, for any reason, the cell or some outside influence should increase the concentration of solutes outside the cell, water would move out of the cell again toward the greater solute concentration. Since with respect to water diffusion potentials, water always moves toward the greater solute concentration, by convention, scientists sometimes say that solutes "osmotically draw water into" a compartment. The water that may move into a compartment as a result of the solutes present exerts water pressure. Therefore, the solute concentration which may result in water moving into a compartment is said to exert an osmotic pressure, the greater the solute concentration, the greater the osmotic pressure in a compartment.

Active Transport.   General concept. Things tend to fall down not up. However, a machine, like an elevator, can take in energy and focus it on carrying a thing upwards and then come down again to carry up something else. Repetition of this cycle actively transports many people from lower to higher floors in a building. Of course, elevators do not just fall down to the bottom floor; that would be too hard on the elevator and on the people inside it. Elevators use energy to transport people from upper to lower floors as well as vice versa. Cell membranes have in them machines analogous to elevators that transport substances from one side of the membrane to the other. These membrane machines are called pumps or active transport mechanisms, see figure 6.16. Usually these pumps transport substances against a diffusion potential, but sometimes they actively transport substances in the same direction as the diffusion potential just as elevators transport people down as well as up. For example, "glucose pumps" in the cells lining the small intestine of humans actively carry glucose from an area of higher concentration in the intestine to an area of lower concentration inside the cells lining the intestine.

Membrane Sodium-Potassium Pump.   Virtually all cells have transport pumps which actively transport sodium ions to the outside of the membrane and actively transport potassium ions from the outside to the inside of the cell membrane. The operation of these membrane pumps exemplifies the cell as a machine. Energy packets released during the breakdown of glucose cause ADP + phosphate to form ATP. Energy is released when ATP breaks down into ADP + phosphate. In some manner, this energy results in a special membrane protein to transport a sodium ion to the outside. This protein may, if necessary, pick up a potassium ion and transport it to the inside of the cell. Repetition of the cycling of the ADP/ATP energy couplers between low and high energy states and the flip-flop of the membrane proteins accomplish the task of removing sodium ions from and taking potassium ions into the cell, as shown in figure 6.16. These pumps must operate continually because diffusion potentials continually cause sodium ions to move into the cell and potassium ions to move out of the cell. When the cell dies, the membrane machines stop and the cell comes into equilibrium with its environment. The sodium and potassium diffusion potentials become zero.

Negative Feedback Regulation of Cell Volume

Regulation of cell volume exemplifies the cell as a homeostatic machine. The sodium-potassium pumps are the major effectors that influence the solute concentration inside a cell. A *decrease* in rate of the sodium-potassium pumps results in some sodium ions that diffuse into the cell to stay there. This results in an increase in solute concentration in the cell. Water then is osmotically drawn in toward the higher solute concentration, and the extra water results in an increase in cell volume. Just the opposite occurs when there is an increase in rate of the sodium-potassium pumps. Thus, decrease in rate of sodium-potassium pumps causes an increase in solute concentration in the cell which causes a net movement of water into the cell (osmosis) which increases cell volume. An increase in rate of sodium-potassium pumps causes a decrease in solute concentration in the cell which causes a net movement of water out of the cell toward greater solute concentration outside (osmosis) which decreases cell volume.

The cell has "something like" a sense receptor, evaluation unit and a command unit. The solute concentration is "recognized," "evaluated," and a command is sent to the appropriate pump. If the sodium ion concentration is above the preset value, the rate of the sodium-potassium pump is increased leading to a decrease in sodium ion concentration inside the cell. This causes movement of water out of the cell resulting in a decrease in cell volume. If the sodium ion concentration is below the preset value, the rate of the sodium-potassium pump is decreased leading to an increase in sodium ion concentration inside the cell due to the sodium diffusion potential. This causes movement of water into the cell (osmosis) resulting in an increase in cell volume. Thus, internal sodium ion concentration, internal water content, and cell volume oscillate about preset values due to negative feedback regulation.

Summary of Negative Feedback Regulation of Cell Volume

1. Regulation of $[Na]_i$ results in regulation of cell volume.
    a. Formal understanding of this:
        1) If $[Na]_i$ > preset value, the cell volume increases to > preset value
        2) If $[Na]_i$ < preset value, the cell volume decreases to < preset value
    b. Metaphorical analogical understanding of this: [Na] may be thought of as [heat energy] which is temperature and relative to this analogy, cell volume may be thought of as cell temperature.
        1) If heat energy flows into a cell causing the cell temperature to rise then $[heat\ energy]_i$ > preset value so that cell temperature increases to > preset value
        2) If heat energy flows out of cell causing the cell temperature to fall then $[heat\ energy]_i$ < preset value so that cell temperature decreases to < preset value
2. Negative feedback regulation of cell volume in terms of negative feedback regulation of $[Na]_i$.
    a. Formal understanding of this: SR (sense receptor) records the $[Na]_i$ and the EU (evaluation unit) compares $[Na]_i$ to the preset value and then the CU gives the appropriate command which is
        1) If $[Na]_i$ > preset value, then turn ON the Na pump
        2) If $[Na]_i$ < preset value, then turn OFF the Na pump
        3) The change in $[Na]_i$ causes a change in cell volume which starts a new negative regulation
    b. Metaphorical analogical understanding of this: the Na pump may be thought of as an air conditioning unit. SR records the internal cell temperature analogous to cell volume and the EU compares this to the preset value and then the CU gives the appropriate command which is

1) If cell temperature (analogous to cell volume) > preset value, then turn ON the air conditioning unit (analogous to the Na pump)
2) If cell temperature (analogous to cell volume) < preset value, then turn OFF the air conditioning unit (analogous to the Na pump)

Regulation of Intracellular Chemical Reactions

Many hundreds of different chemical reactions go on within the cell. Many of these reactions occur in the same general area within the cell and therefore tend to interfere with and in some cases oppose one another. The cell avoids chaos by means of cell compartmentalization via membranes and by protein regulator molecules.

Protein Regulator Molecules

Proteins have three-dimensional configurations which sometimes allows a part of the protein to conform to the shape of another molecule so that the protein and the other molecule fit together like two pieces of a puzzle, figure 6.19. As a result of this characteristic, proteins have *specificity*; that is, a particular protein will bind only to those molecules with a certain shape. The molecule to which a protein can bind is called a substrate. In binding to a substrate, the protein alters the way the substrate molecules will interact with other molecules. Overall, this is how some proteins regulate molecular events within cells and between cells. Enzymes are a type of protein regulator molecule found in all cells.

Enzymes

Enzymes are protein catalysts. A catalyst is a substance that increases the rate of a chemical reaction without altering the end results of the reaction. In order to picture what this means, we need to know three things about chemical reactions. First of all, modern chemical theory tells us what chemical reactions can occur. For example, hydrogen and oxygen can react to produce water, but at normal temperature and pressure, this reaction occurs so slowly that for all practical purposes it does not occur even though it can occur. In contrast, hydrogen cannot react with atoms like sodium and potassium.

Secondly, any atom consists of a positively charged nucleus which is balanced by one or more electrons that surround the nucleus. This means all atoms and molecules have "negative surfaces." Since any two negative particles will repel one another, two atoms or molecules will tend to repel one another and thus not move close enough to one another to chemically interact even though such a reaction is permissible, see figure 6.17. Only those atoms or molecules with a large amount of kinetic energy (i.e., high velocity) can overcome this repulsion and approach one another closely enough for a chemical reaction to occur. This minimum amount of energy two molecules must have in order to overcome repulsion sufficiently to allow a chemical interaction between them is called activation energy.

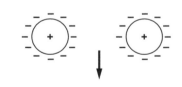

FIGURE 6.17

Atoms or molecules repel one another so a chemical reaction will not occur.

Third, all chemical reactions involve something like climbing a hill: one can climb the hill all at once or do it in small stages. Whichever path one uses, the end result is the same, which is to be on top of the hill, see figure 6.18. Likewise, a chemical reaction can occur in one step with a large activation energy or it may occur in smaller steps each with a much lower activation energy.

In any population of molecules, a few molecules will have very high kinetic energies (high velocities) and thus be able to carry out chemical reactions with high activation energies. Another few molecules will have very low kinetic energies, but most molecules will cluster around some intermediate value. At room temperature, this intermediate value is lower than the activation energies for many

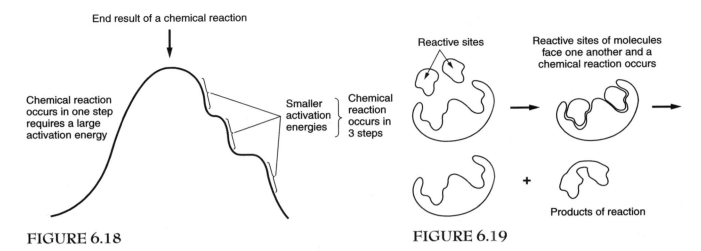

FIGURE 6.18   FIGURE 6.19

types of chemical reactions. A catalyst causes a chemical reaction to occur in steps, each with a lower activation energy. In effect, the catalyst lowers the amount of energy required for a particular reaction to occur. As a result, the intermediate type molecules in a population will have enough energy to enter each of these steps. Therefore, many more molecules will enter the chemical reaction sequence with the result that the reaction will occur much more rapidly. The catalyst has no effect on the end products of the reaction; the catalyst only affects the rate at which these substances are produced. For example, ATP will break down on its own to ADP + phosphate, but a catalyst will cause this same reaction to occur much more rapidly.

Enzymes are the best catalysts in nature. Not only do they lower activation energies for specific reactions, enzymes bind the molecules that will react with one another in such a way that they can more readily interact. Each of the reacting molecules has a specific site that will participate in the reaction. If these molecules were to collide with one another at random, their reaction sites may then face the wrong way. No reaction would occur even though the molecules had enough energy to collide. An enzyme holds the two molecules so that the reactive sites face one another, see figure 6.19.

Enzymes are regulator molecules for two reasons. First of all, virtually all reactions that occur within a cell have fairly high activation energies when no catalyst is present. At room temperature or below, these reactions would not occur at all without a catalyst. Consequently, the presence or absence of an enzyme determines whether a particular chemical reaction will or will not occur. Secondly, the specificity of enzymes enables them to select which chemical reactions will occur. In a cell, a particular molecule may participate in many different types of chemical reactions. For example, glucose may react with phosphate or with another glucose or with an amino acid. Which of these reactions will occur depends on the enzyme present. One enzyme will specifically bind glucose and an amino acid so that these two molecules rather than any of the other molecules will interact.

In a general way, we can see how a cell carries out negative feedback regulation of concentration of various substances. The cell increases or decreases the concentration of a particular enzyme. This will increase or decrease the rate of a particular chemical reaction which in turn will increase or decrease the concentration of a specific substance within the cell.

Cell as a Homeostatic Factory

Quest for the Secret of Life

The "secret of life" relates to a poetic frame of mind in which one could say that the network of living organisms in the world exhibits mindfulness. At various levels of organization, life "knows" how to circumvent the universal trend toward chaos in order to maintain life. Individual cells regulate themselves and coordinate physical-chemical processes "in order to" stay alive. A population of single-celled organisms transcends death by means of reproduction of single cells. The same mindfulness is even more obvious in multicellular organisms. We can be sure that an acorn, if it grows at all, will become an oak tree rather than a dandelion, and a human ovum will become a humanlike creature rather than a lizard or a giraffe. It almost seems that these are purposeful designs which dominate the chance events that occur to a developing organism or that occur to any of its cellular components. The details of the plan are never specified in advance. Instead they emerge from the successive interactions of the organism with a somewhat randomly changing environment. However, the overall plan is unfailingly passed on from one generation to the next. No amount of environmental manipulation will turn a pumpkin seed into a human or a frog ovum into a pig.

Living things are composed of interacting molecules. How can nonliving molecules produce the cellular life properties of homeostasis, reproduction and differentiation which occur during development? The solution of this problem, indeed, would seem to give one of the most fundamental insights about life. In 1953 James Watson and Francis Crick published a model for the structure of DNA which revolutionized the biological approach to understanding life processes. The model in conjunction with biochemistry provided a basis for understanding how cellular homeostasis, reproduction and differentiation could occur at the molecular level.

From 1930 to 1950 several lines of research by people who fundamentally disagreed with one another began to form an outline that, once appreciated, could help produce an understanding of the secret of life. One group worked out techniques (applying x-ray diffraction analysis) for understanding the three-dimensional structure of proteins. Not only do proteins consist of very specific linear sequences of amino acids, but these macromolecules can also have a three-dimensional size and shape. In fact, as we know, it is a protein's characteristic shape that gives it specific biological properties. This is especially true of enzymes. Proteins are like miniature crystals except they lack the pattern that monotonously repeats itself as is true of inorganic crystals; so proteins are called aperiodic crystals (nonrepeating crystals). Some scientists speculated that these unique aperiodic crystals held the secret to life. One of the pioneers in quantum mechanics, Niels Bohr, argued against there being the same sort of biological unpredictability for proteins (his term was "indeterminacy") as was found for subatomic particles. Bohr reasoned that life processes are too well coordinated to be explained on the basis of statistical quantifications of indeterminate chemical interactions. Bohr's student, Max Delbruck, was influenced to look elsewhere than classical and quantum mechanics for the physical basis of life. Based on his studies of x-ray-induced mutations in phage (a virus that invades bacteria), Delbruck proposed that chromosomes are aperiodic crystals. The collection of genes in these aperiodic crystals somehow contain and implement the master plan for life processes.

Meanwhile, many biologists firmly held that life could only be understood in terms of biochemical interactions. Archibald Garrod was able to show that the disease phenylketonuria (PKU) is a result of a gene mutation. Furthermore, he showed that this disorder is due to the absence of a particular enzyme. The connection is obvious: genes determine which enzyme will be produced and enzymes coordinate the life processes. Genes are the executives with master plans, enzymes are the junior executives, and all the other chemicals are the community of molecules keeping the organism successful, i.e., alive. These results were published in 1914; so, of course, no one paid any attention to them—Garrod was too far ahead of his time. But in the 1940s, George Beadle and others showed that, indeed, a change at a single site in Neurospora (a type of fungus) caused an enzyme not to be produced. In the 1950s, a sequence of studies demonstrated that the sickle cell trait results from a mutation of one gene which changes only one amino acid in a long-chain, hemoglobin molecule. Garrod's suggestion was correct after all.

Another quantum mechanics physicist, Erwin Schrodinger, wrote a very influential book, *What Is Life?*[2], first published in 1944. Schrodinger speculated but with a very convincing synthesis of the then available physical and biological data, that genes are a group of atoms, perhaps a thousand of them, located on chromosomes. Gene changes result from an atom's random change in energy level,

---

[2]Erwin Schrodinger, *What Is Life?* (New York: The Macmillan Co., 1947).

which occurs under the "right" conditions. Thus, gene changes are fairly rare events. But these random events are the basis for natural selection of modifications of a species.

In the last chapter of his book Schrodinger goes on to say:

> . . . from all we have learnt about the structure of matter, we must be prepared to find it [the aperiodic structure of genes] working in a manner that cannot be reduced to the ordinary laws of physics. And that not on the ground that there is any 'new force' or what not, directing the behavior of the single atoms within a living organism, but because the construction is different from anything we have yet tested in the physical laboratory.[3]

The obvious inference to be drawn from this is that the secret of life is to be found in the arrangement of the 1,000 or so atoms forming individual genes and in new physical laws which govern the functioning of these very special aperiodic crystals.

By 1952, several lines of evidence had indicated that the chromosomal material containing the genetic master plans is none other than DNA. However, the biochemical mechanists, rightly proud of their accomplishments, would have nothing to do with the wild speculations of physicists, with their new laws of nature. It all sounded so mystical, or at least metaphysical, which is almost as bad. The sophisticated physicists, in turn, thought the biochemists were too caught up with simplistic explanations. Life is much more complex than just chemistry, but then what can you expect from people who mess around with test tubes and such instead of thinking about the deep realities of life?

The Watson-Crick Model of DNA

James D. Watson, an American cell biologist, and Francis Crick, an English physicist, were both inspired by Schrodinger's book during their graduate school days. These two young scientists naturally were drawn to one another by their mutual and unconventional respect for Schrodinger's speculations. They both sought for the secret of life in the structure of DNA in chromosomes. Watson brought to the collaboration a familiarity with and regard for the biochemical studies of gene expression, and Crick brought a keen and quick talent for physical abstraction and a familiarity with techniques of analyzing aperiodic crystals. The key to understanding life lay in simultaneously solving three problems: (1) How does a molecule like DNA store a master plan? (2) Since DNA makes up genes and genes duplicate during cell division, DNA must duplicate also, but how can a molecule duplicate itself? (3) How does DNA cause proteins to be made?

DNA Information.   It is fairly easy to imagine a possible solution to the first question. The DNA molecule stores a master plan just like the arbitrary symbols on this page store meaningful statements. Particular sets of letters represent particular words. A sequence of words represents a sentence. Various other arbitrary symbols such as periods separate one sentence from another and help clarify the information in a particular sentence. Watson and Crick knew that nucleic acids such as DNA consist of nucleotides that contain phosphate groups, 5-carbon sugars (riboses), and two types of nitrogen bases, purine and pyrimidine. In DNA there are two purine derivatives, adenine and guanine, and two pyrimidine derivatives, thymine and cytosine. Hereafter, these four molecules will be represented by A, G, T, and C respectively, see figure 6.20a and 6.20b. These *nucleotides* represented by the letters A, G, C, and T are the letters of the genetic language now called the genetic code. A particular sequence of these letters is a genetic word, and a sequence of these words represents a statement about some life process. The genetic (molecular) statements taken together (the chromosomes) contain the master plan for the life processes of a particular organism.

Nucleotides are the subunits of nucleic acid. As shown in figure 6.20b, a nucleotide consists of a nitrogen containing base, a 5-carbon sugar and one or more phosphate groups. The phosphate groups are normally joined to the C5 hydroxyl of the ribose or desoxyribose sugar. Mono-, du-, and triphosphates are common.

---

[3]Ibid., p. 000.

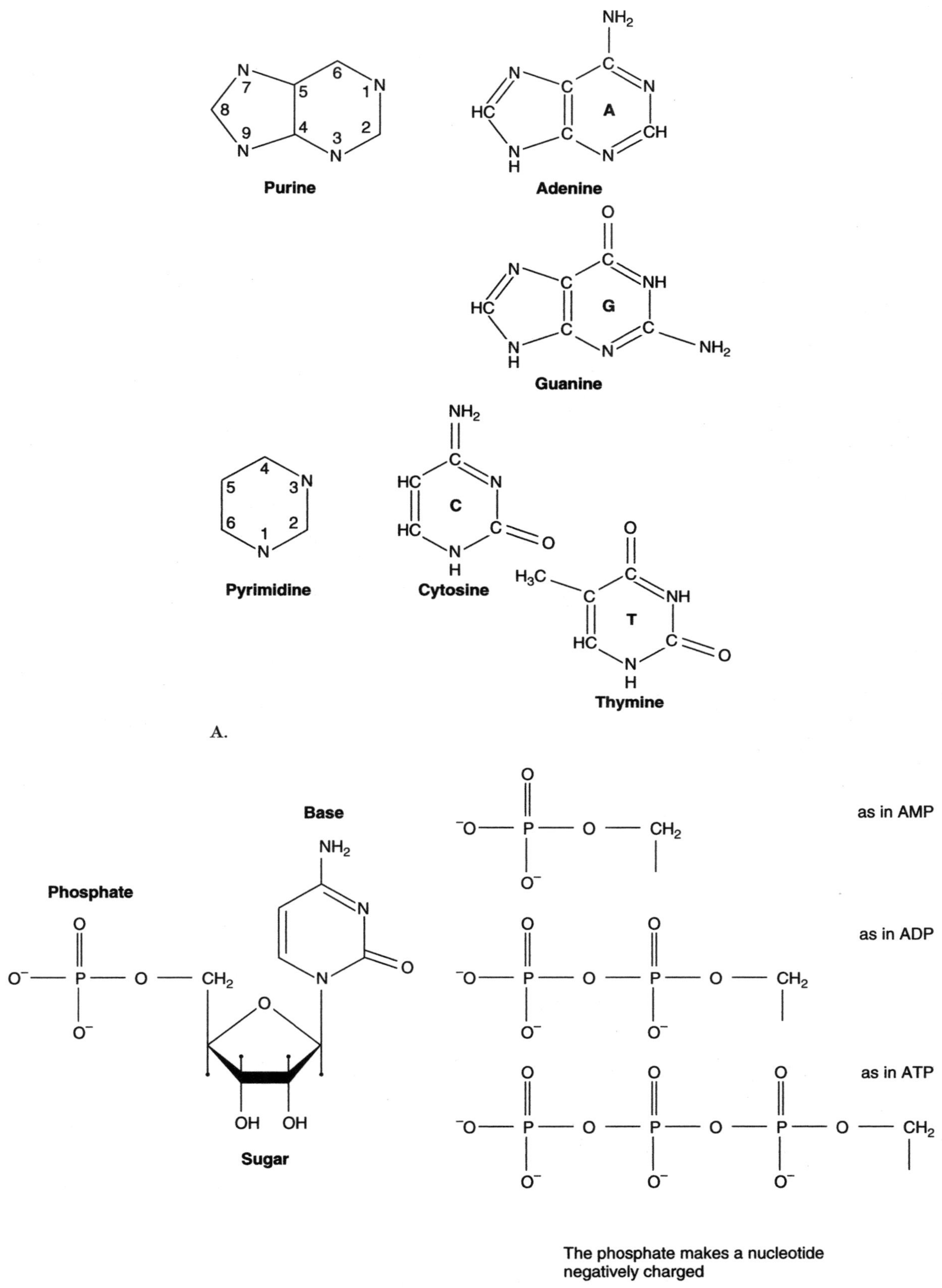
Purine
Adenine
Guanine
Pyrimidine
Cytosine
Thymine
A.
Base
Phosphate
Sugar
as in AMP
as in ADP
as in ATP
The phosphate makes a nucleotide negatively charged

B.

FIGURE 6.20

DNA Replication. It was not so easy to imagine a solution to the problem of DNA duplication. The first mechanism that may occur to one is that of a cookie cutter. A star cookie (produced by a cookie cutter) is not a copy of the star cookie cutter. So this analogy won't help—or will it? Often the solution to a problem involves thinking in unusual ways. The problem, of course, is that instead of making a lot of star cookies, you wanted to make a lot of star cookie cutters. One method is to form metal around a solid star. In this case, the star serves as a template, or form, for making the cookie cutter just as the cookie cutter serves as a template for making a star cookie. The cookie cutter and the star are complementary to one another. Thus, let us represent the complementary pair, star cookie-star cookie cutter, as K-K'. K is the complement of K' and conversely, K' is the complement of K. Furthermore, K can serve as a template for making K', and K' can serve as a template for making K. Voila! Eureka! (or whatever scientists say when the light bulb goes on.) Now it is easy to imagine how such a pair could duplicate itself. The K and K' separate; K is used as a template to make another K' which then pairs with the original K to produce $(KK')^1$. In the meantime, K' is used as a template to make another K which then pairs with the original K' to produce $(KK')^2$. The $(KK')^1$ is identical to $(KK')^2$.

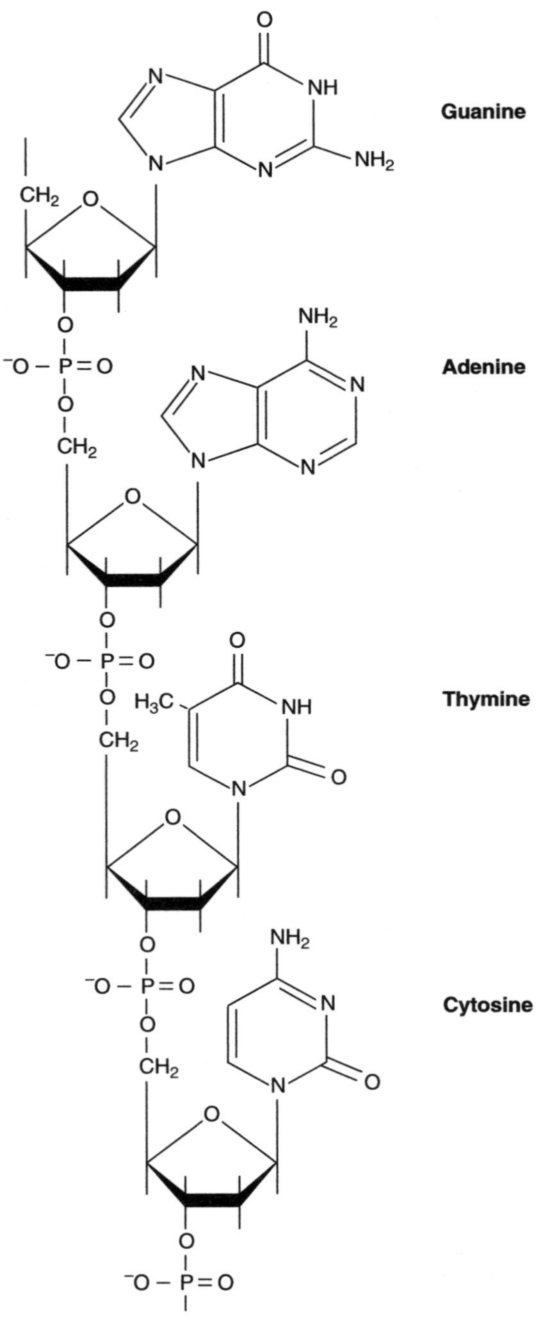

This idea of pairs of complements, each serving as a template for making the other, is the key idea of the Watson-Crick model. Since it was known that the DNA molecule consists of at least two complementary strands of nucleotides, we can see in general terms how Watson and Crick visualized duplication of such a two-strand molecule. First of all we replace the term duplication with replication, since by definition a replica is like one's footprint; it is a "replica" that is a complementary representation of the foot. DNA duplication occurs by the two component strands of nucleotides separating and each strand serves as a template for forming its "replica." Now let's look at some of the molecular details.

Watson and Crick knew that nucleotides are connected to one another by a bond between the phosphate group of one nucleotide and a sugar of another, see figure 6.21. They also knew that nucleic acids consisted of at least two chains of nucleotides. But the missing pieces of information eventually would lead these scientists to propose complementary base pairing between purines and pyrimidines. Watson and Crick gave John Griffith, a young Cambridge mathematician, data about the three-dimensional structure of the four types of nitrogen bases. Griffith calculated which base pairs would form the most stable chemical (hydrogen) bonds between them and therefore would be the most likely pairs to occur. These are adenine-thymine (A-T) and guanine-cytosine (G-C), see figure 6.22.

A chance meeting in June of 1952 between Watson and the Austrian refugee biochemist, Erwin Chargaff, led to the key idea of the DNA model. Just as in evolution of nonliving and living matter, in human affairs

FIGURE 6.21 Sequence of nucleotides in a nucleic acid, such as DNA.

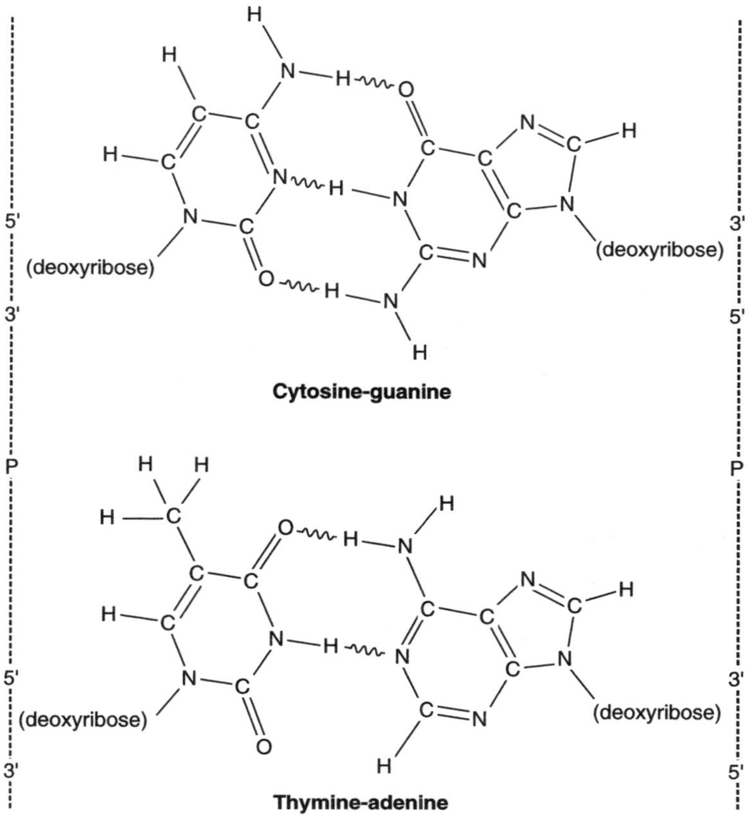

FIGURE 6.22   Nitrogen base pairing such as C-G and T-A, in a segment of DNA.

chance events under just the right circumstances may lead to new insights. In a casual conversation with Watson, Chargaff referred to a paper he had written that made three points. First, the composition of DNA in terms of the percentage of each of the four bases varies from one species to another. DNA composition, we now realize, is a fundamental characteristic of a species. Second, different types of cells from a particular species nevertheless have the same relative proportions of the four bases. This is just a more explicit aspect of the first point. Human tissues, for example, have 28% adenine, 28% thymine, 19% guanine, and 19% cytosine. Third, taking human tissue as an example, there is a one-to-one ratio of adenine to thymine (A-T) and guanine to cytosine (G-C).

Watson and Crick immediately began trying to build a model of DNA with the preceding nucleotide base pairing. By the winter of 1953 they had evidence from an aperiodic crystal analysis of DNA done by Wilkins, which suggested that DNA consists of only two strands of nucleotides. These are arranged like two pipe cleaners wound around a pole. The phosphate groups face outward and the nitrogen bases face inward. At this time, however, Watson and Crick were using somewhat incorrect structures of the purines and pyrimidines. With the wrong form of the bases it was nearly impossible to imagine how the base from one nucleotide strand could point inward and mesh with a complementary base of the other strand. The paired strands seemed to bulge awkwardly because of this poor complementary base fit. However, Wilkins' data indicated that the helical structure formed by the two strands had a constant diameter. Then in February 1953, Jerry Donohue, an American chemist who shared an office with Watson and Crick, pointed out that they were using an incorrect structure for the bases. The alternative structure he suggested revealed that the complementary bases could fit nicely together by hydrogen bonding. Not only that, but the A-T pair had the same molecular shape and length as the G-C pair. Thus, such a base pairing also would explain the constant diameter of the double helix structure of DNA. By April of 1953 Watson and Crick published a short (two pages) article in *Nature*, describing a proposed structure for DNA, see figure 6.23a and b. The model

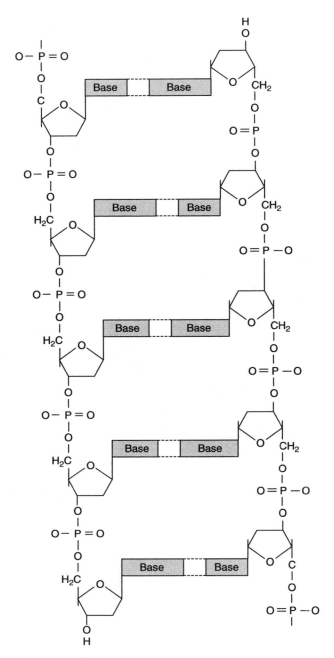

FIGURE 6.23a   Schematic of a portion of a DNA molecule.

fit very well with the data on DNA structure from Wilkins' lab. More important, the model suggested how DNA could store information and how it could duplicate itself by allowing each strand of the DNA molecule to *replicate* itself, see figure 6.24. Furthermore, the complementary base pairing provided a good starting point for deriving a model for the synthesis of a particular protein from a particular DNA strand.

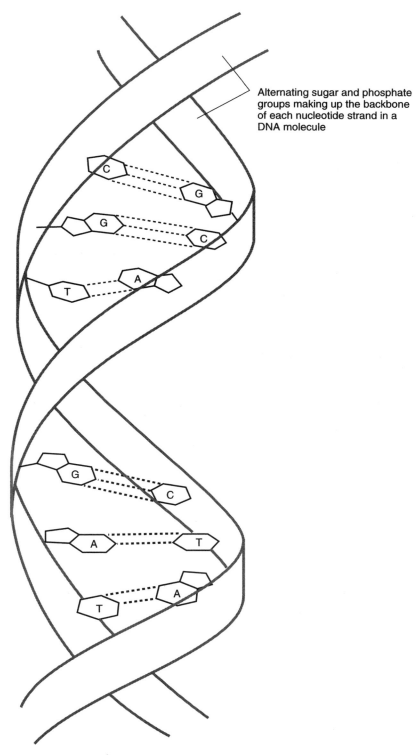

Alternating sugar and phosphate groups making up the backbone of each nucleotide strand in a DNA molecule

FIGURE 6.23b   Schematic of a portion of a DNA molecule.

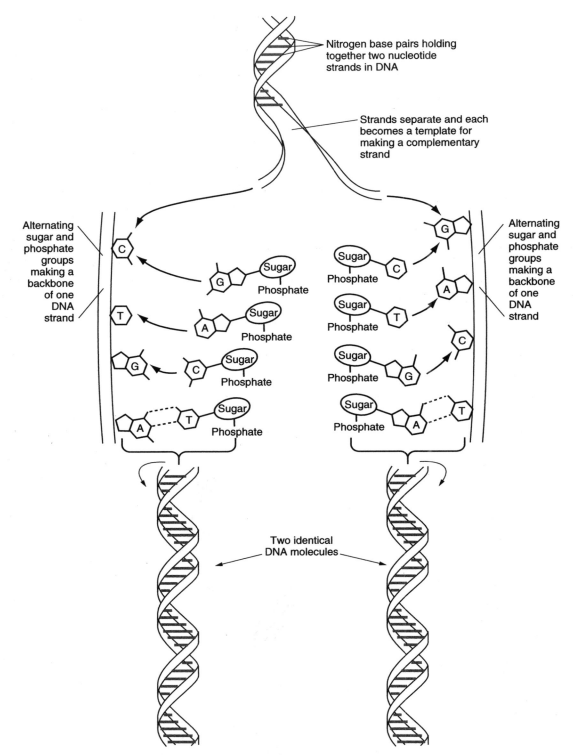

FIGURE 6.24   Replication of a DNA molecule.

Molecular Information Communication

DNA replication exemplifies operationally definable information communication defined as: a source-receptor interaction that generates meaning where the meaning is a nameable event that can be operationally defined. DNA replication occurs in the nucleus of a cell that is preparing to divide into two cells each containing identical DNA information. Such a cell first synthesizes many molecules of each of the four kinds of nucleotides: A, T, C, and G. Then regulator molecules (enzymes) bring about the separation of the two nucleotide strands that make up each DNA molecule. Enzymes facilitate and thus regulate each separated nucleotide strand becoming a template on which is formed its replica. Thus, for example, as shown in figure 6.25, a newly synthesized T nucleotide—represented as T'—is directed to pair with (and form hydrogen bonds with) the A nucleotide which is one member of the A-G-T-C nucleotide template strand. In this molecular communication T' may be thought of as the source, A as the receptor, and A-T' complementary pairing as generating meaning which is the first stage of a potential for T' to chemically bond with a newly synthesized A nucleotide represented as A'. When A' communicates with T on the same template by pair bonding with it, the meaning generated by this communication is the completion of the potential for T' and A' to chemically bond to one another and to begin to form the replica of the template. Another aspect of this generated meaning is the first stage of the potential for A' to chemically bond with a newly synthesized C nucleotide represented as C'. If appropriate regulator molecules and energy

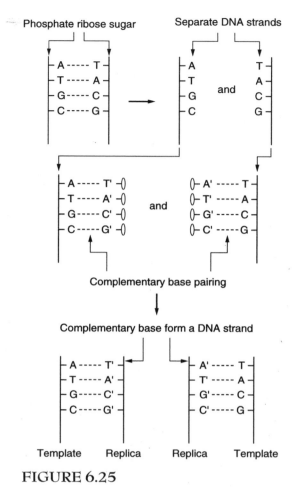

FIGURE 6.25

sources are available in the cell, then another molecular communication occurs: the T' and A' bond with one another and generate meaning which is expressed as the T'—A' bond that is a component of an emerging new DNA molecule. This meaning representing information of the emerging replica of a DNA strand is operationally defined by the procedures for determining the presence of a new DNA molecule containing one nucleotide strand from the original DNA molecule and the replica of this original strand. The two newly formed DNA molecules resulting from the duplication of the original DNA molecule result from each original strand replicating itself; so this DNA duplication process is called DNA replication.

Protein Synthesis

Overview of Protein Synthesis.   The theory of complementary base pairing that occurs in DNA replication provided a good starting point for deriving a model for the synthesis of a particular protein from a particular DNA strand. The details of how DNA directs the synthesis of proteins was worked out between 1953 and 1963. In a manner similar to its own replication, DNA directs the synthesis of three types of RNA. These three types of RNA in turn direct the synthesis of a particular protein; that is, DNA → RNA → protein. RNA differs structurally from DNA in three ways. First, the sugar, ribose, has one more oxygen atom than the deoxyribose of DNA (an OH group replaces H). Second, RNA also contains four nitrogen bases but substitutes uracil (U) for thymine; uracil will form A-U pairs which are equivalent to the A-T pairs found in DNA. Third, RNA consists of only one strand of nucleotides.

The key ideas for the scheme of protein synthesis are as follows. First, one of the DNA strands serves as a template for making a complementary sequence of nucleotides in the RNA molecule. An information sentence in a segment of DNA thus is restated in an RNA molecule, which then moves out of the nucleus into the cytoplasm, see figure 6.26. This formation of RNA on a DNA template is called transcription. The information passed from DNA to RNA is called a message, and the RNA molecule containing it is messenger RNA (mRNA).

Second, several different small RNA molecules in the cytoplasm, called *transfer RNA (tRNA)*, "recognize" specific words in the mRNA message and at the same time bind to specific types of amino acids. The words in the mRNA consist of sequences of three bases and are called *codons*. A particular tRNA molecule recognizes a codon by means of a sequence of three complementary bases that bond to the codon. This sequence of three bases in the tRNA, therefore, is called an *anticodon*. Each type of tRNA characterized by its anticodon specifically binds to only one type of amino acid, see figure 6.27. The collection of tRNAs in the cytoplasm is like a dictionary. In effect, each tRNA defines a particular word (codon) in mRNA to mean a particular amino acid. Of course, tRNA is aided by ATP as an energy source for binding of the tRNA to an amino acid and by enzymes to guide the process.

Third, a dictionary is meaningless unless there is someone or something to use it to translate a sentence, that is, to put the defined words into proper order to form a sentence. Particles in the cytoplasm called ribosomes are involved with protein synthesis. Ribosomes are also composed of RNA and are represented as rRNA. The ribosomes ensure that the tRNA molecules bound to their particular amino acids line up in the order specified by the message in the mRNA. With the aid of ATP as an energy source and of specific enzymes, peptide bonds are formed between the properly arranged amino acids. The resulting polypeptide chain breaks loose to become a protein or one of the strands of a complex protein. Thus, the interactions among mRNA, tRNA, and ribosomes (rRNA) have translated the message obtained from DNA into a specific polypeptide chain. This molecular process is therefore called translation, as shown in figures 6.28 and 6.29.

Genetic Code.   As already indicated, the alphabet of the DNA language consists of the four letters A, T, G, and C representing the four types of nitrogen bases in any DNA. All the various proteins consist of only 20 different amino acids, although a particular protein may not contain all of these. There must be enough DNA words to specify "start a message" or "stop a message" and to specify the 20 different amino acids. If the DNA words consisted of two letters, then there would be 16 different words: (AA, AG, AT, AC; GG, GA, GT, GC; TT, TA, TG, TC; and CC, CA, CG, CT). If the word consisted of three letters, then there would be 64 different combinations. Thus, a four-letter alphabet and one or two letter words would not be sufficient to code for the synthesis of proteins, but three-letter words would do the trick. In 1963 Severo Ochoa's group and Marshall Mirenberg's group independently cracked the genetic code. The DNA words did indeed consist of three letters. These researchers worked out the molecular translation of each three-letter combination in terms of 64 different tRNA molecules. A triplet in DNA is transcribed by three-letter codon in mRNA, which bonds to the complementary three-letter anticodon in tRNA. Ribosomes, with the aid of mRNA and tRNA, translate the sequence of words in DNA into a protein, see figures 6.28 and 6.29.

"The Central Dogma" of Molecular Biology.   The Watson-Crick DNA model and its further elaboration has provided a unifying framework in biology. In the early 1960s the model was half-jokingly refered to as "the central dogma" of molecular biology. We now realize that there are exceptions to this "dogma" and in general the molecular regulation of cell functions is more complex than as described by this theory. The core idea of the model is that the DNA in the nucleus contains information that determines all the cell's structural properties and functions. The way in which a portion of this DNA information is expressed as a cell function is rather complex and may be likened to the following analogy. Suppose for a Christmas holiday you wanted to prepare a special medieval French dish. First you would go to the public library and transcribe into longhand the printed French recipe. Next, assuming you do not read French, you would have the French words translated into their English equivalents. Then you would arrange these English words into sentences that would represent an English translation of the French recipe. This English translation would give you the directions for preparing the desired French dish for your party. On the day of your party, assuming all

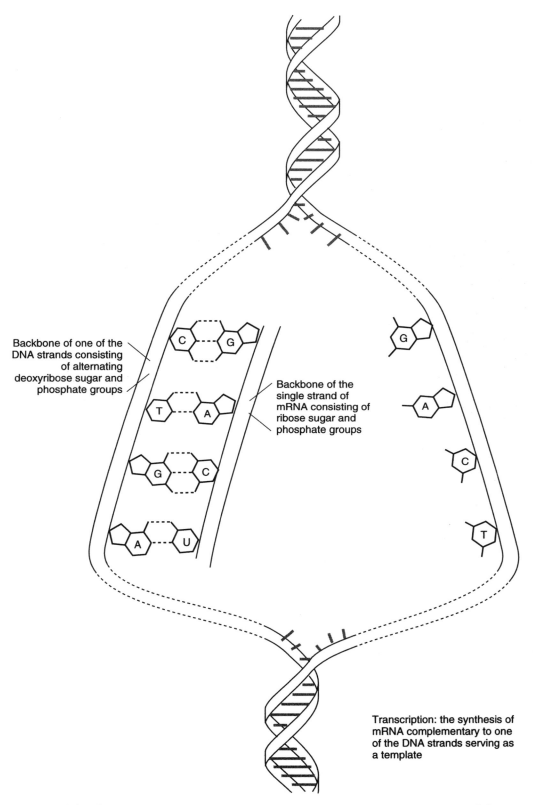

Backbone of one of the DNA strands consisting of alternating deoxyribose sugar and phosphate groups

Backbone of the single strand of mRNA consisting of ribose sugar and phosphate groups

Transcription: the synthesis of mRNA complementary to one of the DNA strands serving as a template

FIGURE 6.26   Transcription: The synthesis of mRNA complementary to one of the DNA strands serving as a template.

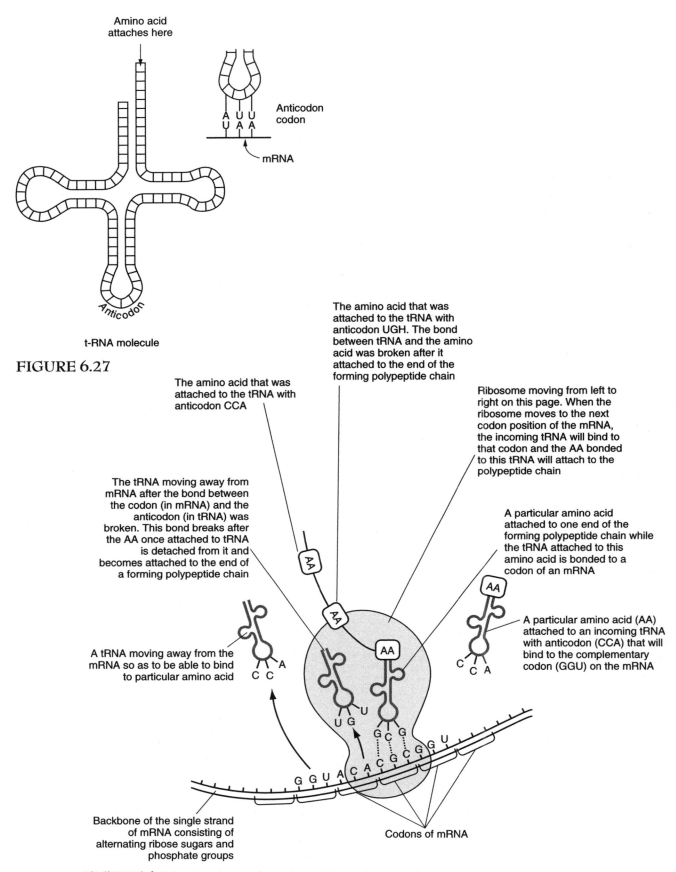

Amino acid attaches here

Anticodon
codon

mRNA

Anticodon

t-RNA molecule

FIGURE 6.27

The amino acid that was attached to the tRNA with anticodon UGH. The bond between tRNA and the amino acid was broken after it attached to the end of the forming polypeptide chain

The amino acid that was attached to the tRNA with anticodon CCA

Ribosome moving from left to right on this page. When the ribosome moves to the next codon position of the mRNA, the incoming tRNA will bind to that codon and the AA bonded to this tRNA will attach to the polypeptide chain

The tRNA moving away from mRNA after the bond between the codon (in mRNA) and the anticodon (in tRNA) was broken. This bond breaks after the AA once attached to tRNA is detached from it and becomes attached to the end of a forming polypeptide chain

A particular amino acid attached to one end of the forming polypeptide chain while the tRNA attached to this amino acid is bonded to a codon of an mRNA

A tRNA moving away from the mRNA so as to be able to bind to particular amino acid

A particular amino acid (AA) attached to an incoming tRNA with anticodon (CCA) that will bind to the complementary codon (GGU) on the mRNA

Backbone of the single strand of mRNA consisting of alternating ribose sugars and phosphate groups

Codons of mRNA

FIGURE 6.28   Synthesis of a polypeptide chain on a ribosome.

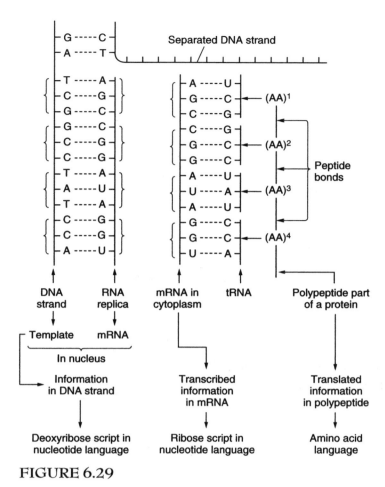

FIGURE 6.29

goes well, the French dish would be the concrete expression of the information contained in the esoteric French cookbook at your public library.

The information of mRNA gotten from one strand of DNA used as a template is analogous to "transcribing into longhand the printed French recipe." The various tRNA molecules bound to specific amino acids recognize codons in mRNA (the anticodon in a tRNA molecule binds by complementary base pairing to a codon in the mRNA). This is analogous to "translating the French words into their English equivalents;" i.e., a codon is translated to mean a particular amino acid. Then the mRNA, the various tRNA, and the ribosomes interact to arrange amino acids into a polypeptide chain, thus forming a particular protein. This is analogous to arranging the English words into sentences that would represent an English translation of the French recipe. In effect, nucleic acid language is translated into protein language. If the protein is an enzyme, it will facilitate the occurrence of a particular chemical reaction in the cell. This enzyme is analogous to the directions for preparing the French dish for your party, and the particular chemical reaction in the cell is analogous to the French dish that is produced, as shown in figure 6.30.

The "central dogma" may be summarized by five aspects of expression of genetic information in a cell.

1. Transcription:
   a. Using the "nucleotide alphabet," the cell nucleus makes mRNA which is a replica of information in a portion of one strand of a DNA molecule.
   b. The information transferred to mRNA is called a *message* and mRNA is called *messenger RNA*.

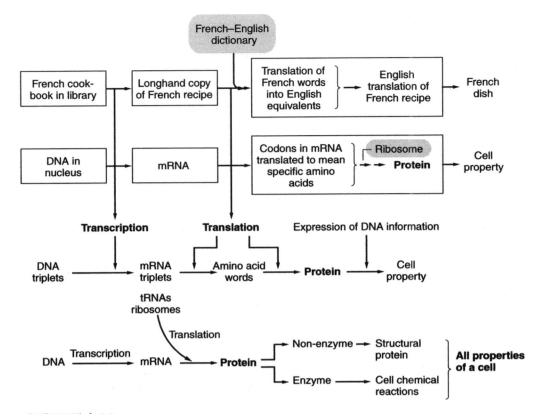

FIGURE 6.30

2. Translation:
   a. Using the amino acid alphabet, the cell makes a portion of a protein which is a sequence of amino acids representing the same information as found in mRNA.
   b. The message in mRNA is translated into an amino acid sequence in a protein.
3. The newly synthesized protein fits into the structural-functional pattern of the cell.
4. As an aspect of the whole cell, the protein participates via molecular communication in producing a particular cell structure or cell function.
5. Cell structure or cell function is an expression of genetic information in a DNA molecule.

Photosynthesis

All cells contain sugars, lipids, and proteins which are available for cellular oxidation. A large portion of the energy released from cellular oxidations is immediately used up in various life activities, but some energy always is recaptured by the organism's continual synthesis of more sugars, lipids, and proteins. However, all cells spend more energy than they save. This means that the total amount of organic chemical energy on earth tends to decrease. Of course, if this actually happened, all life forms would disappear. Life on earth not only goes on but flourishes because some plant cells absorb solar energy, convert it into chemical energy and then use this energy to synthesize organic compounds. Thus, by means of this process which is called photosynthesis, organic chemical energy is continually replenished on earth and directly or indirectly made available to all life forms. Plants provide organic chemical energy to themselves and to animals that eat plants. These animals may be eaten by still other animals and thereby provide organic chemical energy to them. Thus, because of photosynthesis, the sun is the primary source of energy for all life forms, and plants are the primary sources of organic compounds.

General Description of Photosynthesis

Photosynthesis is the process by which plants use packets of light energy called photons to synthesize organic molecules from inorganic molecules. Of course, the sequence of reactions for the synthesis of each type of organic compound will be somewhat different. However, all complete photosynthetic reaction sequences have one common characteristic. Photons provide energy for the transfer of electrons and positive hydrogen ions from some inorganic molecule to the carbon in carbon dioxide. The net result of this process is that several carbon atoms are combined to form an organic molecule. Because of the simultaneous formation of carbon-hydrogen bonds in some carbon atoms, the organic molecule contains a great deal of chemical energy. Thus, as a result of causing electrons and positive hydrogen ions to leave some inorganic molecules and attach to the carbon in carbon dioxide, light energy is transformed into chemical energy of organic compounds.

In green plants, water is the source of electrons and positive hydrogen ions for photosynthesis. As electrons are removed, the oxygen in water becomes a neutral oxygen atom, and the hydrogen atoms in water become positive hydrogen ions. That is, the primary aspect of photosynthesis in green plants is that light energy is focused on breaking the two oxygen-hydrogen bonds in a water molecule. As a result of both hydrogen atoms breaking away from the oxygen atom, the oxygen atom becomes an independent, electrically neutral atom. Two such oxygen atoms spontaneously react to form an oxygen molecule, O_2. All photosynthesizing green plants continually release oxygen gas into the atmosphere. In dramatic contrast, the two hydrogen atoms in a water molecule that break away from the oxygen atom do not become independent, neutral atoms. Rather, the electron of each hydrogen atom goes to a high energy state in pigment molecules in a membrane system which is a network of stacks of flattened discs called thylakoid membrane within chloroplasts. Each of these hydrogen atoms devoid of its electron now is an independent, positively charged hydrogen ion, i.e., a proton.

Some of the high energy electrons generated by absorption of light energy participate in a sequence of reactions that produces carbon-hydrogen bonds, i.e., C-H, in the carbon of carbon dioxide. Each carbon bonded to a hydrogen is used to synthesize a glucose molecule. This sequence of reactions to produce a glucose molecule requires energy from breakdown of ATP. High energy electrons produced by absorbed light energy go through a series of electron transfers similar to what occurs in mitochondria. Such transfers of electrons from higher to lower energy levels are coupled with driving the chemical reaction of forming ATP; that is,

$$ADP + Phosphate + energy\ from\ electron\ transfer \Rightarrow ATP$$

Overall absorbed light energy produces three results: (1) $H_2O \Rightarrow O_2 + H^+$ + high energy electrons; (2) ADP + phosphate + energy from electron transfer $\Rightarrow$ ATP and (3) High energy electrons + ATP breakdown drives the chemical reaction: $H_2O + CO_2 \Rightarrow$ glucose.

The first two results occur in the thylakoid membrane system in chloroplasts. These reactions called light reactions convert light energy into chemical energy that can be used to synthesize glucose. The various reactions that use chemical energy generated by the light reactions to produce glucose are called dark reactions. The term dark refers to the fact that these reactions can occur in the dark, that is, they can occur even when the plant is not absorbing light energy. Light reactions convert absorbed light energy into "useable" chemical energy, i.e., high energy electrons and ATP. Dark reactions use chemical energy from the light reactions to drive the overall process of converting $H_2O + CO_2$ into glucose.

As described in Chapter 5, the system of ADP and phosphate molecules is an energy coupler in all cells for focusing incoming energy to accomplish various cellular tasks. The task of forming a C-H in the carbon of CO_2 requires two different energy couplers. One is the ADP/phosphate system, see Chapter 5. The second is the $NADP^+/H^+$ system. When this system is at a higher entropy (lower order) and lower energy level, $NADP^+$ and H^+ are independent of one another. These independent components tend not to interact because they both are positively charged. Energy derived from excited electrons produced in the light reactions activates $NADP^+$ so that it attracts and then bonds with H^+ to produce NADPH. Now this coupler system is at a lower entropy (higher order) and

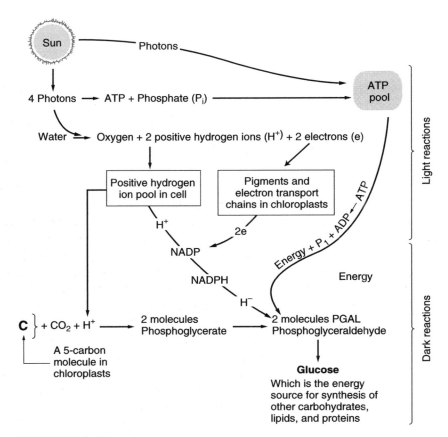

FIGURE 6.31

higher energy level. The increased order is due to the formerly independent components being bonded to one another. Also, because of this bond with H, the NADP has stored chemical energy, i.e., energy from the excited electrons produced by absorption of light is converted into chemical energy. Energy from the breakdown of ATP to ADP + phosphate enables NADPH to lose H as a negative hydrogen ion, H^-, that is, a hydrogen atom with two electrons rather than neutral hydrogen with one electron or H^+ with no electrons. This H^- ion is bonded to phosphoglycerate thus converting it into phosphoglyceraldehyde (PGAL). Two PGAL molecules combine to form one glucose molecule. The coupler system in the form of NADPH now has gone back to a higher entropy, lower energy level as $NADP^+$ plus H^+ that comes in from a H^+ pool in the plant cell, see figure 6.31.

In order to continue to synthesize glucose the chloroplast must continually regenerate more PGAL. This regeneration of PGAL occurs in a process called CO_2 fixation (also called the Calvin cycle). Some of the energy from the breakdown of ATP to ADP + phosphate "energizes" this phosphate so that it combines with a 5-carbon molecule with a phosphate at one end to produce a 5-carbon molecule with a phosphate group at each end. One of the two phosphates of this molecule is a "high energy phosphate." The carbon bonded to this "high energy phosphate" loses this phosphate and the energy released in this process is used to form a bond between the carbon that lost the phosphate and the carbon in CO_2. As a result, the 5-carbon molecule with a phosphate group at one end + CO_2 becomes a 6-carbon molecule with a phosphate at each end. This 6-carbon molecule + another phosphate (from a phosphate pool in the plant cell) breaks in half to form two 3-carbon molecules each with a phosphate at one end. This 3-carbon molecule is called phosphoglycerate, see figures 6.31 and 6.33. Thus, in producing PGAL, CO_2 fixation accomplishes two tasks: (1) some PGAL is used to synthesize glucose and (2) the remaining PGAL is used to produce a 5-carbon molecule that can eventually react with more CO_2 and start a new cycle that regenerates PGAL, see figure 6.33.

In summary, photosynthesis consists of two interconnected series of reactions: light reactions and dark reactions. During the light reactions, the energy from absorbed photons is used to transfer two electrons and one of the positive hydrogen ions from water to $NADP^+$ which thereby becomes NADPH. As a result of losing two hydrogen atoms, each oxygen atom of two water molecules combines to form an oxygen gas molecule which diffuses into the atmosphere. Some absorbed photon energy also is used to convert ADP plus phosphate into ATP. The dark reactions, occurring simultaneously with the light reactions, combine CO_2 with a 5-carbon molecule to produce two 3-carbon molecules called phosphoglycerate. The ATP and NADPH produced in the light reactions convert phosphyglycerate into phosphoglyceraldehyde (PGAL). Two PGAL molecules combine to produce one glucose molecule. Some PGAL molecules, instead of being converted into glucose, participate in CO_2 fixation that generates more PGAL.

Biochemical Details

Converting Light Energy to High Energy Electrons. Light Energy.   Reactions like the ones that occur in the hydrogen bomb continuously occur within the sun. These violent reactions release energy in the form of light and heat energy which is radiated away from the sun in all directions. Green plants on earth use this light as a source of activation energy for photosynthesis. But what are the properties of light that make this possible? First of all, there is still much controversy as to how this energy travels from the sun through empty space to the earth. Current theories suggest that a ray of light is both a train of waves and a stream of particles. This, of course, is a very difficult concept which depends on a lot of physics and mathematics to verify. For our purposes, however, we can view light as consisting of packets of energy called photons, which differ in the amounts of energy they possess. These photons can be viewed as traveling in continuous wave patterns called electromagnetic waves. The light rays coming from the sun range in wavelength from very short (x-rays) to very long (radio waves). The amount of energy in a light wave is related to its wavelength: the shorter the wavelength, the greater the energy. Therefore, the shorter the wavelength of a light wave, the greater the energy of its constituent photons. Now when light strikes an object, some of the photons are absorbed, some pass through the object (possibly), and some are reflected. Absorbed light is what is used to activate photosynthesis. Reflected light is what gives an object its color. The degree to which light passes through an object determines whether the object is transparent, translucent, or opaque and gives its color as well.

Chlorophyll Absorption of Light Energy.   The cells of green plants contain special plastids called chloroplasts which get their name from the light sensitive green pigment, chlorophyll, they contain. Pigment, mainly chlorophyll, located in unstacked and stacked membrane discs within the chloroplasts has a molecular structure that readily enables it to absorb photons from visible light rays as opposed to x-rays, ultraviolet rays, infrared waves or radio waves. In addition, other pigments besides chlorophyll within the chloroplasts absorb light but to a much less degree and pass its energy to chlorophyll.

Chlorophyll molecules have some electrons that are associated with the molecule as a whole rather than being limited in location to a specific chemical bond. When chlorophyll molecules have not yet absorbed light, these electrons are in the lowest possible energy levels available to them. However, there are unoccupied higher energy levels in chlorophyll that differ from the lowest ones by the amount of energy contained in the photons of red and violet light. As a result, chlorophyll absorbs red and violet light from the sun's rays resulting in the movement of these electrons to higher energy levels. Chlorophyll's green color results from the fact that it reflects and transmits the green wavelength of light from the middle part of the light spectrum. In solution chlorophyll reflects only a small amount of green light. In a plant the green light transmitted by the chlorophyll is reflected back by other surfaces within a leaf or stem to make the plant appear green.

Photocenters.   The various pigments in chloroplasts are called photosynthetic pigments. These pigments are organized into light absorbing systems called photocenters. Each photocenter consists of many pigments that only absorb light energy and a chlorophyll molecule that acts as a reaction center. The light absorbing pigments transfer the energy of their excited electrons to electrons at a lower energy level in the reaction centers. The resulting high energy electrons in each reaction center

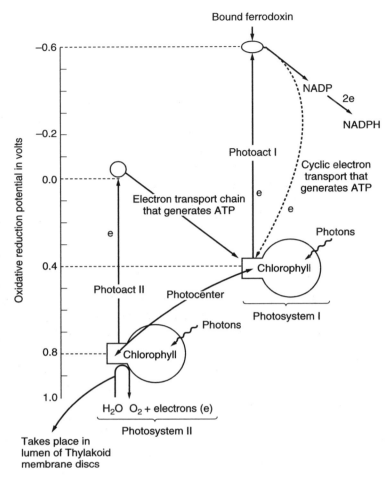

FIGURE 6.32   Energy pathway of the light reactions.

move to an acceptor molecule in an electron transport chain. These electrons are like baseballs thrown out by a machine and the acceptor molecules are like baseball gloves designed to catch the baseballs. These high energy electrons now flow through the electron transport chain back to low energy levels in the pigments of the photocenters.

Linear Electron Flow in the Light Reactions.   Light energy is absorbed simultaneously in two systems in green plant cells, called photosystem I and II respectively, see figure 6.32. In each system photons are absorbed by pigments that are connected to specialized pigment molecules called reaction centers (these once were called trapping centers). Absorbed light energy collected by a reaction center causes one of its electrons to jump to a high energy level in an electron acceptor molecule, mentioned earlier. Because this molecule now has an electron which it can readily donate to a lower energy level in some other molecule, it is said to have chemical energy. The conversion of light energy into chemical energy at a reaction center is called a photoact, see figure 6.32.

Overall there is electron flow from photosystem II to photosystem I which then is transferred to $NADP^+$ plus H^+ to form NADPH, see figures 6.31 and 6.32. In effect electrons flow from photosystem II to photosystem I and then into $NADP^+$ to form NADPH. The photoact that occurs in photosystem I makes this sequence of flow from II to I possible. In generating high energy electrons the photoact I also causes its reaction centers to become positive. As a result, high energy electrons generated in photosystem II flow through an electron transport chain to the positively charged reaction centers in photosystem I. Thus, the electrons lost by photosystem I in converting $NADP^+$ to NADPH are replaced by electrons generated by photoact II in photosystem II, see figure 6.32. This means that photosystem II is continually losing electrons and would very quickly become exhausted, i.e., the system would no longer be able to replenish the electrons lost by photosystem I. What prevents this from happening is that the energy generated by photoact II enables H_2O to break down into O_2 plus 2 H^+ plus 2 electrons to replenish the electrons lost by photosystem II to photosystem I, see figure 6.32.

Breakdown of Water Coupled to Absorption of Light. The process by which absorbed light brings about this breakdown of water is rather complex. Overall the process occurs this way. As indicated earlier, photoact II generates high energy electrons that begin to flow down an electron transport chain analogous to that found in mitochondria. However, instead of being coupled to converting ADP + phosphate to ATP, some of the released energy associated with this flow of electrons is coupled to converting $H_2O \Rightarrow O_2 + 2\,H^+ = 2\,e$.

Absorbed Light Coupled to Generating ATP. Photoact II generates high energy electrons that flow down an electron transport chain to photosystem I. This electron flow generates chemical energy potentials that drive two energy requiring chemical processes. One, as described earlier, is the breakdown of water into $O_2 + 2\,H^+ + 2\,e$. The other process is the synthesis of ATP from ADP + phosphate. As occurs in mitochondria, the electron transport sets up a chemical potential (a proton gradient). The chemi-osmotic theory—not described in this text—explains how this potential drives the synthesis of ATP. Photoact I transports "excited" electrons to a molecule called ferrodoxen. From here the high energy electrons flow into one of two pathways. In one pathway two electrons plus H^+ go to $NADP^+$ converting it into NADPH. In the other pathway the electrons go through a series of transfers down an electron transport chain to a ground state, lower energy level in photosystem I. In this cyclic pathway electrons go from a lower energy to a higher energy level in ferrodoxin and back to a lower level in chlorophyll in photosystem I. As in the electron transport in photosystem II, this electron transport is coupled with driving the synthesis of ATP from ADP + phosphate. Overall, photosystem II leads to the breakdown of H_2O into $O_2 + 2\,H^+ + 2\,e$, and to the synthesis of ATP. Photoact I leads to electrons plus H^+ joining $NADP^+$ to produce NADPH and to the synthesis of ATP.

Dark Reactions. These reactions are called "dark" because they do not directly depend on the absorption of light. The reactions occur when ATP, NADPH, and CO_2 are available. Of course indirectly these reactions depend on the absorption of light in that the light reactions generate ATP and NADPH that are used in the dark reactions. Overall the dark reaction is a cyclic process, and like any such process one can describe it from any starting point in the cycle. Let us start with the reaction involving CO_2 fixation. Six 5-carbon molecules derived from six other 5-carbon molecules, each called Ribulose di phosphate, combine with six CO_2 molecules to produce six 6-carbon molecules. Each of these 6-carbon molecules breaks down into two 3-carbon molecules called phosphoglycerate. Thus, six CO_2 plus six 5-carbon molecules produce twelve molecules of phosphoglycerate. Each of these molecules combines with two negative hydrogen ions to produce twelve molecules of PGAL. Two of these twelve PGAL molecules combine to produce one glucose molecule. The remaining ten molecules of PGAL are converted into six molecules of Ribulose di phosphate. Six of these 5-carbon molecules are converted into another six 5-carbon molecules that react with six CO_2 molecules. This marks the beginning of a new cycle, see figure 6.33.

In general, glucose usually is not the final product of photosynthesis. In many circumstances most of the fixed CO_2 shows up in sucrose that is synthesized in chloroplasts. Most of the CO_2 fixed by plants is used to produce organic compounds that sustain plant life. The fixed carbon is converted to CO_2 during respiration of these organic molecules and thus is returned as a gas to the atmosphere without passing through animals as food. However, under some circumstances, the plant will convert thousands of PGAL molecules into starch which then is a source of food for animals. Under some circumstances such as when the plant is growing rapidly, some PGAL molecules diffuse into the cytoplasm. There they are used as the starting points for the synthesis of other organic molecules such as lipids and proteins. Sometimes only 30% of the fixed carbon shows up as carbohydrates. There always will be some fixed carbon that participates in the synthesis of non-carbohydrates. This synthesis often occurs in the cytoplasm and may use ATP and NADPH produced by cellular respiration. Therefore, we cannot draw a sharp distinction between products of photosynthesis and of other types of biosynthesis in plant cells.

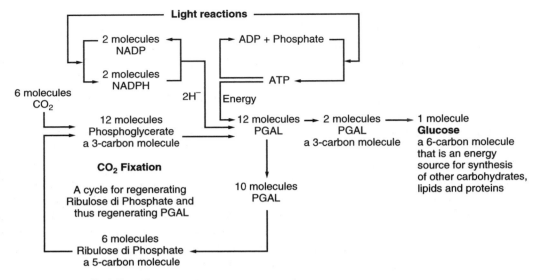

FIGURE 6.33

Tearout #1

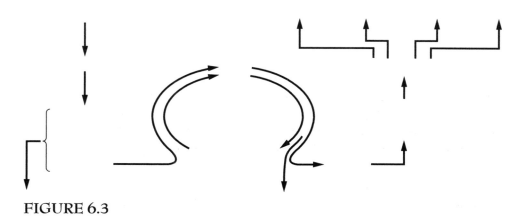

FIGURE 6.3

Directions:

1. Using figure 6.3 as a reference, fill in all labels.

2. Describe how the population of ADP and phosphate molecules operates as an energy coupler in a cell.

Tearout #2

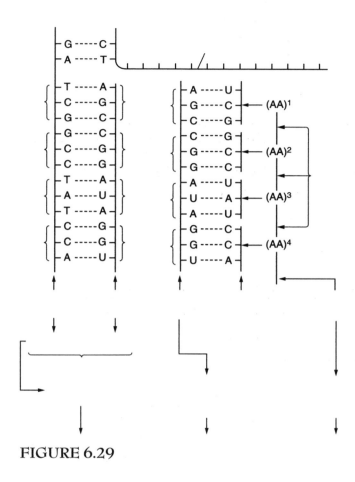

FIGURE 6.29

Directions:

1. Using figure 6.29 as a reference, fill in all labels.

Human Homeostatic Body Systems

Summary of Homeostatic Body Plan

As described in Chapter 4, homeostatic refers to the ability of any system to make adjustments so that a particular quantitative characteristic of the system, e.g., temperature, will stay near (fluctuate around) some steady state value, e.g., the human's steady state body temperature is 98.6°F. The animal homeostatic body plan includes the ideas of: internal milieu, blood-internal milieu interaction, and coordinated organ systems.

Internal Milieu

Single-celled organisms were the first forms of life to appear on earth. Random association of single cells into colonies had the adaptive advantage of size. Another competing, single-celled organism which can "eat" (engulf) cells could not subdue a colony. Cell colonies led to the emergence of bona fide individual organisms consisting of many cells. In addition to increased size, such organisms were able to cope with the challenges of the environment more efficiently by means of cell specialization. Most of the cells, called vegetative cells, deal with getting chemicals and energy from the environment in order to stay alive. A few cells became specialized for reproduction of the multicellular organism. The vegetative cells take care of the chemical and energy needs of the reproductive cells.

The multicellular adaptation resulted in organisms having an internal environment, also called internal milieu, as well as an external environment. All the surface cells of the multicellular organism provided the boundary between the individual and the external environment. The spaces between the cells and the fluid filling these spaces, called tissue fluid, are the *internal* environment, the internal milieu, of an organism. Very often this fluid filled space, a microfilm around each cell, can only be seen with the aid of an electron microscope. In simple organisms, the internal milieu is only one cell removed from the external environment from which nutrients can move in and to which wastes can move out. However, with the increase in multicellularity, the internal milieu becomes more distinct and isolated from the external environment. This means that for many cells of the organism, the only environment they deal with in their whole lives is the fluid that surrounds them. If these cells are to continue to live, the whole organism must regulate the interconnected fluid internal milieu to make it suitable for cellular life.

Blood-Internal Milieu Interaction

Individual animal cells are like car engines; they take in oxygen and glucose (analogous to gasoline) and chemically combine them to produce energy to carry out life processes plus wastes, thermal energy, carbon dioxide and water that are excreted into the internal milieu. Animal cells also take up many other nutrients. After awhile the internal milieu becomes depleted of nutrients and polluted by wastes so that it no longer is suitable for animal cells to continue to survive. The only hope for the continued survival of multicellular animals is some mechanism for the renewal of the internal milieu making it suitable again for these cells to survive. Continually circulating blood provides this mechanism. Namely, as blood from small arteries (arterioles) flows through capillaries which are bathed in the internal milieu just as all cells are, the flowing blood takes up excess heat energy, water, carbon dioxide and other wastes from the internal milieu and simultaneously resupplies the internal milieu with oxygen and glucose. Thus, the internal milieu is renewed by the blood in capillaries flowing past it, but now the blood is polluted by the wastes it picks up and depleted of oxygen and glucose.

Coordinated Organ Systems

In smaller organisms such as the earthworm, the circulating blood is continually depleted of nutrients and polluted by wastes as a result of interactions with the internal milieu. Nevertheless, the blood is continually renewed sufficiently rapidly because of its proximity to the digestive tube and to the external environment. In larger organisms, especially those that are more active, the blood must be renewed in a more efficient manner. Animals with backbones, such as vertebrates, have developed organ systems that keep properties of blood close to preset values (by means of negative feedback regulation). For example, the respiratory system regulates oxygen and carbon dioxide concentrations; the digestive system regulates the concentration of glucose along with many other nutrients; the kidney removes wastes and poisons and regulates concentrations of water, various ions, and other substances. These homeostatic mechanisms for renewing the blood lead to the efficient continuous renewal of the tissue fluid. Cells are able to survive in vertebrates because the circulatory system has become elaborate enough to help coordinate the activities of all organs that indirectly modify the internal milieu by modifying the blood or that directly modify the internal milieu; for example, the actions of the lymphatic and immune systems. In less complex animals (invertebrates), some of these organs, e.g., excretory organs, operate independent of the circulating blood. In vertebrates, blood moving in a closed circulatory system serves as an intermediary between the external environment of the animal and the internal milieu of its cells.

Various organ systems, e.g., respiratory system, digestive system, kidney, continually interact with the external environment to *renew* the blood that is polluted by wastes and depleted of nutrients, for example, oxygen and glucose. The *renewed* blood as a result of being kept flowing past the tissue fluid (internal milieu) renews it, and thus, this tissue fluid is continually maintained suitable for the cells surrounded by it to stay alive, see figures 7.1 and 7.2. A portion of the central nervous system (the hypothalamus) regu-

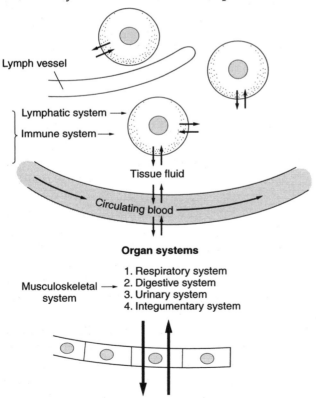

Lymph vessel

Lymphatic system →
Immune system →

Tissue fluid

Circulating blood

Organ systems

Musculoskeletal →
system

1. Respiratory system
2. Digestive system
3. Urinary system
4. Integumentary system

External environment

FIGURE 7.1

lates the flow of blood to all tissues in the body, and it also regulates all other organ systems so that the properties of the blood, e.g., temperature, concentration and volume of water, concentration of oxygen, carbon dioxide, and glucose, are kept near constant values. As a result, the tissue fluid is maintained suitable for cells to stay alive. This overall pattern of interactions among tissue fluid, circulating blood, other organ systems, and a regulatory portion of the central nervous system is called homeostatic body plan.

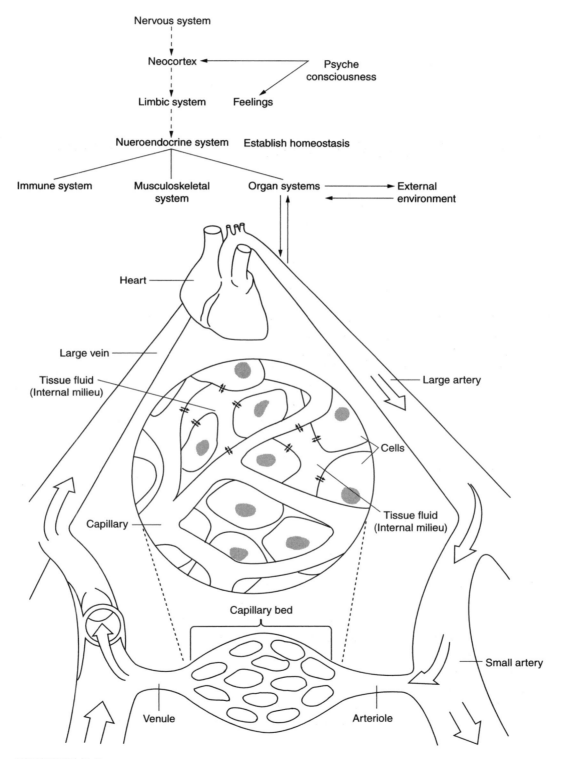

FIGURE 7.2

In summary, the human body consists of trillions of cells each surrounded by tissue fluid which supports cellular life. Each cell continually pollutes the tissue fluid and depletes it of nutrients. Circulating blood continually reestablishes the life supporting properties of tissue fluid, but in doing this, the blood is polluted and depleted of nutrients. Four organ systems, (1) the integumentary system (skin), (2) the respiratory system, (3) the digestive system, and (4) the urinary system, carry out exchanges between the external environment and the blood resulting in the reestablishment of the blood's life supporting properties. The lymphatic system removes excess water and debris from the tissue fluid, and the immune system in conjunction with the lymphatic and circulatory systems removes foreign substances, e.g., bacteria, viruses, and poisons, see figure 7.1.

The neuroendocrine system interacts with all other organ systems to establish homeostasis in which various properties of the body are kept at near constant values. A portion of the brain called the limbic system is associated with the experience and expression of feelings. Feeling is the connecting link between human consciousness and animal consciousness which is manifested through animal feelings. Animal feeling, in turn, is the connecting link between the animal psyche and the nonconscious body. Thus, human feeling is the connecting link between the human psyche and the nonconscious human body, see figure 7.2.

Human Circulatory System

Blood Flow Through the Body

The heart both initiates and directs blood flow through the body via a two-stage cycle. The heartbeat marks the first stage, called systole, where the heart contracts, pushing blood from the atria into the ventricles and then from the ventricles out into the aorta and pulmonary trunk. After the ventricles have emptied, the second stage, the diastole, begins where blood returning from the lungs enters the left atrium via the pulmonary veins, and blood from the rest of the body enters the right atrium via the superior and inferior vena cava. During diastole, as blood collects in the atria, the atrioventricular valves are pushed open so that the ventricles also begin to fill. The pulmonic and aortic valves also are open at the beginning of diastole, but as blood in these arteries flows back toward the empty ventricles, the cusps of the valves are forced backward and thus close, preventing blood from moving back into the heart. Systole begins when both atria contract which, in turn, completes the filling of the ventricles. A few milliseconds later, the ventricles contract pumping blood to all four openings. The aortic and pulmonic valves are forced open, and the atrioventricular valves are forced closed, see figure 7.3.

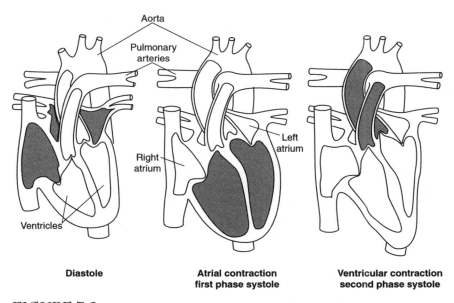

Aorta

Pulmonary arteries

Left atrium

Right atrium

Ventricles

Diastole

Atrial contraction first phase systole

Ventricular contraction second phase systole

FIGURE 7.3

The following "landmarks" indicate one complete cycle of blood circulating in the body: (1) the superior and inferior vena cava, (2) the right atrium, (3) the tricuspid valve (right atrioventricular valve), (4) the right ventricle, (5) the semilunar valve (pulmonic valve), (6) the pulmonary trunk, (7) the right and left pulmonary arteries, (8) the capillary beds in the lungs, (9) the pulmonary veins, (10) the left atrium, (11) the mitral valve (left atrioventricular valve), (12) the left ventricle, (13) the semilunar valve (aortic valve), (14) aorta, and (15) the capillary beds throughout the body which eventually empty into the superior and inferior vena cava, see figures 7.4 and 7.5.

Vascular Basis for Regulating Blood Pressure

Besides carrying blood to and from the heart, some portions of the blood vessels actively participate in modifying blood flow. The flow in the arteries, capillaries and veins is similar to the flow of water in pipes. The pressure difference between two points along the length of a pipe causes water to flow.

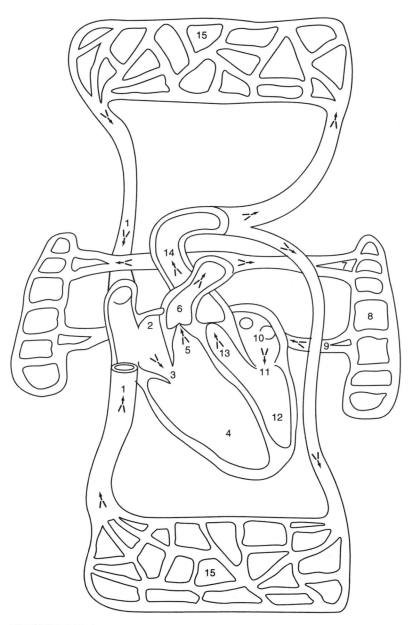

FIGURE 7.4

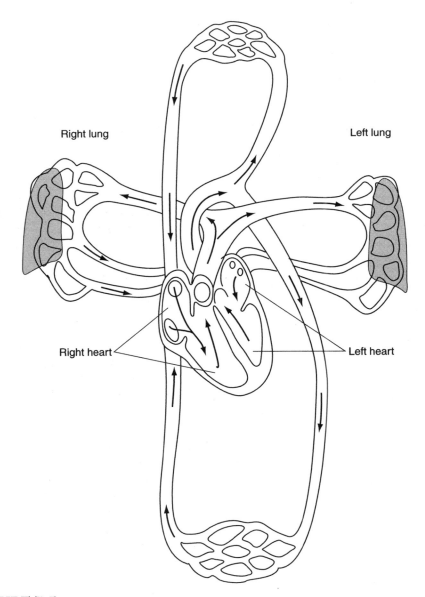

FIGURE 7.5

That is, water flows from a point of higher pressure to a point of lower pressure. In the circulatory system, when the heart pumps blood into the large arteries, there is an increase in pressure due to the increased amount of fluid in the arteries. This is similar to increasing the pressure in one's car (or bike) tire by pumping air into it. Just after a heartbeat the pressure at the beginning of these large arteries is greater than anywhere else in the circulatory system. Correspondingly, the atria, the portion of the heart into which blood flows, has the lowest pressure. As a result of this pressure difference, blood flows away from the ventricles of the heart, through the capillaries, and back to the atria. But the process is more complex than this. Various portions of the blood vessels modify the flow of blood through them due to this pressure difference.

The aorta branches and rebranches into various other arteries. This branching continues and the arteries become progressively smaller as they become continuous with an extensive network of capillaries. The veins exhibit a corresponding branching network. For convenience, these three networks will be referred to as the arterial vascular tree, capillary bed, and the venous vascular tree, respectively.

Arterial Vascular Tree.   The arterial vascular tree plays a dominant role in the circulation of blood. It keeps the blood moving at all times, and it is the primary site of regulation of circulation. When the heart contracts, the pressure in the aorta and the pulmonary arteries is greater than in the ventricles.

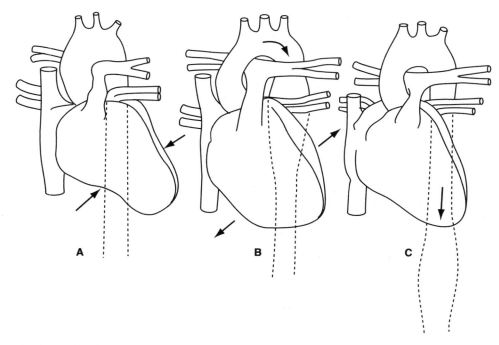

FIGURE 7.6

As a result, some blood flows back toward the heart. This forces the ventricle outlet valves to close. Now if this were all that happened, the blood flow would greatly slow down and eventually stop until the next heartbeat. This "on-again, off-again" type of blood flow would be disadvantageous to the cells of the body. The cells would not receive a steady continuous supply of essential nutrients such as oxygen and sugar.

The large arteries prevent this from happening in the following way. These arteries are composed of elastic tissue, so as blood is pumped into them, they stretch like a rubber band or balloon. When the heart is relaxing and therefore no longer pumping blood into the large arteries, the pressure drops. It would go to zero except now the stretched vessel walls spring back to their initial positions. This decrease in vessel size is analogous to contraction of the ventricles. The blood pressure is not allowed to fall to zero and blood is pushed along even when the heart is relaxing. The pressure in the large arteries when the heart is (in systole) pumping blood into them is called systolic pressure. When the heart is in diastole, the pressure in the stretched large arteries is due to their recoil and is called the diastolic pressure, see figure. 7.6.

The arterioles at the end of the arterial vascular tree have thick walls composed of smooth muscle. Remember that the capillaries are very delicate tubes. Blood flowing into them at high pressure would rupture them. The thick walled arterioles absorb the high pressure of blood flowing into them. The arterioles are like a room packed with people leading into a narrow hallway allowing only one person out at a time. As more people are pushed into the room, the ones there are squeezed together, and only a few are pushed out into the hall because of the narrow exit. Likewise, as blood under relatively high pressure comes to the arterioles, the blood pressure remains high because the narrow inner diameter of the arteriole offers resistance to blood flow through it. If the arteriole had only one outlet, then the blood would flow out of it at high velocity—like a garden hose with the nozzle greatly reduced. However, each arteriole empties into tubes leading to a capillary network. As a result, the rate of blood flow into a capillary network is fairly high but the velocity of blood flowing through each outlet from an arteriole is quite low. That is, the high rate of blood flow into capillaries is due to many outlets from each arteriole rather than due to a high velocity of flow of blood.

It can be shown that for a particular pressure difference, the volume of fluid (such as blood) flowing through a tube per unit time is directly proportional to the radius of the tube taken to the fourth power. That is, if one increases the radius by two units, one increases flow by 16 units (i.e., 2^4). A small increase in the arteriole radius will bring about a large increase in blood flow. Conversely, a

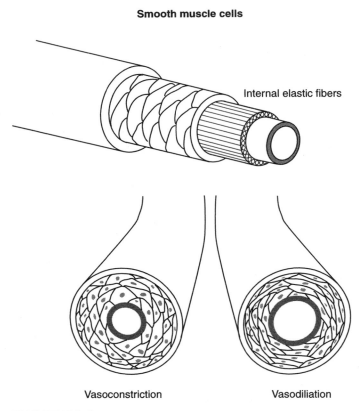

Smooth muscle cells

Internal elastic fibers

Vasoconstriction   Vasodiliation

FIGURE 7.7

small decrease in the radius of an arteriole will bring about a large decrease in flow of blood to the capillaries. When the many smooth muscle fibers in the arteriole contract, they decrease the radius of the vessel; this is called vasoconstriction. If the smooth muscles decrease their constriction, the radius becomes larger; this is called vasodilation, see figure 7.7.

By controlling the degree of contraction of the smooth muscles in the arterioles, the body controls the amount of blood flow to a particular organ. A few examples will indicate the advantages of this feature of the circulatory system. The circulatory system does not have enough blood to provide a rich supply of blood to all the organs in the body at the same time. If the circulatory system did this, then the heart would have to pump much harder. This would be a wasted effort because the organs do not always need a rich supply of blood. Instead, when one eats, the blood supply of the voluntary (skeletal) muscles is decreased by means of arteriolar constriction and the supply to the digestive system is increased by means of arteriolar dilation. When one exercises, the blood supply to voluntary muscles is increased while the blood supply to the digestive, urinary, and reproductive systems is decreased. This is why it is unwise for an individual to do heavy exercise immediately after eating a large meal. Also, when one increases heat energy production by exercising, the arterioles supplying the deeper organs constrict, and those supplying the skin area dilate so that heat energy will more easily be radiated away from the body. When the body temperature starts to decrease, just the opposite happens—the constriction of the surface arterioles tends to conserve body heat energy.

Capillary Beds.   The capillary network is where all the life and death drama at the cellular level occurs. Here the blood gives off oxygen to the cells and picks up carbon dioxide. Here cells pick up nutrients brought by the blood and give off waste products that are carried away by the blood. One might reasonably ask how blood could so easily flow in the capillaries when the pressure there is so low—remembering that flow depends on blood pressure differences. The answer is that the resistance to flow is very low. The capillaries coming from the arterioles branch and rebranch many times. This high degree of branching is equivalent to greatly increasing the capillary tube radius. Though the pressure difference between the arteriole and the venous end of the capillary network is relatively small, there is a smooth, steady flow of blood through the capillaries because of the low resistance to flow.

Venous Vascular Tree.   The venous vascular tree not only conducts blood back to the heart, but it also serves as a blood reservoir. The veins are easily distended, but stimulation of the muscle fibers in their walls—particularly in the larger veins—causes them to constrict. Thus, when we are at rest and do not need a rapid supply of oxygen and nutrients to our tissues, the veins are distended and so a percentage of blood is always found in the venous vascular tree. During activity, the veins constrict and a large percentage of the blood is continuously pumped into the arterial vascular tree. That is, during activity, the recycling of blood occurs faster than during rest to meet the higher demands of the body for oxygen and nutrients.

The blood pressure in the veins is even less than in the capillaries and the venous vascular tree becomes progressively less branched as it funnels blood back to the heart. So how does the blood

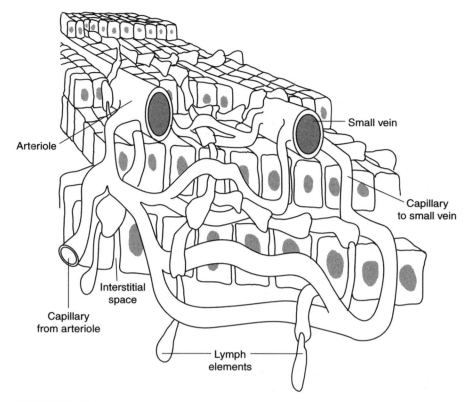

FIGURE 7.8

make it back to the heart? At times it is not easy. If one stands at attention for a long time without flexing one's leg muscles, a significant portion of blood may pool in distended leg veins. The blood pressure tends to fall because not enough blood gets back to the heart. Therefore, not enough blood gets pumped to the brain, and one can faint. The leg region presents the difficulty of blood having to flow back to the heart against the force of gravity. The body has various ways of overcoming this difficulty. For one, the veins have valves which allow blood to flow toward the heart but which close as soon as blood flows backwards. Furthermore, there is muscle tone in all the voluntary muscles throughout the body. This low-level, periodic contraction of the leg muscles surrounding the veins exerts a pump-like action on those veins, forcing the blood up toward the heart.

Human Lymphatic System

The lymphatic system regulates the quantity of water in the tissue fluid. Ordinarily, the albumin in the blood plasma is unable to diffuse across the vessel walls into the tissue fluid. Occasionally, however, some albumin leaks through. If the albumin accumulates in the tissue spaces, it will exert osmotic pressure and cause water to accumulate there also. This excessive accumulation of tissue fluid is known as edema. Edema makes tissues puffy and swollen, as occurs, for example, when one sustains a bruise. The reason why all body tissues eventually do not become puffy and swollen is because the lymphatic system continually drains these spaces of albumin, other proteins, debris, bacteria, foreign matter and excess water that tend to accumulate there.

Lymph Vessels

There is an extensively branched tree of lymph vessels that end blindly in the tissue spaces, see figures 7.8, 7.9 and 7.10. The small terminal branches of this tree are like capillaries. They collect excess tissue fluids along with plasma proteins, foreign particles, parts of dead cells, and other debris and

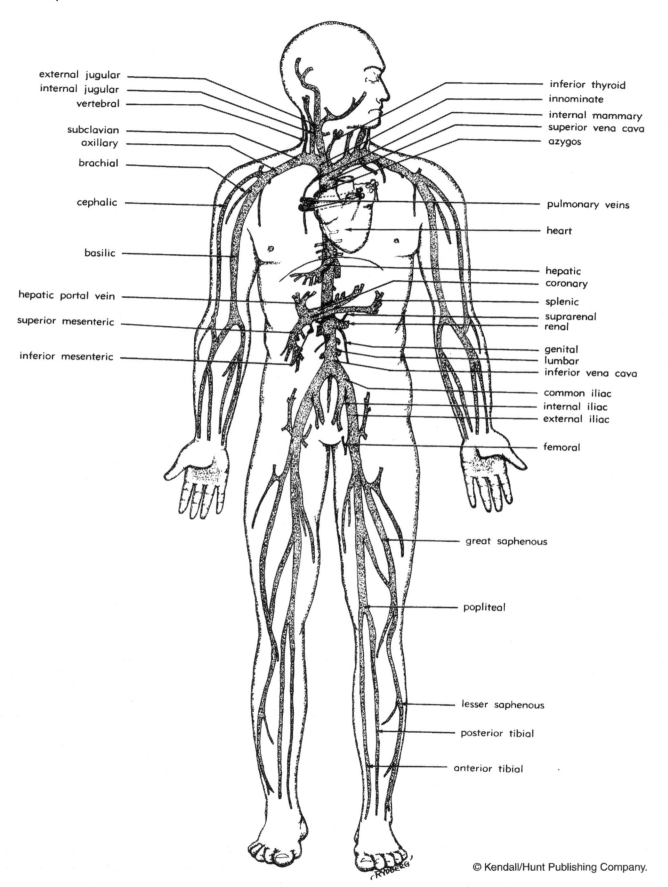

external jugular
internal jugular
vertebral
subclavian
axillary
brachial
cephalic
basilic
hepatic portal vein
superior mesenteric
inferior mesenteric

inferior thyroid
innominate
internal mammary
superior vena cava
azygos
pulmonary veins
heart
hepatic
coronary
splenic
suprarenal
renal
genital
lumbar
inferior vena cava
common iliac
internal iliac
external iliac
femoral
great saphenous
popliteal
lesser saphenous
posterior tibial
anterior tibial

FIGURE 7.9

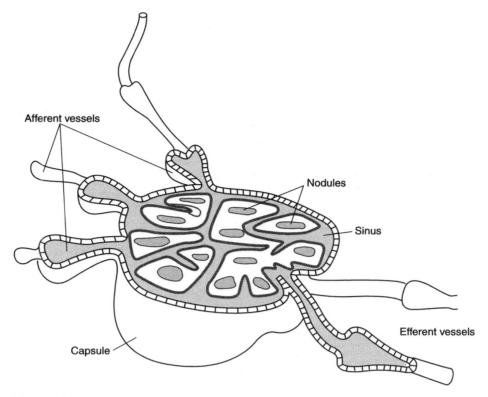

Afferent vessels

Nodules

Sinus

Efferent vessels

Capsule

FIGURE 7.10

transport them into larger lymph vessels, see Figure 7.8. This fluid, called the *lymph,* is transported to the *right lymphatic duct* and the *thoracic duct*. The right lymphatic duct carries lymph from the right side of the head and trunk and empties into the right subclavian vein, the major vein of the right arm. The thoracic duct carries lymph from all other tissues of the body and empties it into the junction of the left internal jugular and left subclavian vein, see figure 7.9. The large lymph vessels are similar in structure to veins of comparable size and, like veins, have valves that allow the lymph to flow in only one direction. Contraction of surrounding muscle tissue, for example, in the arm and legs, brings about lymph flow. In areas where lymph vessels are not surrounded by muscle tissue, such as the kidney or liver, variations in tissue fluid pressure cause lymph flow.

Lymph Nodes

At intervals along the larger lymph vessels, there are nodules, called lymph nodes, that contain lymphocytes and monocytes. The lymphocytes participate in the immune response, see figure 7.10. The monocytes and the cells lining the vessels in these nodes engulf and remove bacteria and particulate matter as the lymph percolates through the nodes. This is why a localized bacterial infection in the throat causes swelling and tenderness of the lymph nodes there. The lymph nodes act like filtering stations cleaning the lymph before it empties into the larger veins nearer the heart. Since foreign bodies and debris in the blood will eventually leak into the tissue spaces, the lymph nodes help "clean" the blood as well as the tissue fluid.

Spleen

The spleen is about the size of a person's fist and is located to the left and somewhat behind the stomach and below the diaphragm but directly above the left kidney. Since the spleen is similar in structure and function to a lymph node, it is considered part of the lymphatic system even though it does not filter lymph. The spleen filters blood that percolates through it and accomplishes three

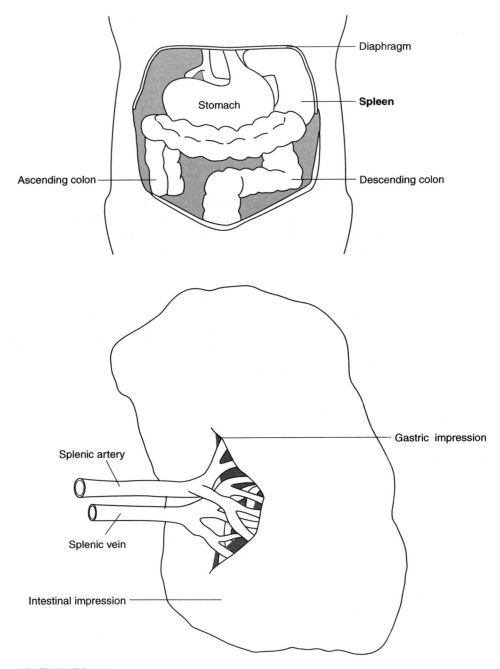

FIGURE 7.11

functions: (1) it destroys old red blood cells and platelets; (2) it forms lymphocytes; and (3) it is a blood reservoir. During times of stress, smooth muscles in the spleen contract forcing more blood out into the circulatory system, thereby increasing its oxygen-carrying capacity.

Regulation of the Amount of Tissue Fluid

The amount of fluid in the tissue spaces is determined by interactions among the circulating blood, tissue fluid, and the lymphatic system. The tissue fluid is in dynamic equilibrium with the blood plasma. That is, any substance that can diffuse across the capillary walls will distribute itself so that its concentration is the same in the tissue fluid as it is in the blood plasma. Ordinarily the amount of water in the tissue fluid remains constant. On a moment to moment basis, some water enters the tissue fluid across the arterial side of the capillary bed, but the same amount of water re-enters the blood across the venous side of the capillary bed, see figure 7.12. Occasionally plasma albumin may leak into the tissue fluid thereby increasing the water concentration due to osmosis (see next section). This excess water is drained away from the tissue spaces by the lymphatic system. The local blood pressure and plasma composition greatly affect the amount of water in the tissue fluid. Since the kidney regulates the concentration of many substances in the blood, it ordinarily helps determine the amount of water in the tissue fluid.

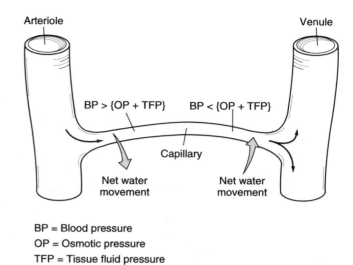

BP = Blood pressure
OP = Osmotic pressure
TFP = Tissue fluid pressure

FIGURE 7.12

Starling's Principle

The dynamics of how the amount of water in tissue fluid is maintained constant is explained by Starling's principle (or the law of the capillaries). The blood pressure or hydrostatic pressure on the arterial side of a capillary bed is greater than that on the venous side. This pressure difference forces blood through the capillaries, but it also tends to force some water with dissolved substances out into the tissue fluid. The small amount of water already in the tissues exerts a small opposing pressure which forces some water back into the capillaries. Osmosis due to high albumin concentration in the plasma also tends to move water from the tissue fluid into the capillaries. Normally the sum of the tissue fluid pressure and the albumin osmotic pressure is less than the blood pressure, and the net effect is that some water is forced into the tissue fluid. On the venous side of the capillary bed, the albumin osmotic pressure has increased because the albumin has become more concentrated due to the net movement of water into the tissue fluid on the arterial side. On the venous side the blood pressure has decreased and the tissue fluid pressure increased. Thus, now the net effect is that water moves from the tissue fluid back into the venous system. The amount of water lost on the arterial side is offset by the amount of water regained on the venous side. As a result of this delicate balance, very little water accumulates in the tissues.

From time to time, some cell debris or plasma proteins leak into the tissues. The lymphatic vessels there collect the excess water and particles, filter them through the lymph nodes and eventually empty them into the venous vascular tree. The lymphatic system is a kind of safety valve for tissue fluid balance and a way of keeping the tissue space free of bacteria, poisons, and foreign particles. Of course, if proteins were allowed to accumulate in the tissue spaces, the proteins would osmotically cause more and more water to move into these spaces.

Tissue Fluid Imbalances

Blood pressure tends to force water from the plasma into the tissue fluid whereas osmotic pressure of the plasma solutes and tissue fluid pressure tends to force water from the tissues into the blood. Ordinarily these opposing forces offset one another so that the amount of water in the tissue spaces remains constant. A change in any one of these forces will upset the balance producing edema or dehydration, the accumulation or decrease respectively of water in the tissue spaces. For example, any increase in blood pressure would cause edema. People with high blood pressure, especially essential hypertension which is sustained high diastolic pressure, will have edema throughout the body. Left heart failure results in the left ventricle not completely emptying with each heart beat. This leads to an increased blood pressure in the pulmonary veins. An individual with this condition will develop pulmonary edema, water in the lung tissue. In a similar way, right heart failure leads to increased pressure in the systemic veins which produces edema throughout the body. Any person who is up much of the day without moving around a lot will retard the return of blood from the legs. The resulting increase in venous pressure in the legs will cause the feet and ankles to swell due to edema. People with poor circulation routinely have swollen feet. Pressure of the growing fetus will impede the venous return in a pregnant woman and cause her legs to swell.

Inflammation is a body response to some foreign substance or local irritant. Inflammation always involves localized vasodilation resulting in the arterioles producing a localized increase in blood pressure. The resulting localized edema is what produces the red swelling in an inflamed area associated with hives due to allergy, or a bee sting, a cut or a bacterial infection.

A decrease in plasma osmotic pressure also causes widespread edema. As a result of an infection in the kidney called glomerular nephritis, albumin (which leaks into the nephron) is lost in the urine. The resulting decrease in osmotic pressure of plasma albumin causes widespread edema. The reason why people who are starving often have large abdomens is that they have an inadequate intake of proteins and therefore a lowered concentration of plasma albumin.

All tissues in the body tend to accumulate some water due to various causes such as plasma albumin leaking into the tissue spaces. Ordinarily the lymphatic system drains this excess water as fast as it accumulates. However, any decreased efficiency of the lymphatic drainage will produce edema. For example, the flow of lymph is aided by contraction of skeletal muscles in the vicinity of the lymph vessels. When a person stops using an arm or leg for a long time or if he is immobilized in a hospital bed, muscle tone decreases. As a result, the lymph flow decreases producing localized or generalized edema. In Africa some people are infected by a parasite that lodges in the lymphatic system preventing tissue water and proteins from being drained off. Massive edema develops in the infected area such as the leg. This condition is called elephantiasis.

Nerve Cell Dialogue

Dialogue is a type of communication that is the basis for a distinction between living and nonliving systems. Communication between any two non-living systems is an interaction between one of these systems called a source and the other system called a receptor. The interaction involves the one-way transfer of information from the source to the receptor. *Note:* the information does not exist in itself; it only exists during the communication between the source and the receptor. In the perspective of modern mechanistic science, *force* is the most elemental communication; all other more complex communications among nonliving organisms can be understood in terms of a network of forces. The information of any "force communication" always is located at a particular point in space and time and has meaning in relation to an observable event at a point which is an event that is part of a sequence of events that culminates in a complex observable event. This complex event, in turn, can be observed by any human with the appropriate sense receptors (eyes and ears, etc.) plus the appropriate detectors, e.g., microscope, voltmeter, radio receiver. The foundation for all inanimate communication is the second law of thermodynamics which describes the one way transfer of information in the universe. In radical contrast, all living things participate in a two-way conversation between an organism and some aspect of its environment. The environmental aspect may be a non-living system or another living system. In general this two-way conversation usually involves the environmental aspect initiating the conversation by a transfer of information to the living organism. This transfer of information can be described by mechanistic science. The organism receiving this information is able to process it and then transfer the processed information to some part of itself or transfer it to the environment thereby changing the environment in some way. Put more simply: the environment "talks" to the organism which receives information, processes it, and then "talks back" to the environment. The environment's "talk" is called a stimulus and the organism's "talking back" is called a response. Thus, the fundamental characteristic of any living thing is that it continually participates in or, under the right conditions, could participate in a *dialogue with the environment*. This dialogue between any living thing and the environment is called a Stimulus-Response Process.

Most fundamentally a cell is alive because it is a homeostatic system. Some cells may not be able to reproduce, that is, participate in cell division, and some cells such as mature red blood cells do not contain any DNA (genetic information). Nevertheless, all living cells can receive information from the environment, process this information, and make a homeostatic response that allows them to continually adapt to the environment. Animals without backbones are similar to plants and fungi in that all these organisms consist of a non-homeostatic community of cells that establish a network of intercellular communications coordinated to achieve survival of the whole organism. Besides the chemical communications found in plants and fungi, animals with backbones (the vertebrates) increase the efficiency of cellular coordination by means of nerve networks.

Nerve cells, called *neurons*, have some stimulus-response processes specialized for representing, transmitting, and processing information which then is expressed as some activity, see figure 7.13.

All of these "neuronal information events" are analogous to processes exemplified by machines and homeostatic systems. An energy coupler (see Chapter 1) translates the information in some environmental potential into a specific activity representing the task of the machine. The homeostatic system regulating room temperature exemplifies all of these information processes (see Chapter 4). The sense receptor represents temperature by a certain quantity of electrical activity (current) which then is transmitted to the evaluation unit. The evaluation unit processes this information by comparing the received quantity of current to a preset quantity of current representing the temperature at which the room is to be kept; for example, the temperature dial adjustment one makes on the thermostat. Based on this comparison, the evaluation unit in effect "selects" ("chooses") one of three possible pathways for the further transmission of electrical activity to a command center: (1) electrical activity turns on the heating unit and turns off the air conditioning unit; (2) electrical activity turns off the heating unit and turns on the air conditioning unit; (3) electrical activity either is not sent or if sent does not effect either the heating unit or the air conditioning unit (see Chapter 4).

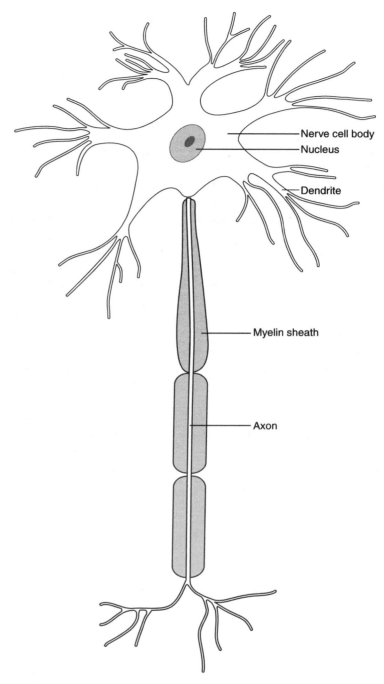

FIGURE 7.13

Intraneuron Communication

A neuron is first and foremost a homeostatic machine that coordinates stimulus-response processes toward its continued survival, but it no longer uses "its energies" to reproduce itself. Neurons never divide unless they become renegades known as neuron cancer cells. Instead neurons use their excess energies for becoming expert cell communicators. As shown in figure 7.13, the nucleus and machinery for carrying out various life processes such as synthesizing proteins lie in the neuron cell body from which emerges one usually long projection called an axon and several other usually much smaller projections called dendrites. Some neurons are surrounded by Schwann cells which are arranged sequentially along the length of the axon. Axons are said to be myelinated nerves when each associated Schwann cell wraps itself around the axon several times, resulting in alternating layers of proteins and lipids that comprise the cell membrane of the Schwann cell. The outermost part of the Schwann cell is called the neurilemma and the alternating layers of proteins and lipids is called the myelin sheath because most of the lipid present is myelin, see figures 7.14 and 7.15. Some axons associatd with Schwann cells are called nonmyelinated nerves because they have a neurilema but no myelin sheath, see figure 7.15.

Here is how a neuron represents and transmits information from one part of itself to another. Normally the inside of any cell is negatively charged with respect to the positively charged

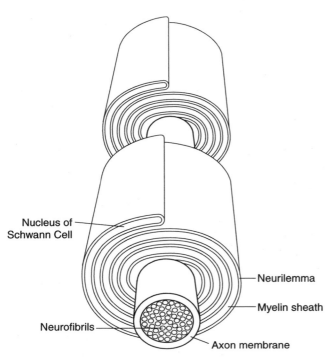

Nucleus of
Schwann Cell

Neurilemma

Myelin sheath

Neurofibrils

Axon membrane

FIGURE 7.14

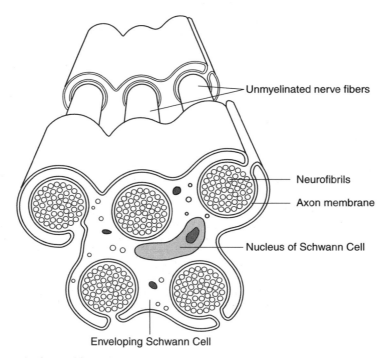

Unmyelinated nerve fibers

Neurofibrils

Axon membrane

Nucleus of Schwann Cell

Enveloping Schwann Cell

FIGURE 7.15

outside of its membrane. The amount of this inner negativity represents the cell membrane electrical potential. When some environmental potential decreases the neuron membrane negativity, the homeostatic cell responds by reestablishing the normal membrane potential. However, sometimes for one reason or another, such as several environmental potentials simultaneously effecting the neuron at a particular membrane site, the decrease in membrane potential leads to a new kind of response. Instead of making a homeostatic adjustment, the neuron responds by reversing the electrical potential at the effected site of the membrane; that is, the inside becomes positive with respect to the negative outside. This highly localized reversed membrane potential is called an action potential, and it represents one or a net set of environmental potentials; i.e., it represents some information present in the environment.

The action potential at site 1 stimulates its immediate neighbor, membrane site 2, to generate its own action potential. This second action potential stimulates its neighbor, site 3, to generate an action potential. In the meantime, membrane site 1 not only stops generating an action potential, it reestablishes the normal membrane potential. Site 3 stimulates site 4 to generate an action potential which stimulates site 5 to produce one and so forth. As new sites are generating action potentials, the preceding ones are reestablishing the normal membrane potential. In this manner, a sequence of stimulus-response processes beginning, say, at the tip on an axon, results in the transmission of information to the neuron cell body and from there perhaps down one or more of its dendrites. This "transmission of an action potential" is called a nerve impulse, see figure 7.16.

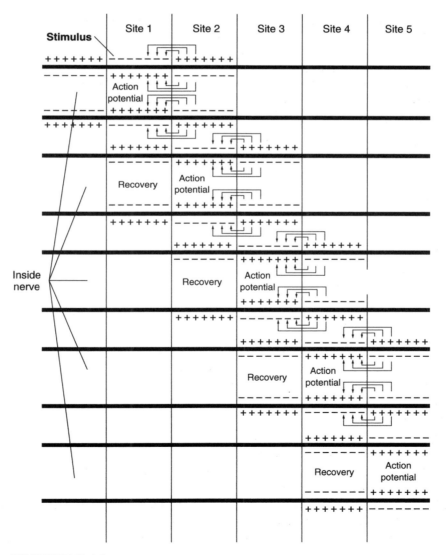

FIGURE 7.16

The action potential is analogous to a word we create in order to represent our experience of some information in the environment. The nerve impulse is analogous to speaking or writing the word thus transmitting it to someone else. Neuronal information processing and expression are carried out by a network of neurons and effectors. The key components for doing this are synapses, and the simplest cellular network is a reflex arc.

Synaptic Communication

A synapse is a point of functional contact specialized for intercellular communication between one neuron and another or between a neuron and an effector cell such as a muscle cell or a gland cell. The synapse is the material representation of an intercellular communication involving at least one neuron, just as a sentence represents a communication between two people. The source of the communication is a specialized section of the membrane, the presynaptic membrane, of a source neuron. The receptor of the communication is another specialized section of the membrane, the postsynaptic membrane, of some other cell. The gap between the pre- and post-synaptic membranes is the synaptic cleft. The meaning of the intercellular communication is an electrical activity or more usually a chemical, the transmitter substance, that moves from the source neuron, across the presynaptic membrane and the synaptic cleft to a receptor in the postsynaptic membrane of the receptor cell. The source neuron continually synthesizes transmitter substances which are packaged in small secretory vesicles that move to the vacinity of the presynaptic membrane, see figures 7.17a and 7.17b.

The following sequence of events accomplishes intercellular excitatory communication in a chemical synapse: (1) an action potential in the area causes many secretory vesicles to attach to the inner side of the presynaptic membrane; (2) the attached vesicles release transmitter molecules into the synaptic cleft; (3) the transmitter molecules diffuse across the synaptic cleft and fit into receptors in the postsynaptic membrane just like a key fits into a lock; (4) each transmitter molecule-receptor interaction results in a small lowering of the normal ("resting") potential of the postsynaptic membrane; (5) if enough of these molecular communications occur due to an incoming barage of action potentials, the postsynaptic membrane generates an action potential which is transmitted to other parts of the cell, and (6) nearby enzymes facilitate the breakdown of the transmitter molecule fitting into a receptor and the breakdown products diffuse back to the presynaptic membrane where they are taken into the source neuron and recombined to make more transmitter molecules. The action potential of the receptor cell eventually may lead to an observable event that is relevant to the organism as a whole. This event is the expression of the meaning of the transmitter substance which in turn is the meaning of the synaptic communication.

The transmitter substance does not have meaning in itself; rather its meaning is influenced by the nature of the receptor in the postsynaptic membrane and the receptor cell and its functional interactions (communications) with other cells. The integration of many molecular communications with respect to a particular synapse is neuron information processing. This foundational idea has led some authors to define a neuron as ". . . a cell connected to other cells by synapses that mediate specific signals involved in behavior."[1] Thus, the synapse rather than the neuron is the fundamental functional unit of the nervous system. According to this viewpoint, neurobiology may be defined as: ". . . the study of the molecular organization of the nerve cell, and the ways that nerve cells are organized, through synapses, into functional circuits that process information and mediate behavior."[2] How neuronal information processing leads to behavior is exemplified by reflex activity of a reflex arc.

Reflex Arc

Background Definitions.   As we have seen, a neuron may respond to a stimulus by generating an electrical event, e.g., an action potential, or a chemical event, e.g., release of a transmitter substance. These electrical or chemical events are called neuron signals. A neuron signal may stimulate a sec-

[1]Shepherd, G. M., *Neurobiology* 2nd ed., (New York: Oxford University Press, 1988) p. 67.
[2]Ibid., p. 5.

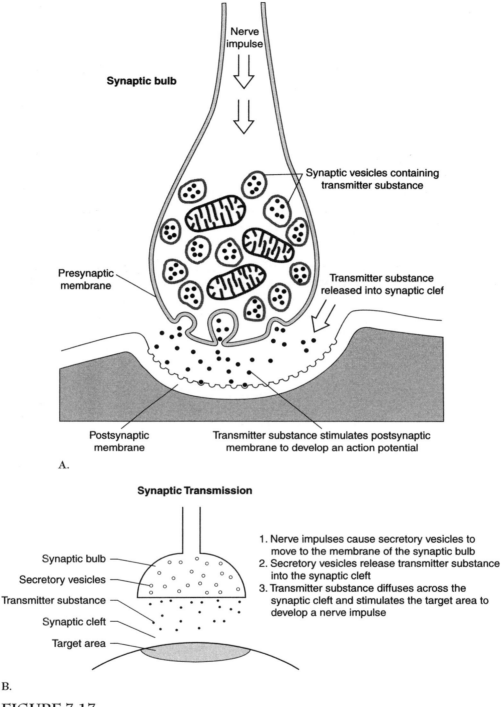

Nerve impulse

Synaptic bulb

Synaptic vesicles containing transmitter substance

Presynaptic membrane

Transmitter substance released into synaptic clef

Postsynaptic membrane

Transmitter substance stimulates postsynaptic membrane to develop an action potential

A.

Synaptic Transmission

Synaptic bulb

Secretory vesicles

Transmitter substance

Synaptic cleft

Target area

1. Nerve impulses cause secretory vesicles to move to the membrane of the synaptic bulb
2. Secretory vesicles release transmitter substance into the synaptic cleft
3. Transmitter substance diffuses across the synaptic cleft and stimulates the target area to develop a nerve impulse

B.

FIGURE 7.17

ond neuron or an effector cell. Correspondingly, we say the first neuron communicates with the second neuron or with the effector cell to form a communication network. These communication networks may consist of many cells of which the component neurons are classified according to their function within the network as follows: (1) sensory neuron, (2) interneuron, and (3) motor neuron. A sensory neuron has an axon that terminates in one or several *sense receptors* which communicate with the external or internal environment of an animal, for example, some pattern in the environment stimulates the sense receptor membrane to generate an action potential that is transmitted up the sensory axon. An interneuron is stimulated by a signal from one neuron and, in turn, stimulates another neuron to generate a neuron signal. A motor neuron communicates with one or more effector cells. The junction between a motor neuron and an effector cell is called a neuroeffector junction or end plate. Several neurons and effectors which as a group receive and transmit signals and are interconnected via synapses make up a neuroeffector network of cells; the component network of neurons is called a neuron circuit. A reflex arc is the simplest neuroeffector network.

Reflex Activity.   A reflex arc consists of: (1) a sensory neuron, (2) one or more interneurons, (3) a motor neuron, and (4) an effector, see figures 7.18a and 7.18b. The sense receptor of a sensory neuron generates an action potential in response to a particular stimulus. This starts a nerve impulse moving up the afferent fiber (sensory fiber) toward the synapse with an interneuron. The interneuron generates signals that stimulate a motor neuron. The motor neuron, in turn, generates signals that travel down the efferent fiber (motor fiber) toward the synapse with an effector. Nerve impulses arriving at the neuroeffector junction stimulate the effector to do something; i.e., muscles contract or glands secrete substances. The muscle contractions or glandular secretions lead to some overt event in or behavior by the animal. These events are the ultimate meanings of neuroeffector communications, see figures 7.18a and 7.18b.

II. Reflex Activity
1. Sense receptors generate an action potential in response to a particular stimulus.
2. The sensory action potential starts a nerve impulse moving up the sensory axon (afferent axon).
3. The sensory nerve impulse, via synaptic transmission, leads to a nerve impulse in the interneuron.
4. The interneuron nerve impulse, via synaptic transmission, leads to a nerve impulse in the motor neuron.
5. The motor nerve impulse, via synaptic transmission across the neuroeffector junction stimulates the effector to do something.
   a. Muscle cells contract
   b. Glands secrete substances

Autonomic Nervous System

The autonomic nervous system (ANS) is that portion of the nervous system that rapidly initiates adjustments to changes or threatened changes of the body's homeostasis. Thus, it is the ANS that regulates blood pressure and blood flow to various tissues and initiates the homeostatic adjustments of the various organ systems. Overall, the functions of the ANS are: (1) to initiate "emergency mechanisms" to preserve the life of the organism or of the species, e.g., the fight, the flight, and sex drives; (2) to initiate repair in any damaged portion of the body; and (3) to maintain homeostasis in various parts of the body and in the body as a whole.

The ANS consists of two divisions, the sympathetic division (SD) and the parasympathetic division (PD). The sympathetic division prepares for or adjusts the body to external stress. For example, when we become frightened, angry, or sexually aroused, the heart rate, blood pressure, and respiratory rate greatly increase. We are much more alert than normal. Our body muscles are tense and ready to respond to some urgent command. These are generalized body responses that prepare us for the intense activities that these strong drives may initiate. A similar type of generalized response occurs when a person is injured. The parasympathetic division regulates localized responses. For example, the digestive glands of the GI tract (gastrointestinal tract) are stimulated to secrete when we become hungry. Despite this difference, the sympathetic and parasympathetic divisions work together in regulating several autonomic reflexes. For example, the PD (Parasympathetic Division) stimulates increased motility or peristalsis of the GI tract, while the SD (Sympathetic Division) causes decreased motility. The reverse responses are found in the heart. The PD decreases the heart rate; the SD increases it.

The PD and SD produce antagonistic responses in the heart, coronary vessels, bronchioles of the lungs, the muscles and glands in the GI tract, the salivary glands, and the eye pupils. The body achieves very fine control of these organs by carefully balancing the antagonistic responses by the PD and the SD. Other autonomic effectors, including the core of the adrenal gland, the sweat glands, the muscles associated with hair, and many vascular beds such as those of the skin and of the voluntary muscles, are regulated only by the SD.

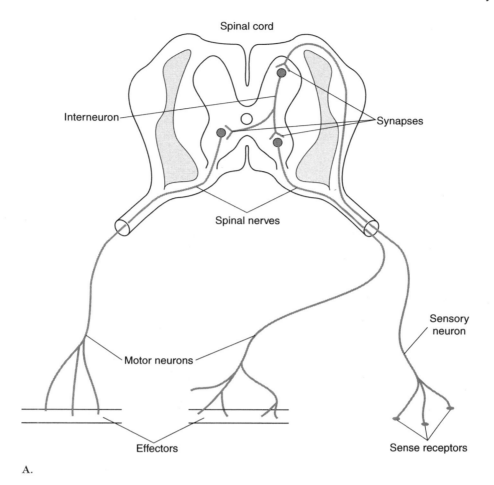

A.

Reflex activity

I. Schematic of a reflex arc in mammals.

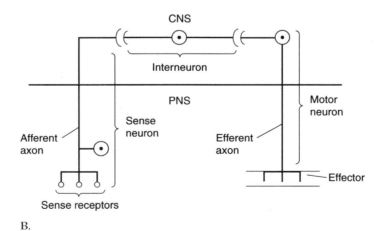

B.

FIGURE 7.18

Anatomical Distinctions

The two major anatomical distinctions between the sympathetic and the parasympathetic divisions of the ANS are illustrated in figures 7.19a and 7.19b. First, the sympathetic efferent axons come from the middle part of the spinal cord while the parasympathetic efferent axons come from the two end regions of the spinal cord. Second, most sympathetic ganglia are located along each side of the spinal cord in the two sympathetic trunks, see figures 7.19a and 7.19b, and in association with major abdominal arteries while

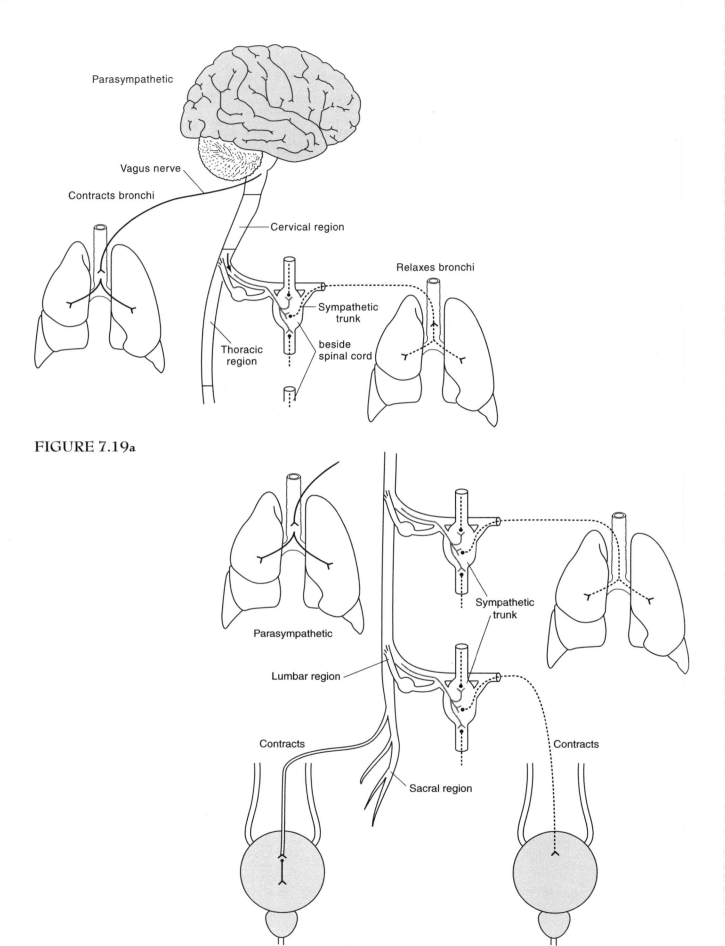

Parasympathetic

Vagus nerve

Contracts bronchi

Cervical region

Relaxes bronchi

Sympathetic trunk

beside spinal cord

Thoracic region

FIGURE 7.19a

Parasympathetic

Sympathetic trunk

Lumbar region

Sympathetic trunk

Contracts

Contracts

Sacral region

FIGURE 7.19b

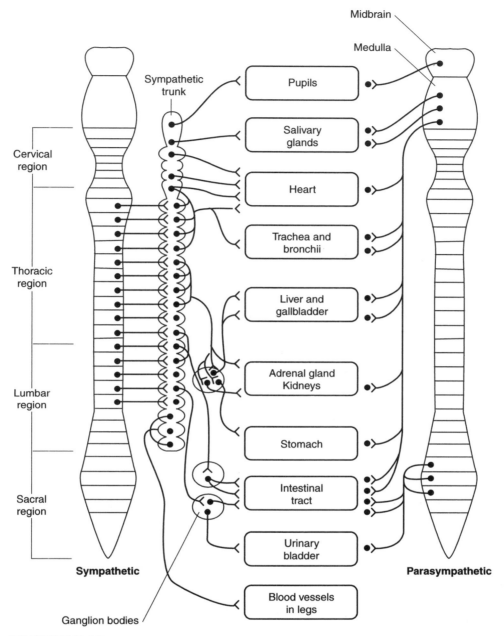

FIGURE 7.19c

the parasympathetic ganglia are located near the organ they control. The PD stimulates constriction of the urinary bladder and inhibits contraction of the muscle around the opening of the bladder. The SD stimulates relaxation of the bladder and contraction of the muscle surrounding the opening of the bladder.

Functional Distinctions

The two divisions of ANS differ in function in the following ways. First, the PD is concerned with local regulation of vital functions under normal conditions; the SD is more concerned with overall regulation of vital functions in emergency situations. Second, the PD normally is constantly active, but the magnitude of activity varies locally. The SD also is constantly active, but without much local variation. Third, the PD and SD are antagonistic to one another in many of the effectors they regulate. Fourth, the PD can be influenced by nerve centers in the brain stem and cerebrum but does not interact with them synergistically, so that the combined effect is not greater than the sum of each acting separately. In contrast, stimulation of higher nervous centers also stimulates the SD. Depression of these centers also depresses the SD. The combined action is synergistic, that is, it is greater than the SD actions plus the actions of the higher centers each acting separately.

The activity of many parts of the body is regulated by antagonistic signals from the parasympathetic and sympathetic divisions. The PD stimulates contraction and the SD stimulates dilation of the pupils of the eyes. The PD stimulates increased flow of bile from the gall bladder and increased GI tract mobility whereas the SD inhibits GI tract mobility. The PD stimulates decreased heart rate; the SD stimulates increased heart rate. The PD causes bronchioles to constrict; the SD causes bronchioles to dilate. The PD stimulates constriction of the urinary bladder and inhibits contraction of the muscle around the opening of the bladder. In contrast, the SD stimulates relaxation of the bladder and contraction of the muscle surrounding the opening to the bladder. In the arterioles in salivary glands, sexual erectile tissue, external genitalia, and the heart, PD stimulation produces vasodilation and SD stimulation produces vasoconstriction.

Visceral sense receptors bring the PD to dominance over the SD in localized areas in the body. The result is spinal cord or brain stem autonomic reflex activity which includes tear flow in crying, increased GI tract mobility, urination, defecation, erection of the penis in males and clitoris in females, and secretion of substances by the pancreases, the salivary glands, the gastric glands in the stomach, and the intestinal glands.

In response to stressors, the SD is stimulated to dominance over the PD. This stress response sometimes inhibits local autonomic reflexes controlled by the PD. For example, a person under stress may be constipated and/or unable to properly digest food because of inhibition of defecation reflex, poor GI tract mobility and reduced secretion of digestive juices. The dominance of the SD causes the pupils to dilate, the hairs to bristle, the bronchioles to dilate, and endocrine glands and sweat glands to secrete. Moreover, blood pressure rises due to increased heart rate and a net vasoconstriction of arterioles. Sympathetic stimulation accompanies emotional responses and is correlated with increased mental activity.

Human Endocrine System

The endocrine system is composed of a group of glands that initiate homeostatic adjustments by increasing or decreasing the rate at which they secrete particular hormones into the circulating blood. Hormones are chemical substances secreted by specific organs into the bloodstream which then carries them to other body sites to exert some kind of action. Glands whose primary function is to secrete hormones into the blood are called endocrine glands.

The concentration of a hormone in the blood affects the rate at which a specific process occurs in a target organ, the specific organ the hormone regulates. If the concentration level changes, the rate of the specific process changes. For example, testosterone, one of the male sex hormones secreted by the testes in men and to a lesser extent secreted by the ovaries in women increases the uptake of amino acid in muscle cells. As a result, this uptake increases the rate of synthesis of proteins such as muscle proteins. Thus, the greater testosterone level in males causes them to have greater muscle mass. A decrease in testosterone level in the blood brings about a decrease in muscle mass. A castrated man not only becomes more effeminate in appearance, his muscles become smaller. A woman with a higher than normal level of testosterone becomes masculine in appearance because of her increased muscle size.

By affecting the rate of specific processes, hormones affect the organism in three ways. First, hormones control the growth and development of certain tissues. For example, at puberty the female sex hormone, estrogen, which is secreted by the ovaries, regulates among other things a pattern of deposition of fat under the skin. The contour of a young girl's body becomes rounder because of these deposits of fat. This shapeliness is particularly pronounced in breasts, buttocks, hips, and thighs. By way of contrast, the relatively low concentration of estrogen in males results in the absence of such fat deposits. Second, hormones maintain some properties of the animal constant. For example, both estrogen and testosterone help regulate the formation of bone. This, in turn, helps maintain the concentration calcium and phosphate in the blood near constant values. Finally, hormones affect emotional patterns. For example, experiments with animals suggest that the androgen hormones (note that testosterone is one type of androgen) is associated with aggressiveness, while estrogen is associated with submissiveness. This pattern is particularly evident in animals showing periodic mating seasons.

The Major Endocrine Glands

The major endocrine glands of mammals are the (1) pituitary, (2) adrenal, (3) thyroid, (4) parathyroid, (5) pancreas, and (6) gonads, i.e., ovaries and testes. These glands are widely separated throughout the body, as can be seen from figure 7.20. Each gland controls processes related to day-to-day survival or reproduction. Nevertheless, these glands operate as a single system, the endocrine system, because of two factors. First, they are in communication with each other by means of the circulating blood. Hormones in the blood operate as messages analogous to nerve impulses. Secondly, the pituitary gland directly controls the adrenals, the thyroid gland, and the gonads, and indirectly controls the pancreas by means of the adrenals. Thus, coordination of the action of the various endocrine glands is done by the pituitary gland.

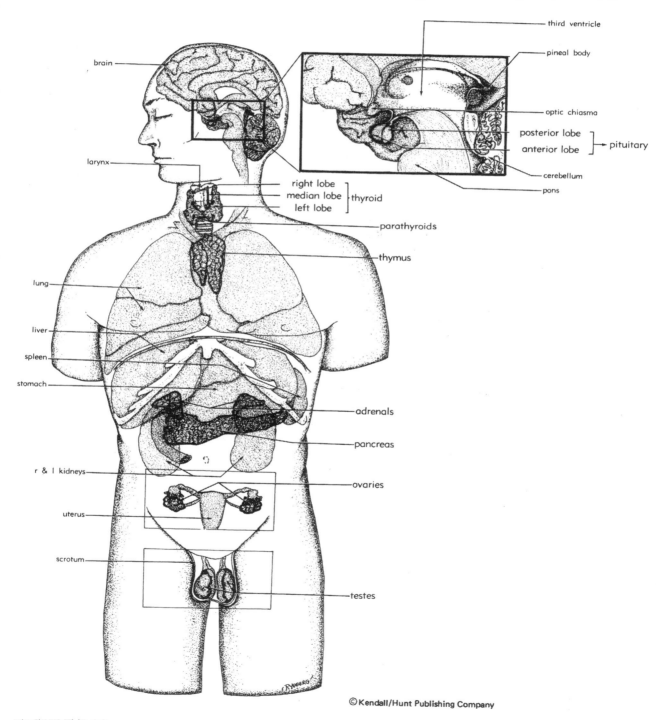

FIGURE 7.20a

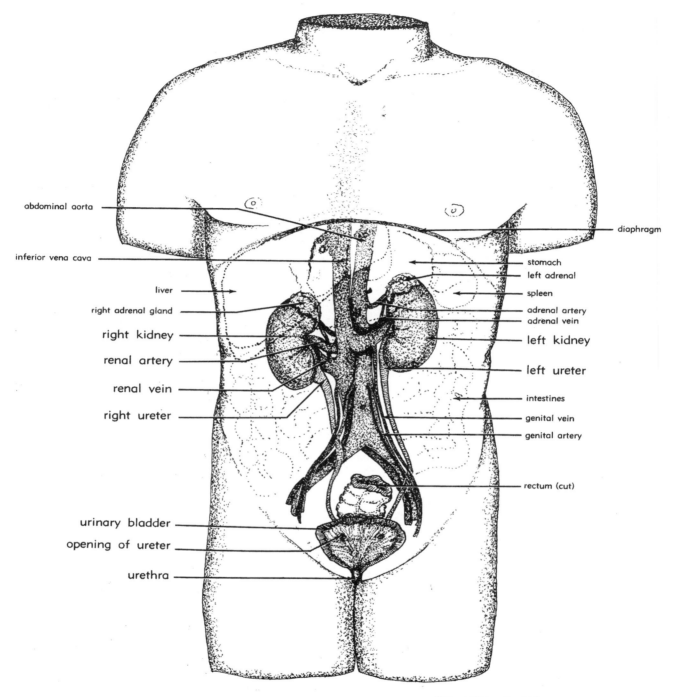

abdominal aorta

inferior vena cava

liver

right adrenal gland

right kidney

renal artery

renal vein

right ureter

urinary bladder

opening of ureter

urethra

diaphragm

stomach

left adrenal

spleen

adrenal artery
adrenal vein

left kidney

left ureter

intestines

genital vein

genital artery

rectum (cut)

FIGURE 7.20b

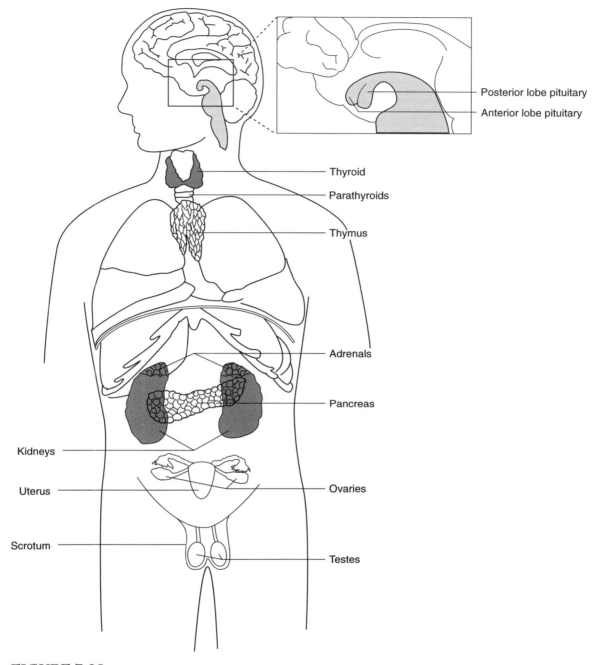

Posterior lobe pituitary

Anterior lobe pituitary

Thyroid

Parathyroids

Thymus

Adrenals

Pancreas

Kidneys

Uterus

Ovaries

Scrotum

Testes

FIGURE 7.20c

The Pituitary Gland

The pituitary gland is often referred to as the master gland because, in addition to secreting hormones which regulate specific body processes, it secretes a special group of hormones, the trophic hormones, which regulate the rate at which other endocrine glands synthesize and secrete their hormones.

The pituitary gland is located immediately below the hypothalamus and behind the nose. The gland consists of two parts: the anterior lobe and the posterior lobe, figure 7.20a, an extension of nervous tissue from the hypothalamus. Nerve cells in the hypothalamus secrete two hormones, antidiuretic hormone (ADH or vasopressin) and oxytocin, which pass down nerve fibers into the posterior lobe where they are stored until secreted into the bloodstream.

The pituitary gland secretes at least nine hormones. Two already mentioned, are secreted by the posterior lobe, and seven are secreted by the anterior lobe. Those secreted by the anterior lobe include the growth hormone (GH), melanocyte-stimulating hormone (MSH), prolactin, and the four trophic hormones: adrenocorticotrophic hormone (ACTH), thyroid-stimulating hormone (TSH), follicle-stimulating hormone (FSH), and luteinizing hormone (LH).

Hormones Secreted by the Anterior Lobe.   The growth hormone stimulates the growth of body tissues, especially bone. This is the most important hormone in the body for regulation of the synthesis of proteins, sugars, and fats. In addition to regulating the turnover of minerals and the body's need for vitamins, the growth hormone stimulates growth of practically all organs and tissues in the body. Its major action is to increase the length of bones during development. The effect of this hormone depends on the circumstances of various body tissues.

Hyperpituitarism is a condition in which higher than normal levels of GH are produced. If this occurs during childhood, giantism results. If hyperpituitarism occurs after childhood growth is complete, the result is a condition known as acromegaly. The long bones and other bones formed from pre-existing cartilage do not grow in length but may thicken. Bones formed by a different mechanism do grow resulting in changes of facial features and an increase in size of hands and feet. Hypopituitarism is a condition in which lower than normal levels of GH are produced. If this occurs during childhood, dwarfism results. If hypopituitarism occurs after normal growth is complete, the result is a condition called Simmonds disease which is characterized by premature ageing.

Prolactin stimulates the growth of the glands of the female breast and is involved with the production and secretion of milk. The melanocyte-stimulating hormone MSH affects the distribution of the pigment, melanin, in the skin of lower vertebrates, some frogs for example, allowing them to change color so as to blend in with the environment. In humans, MSH appears to increase the excitability of neurons in the central nervous system thereby affecting learning and memory processes.

The Trophic Hormones.   Adrenocorticotrophic hormone (ACTH) stimulates the adrenal cortex to secrete glucocorticoids which regulate carbohydrate metabolism. Above normal levels of ACTH also may stimulate the adrenal cortex to secrete mineral corticoids which regulate salt and water retention by the kidneys. The thyroid-stimulating hormone (TSH) stimulates the thyroid gland to produce and secrete thyroid hormones. The gonad trophic hormones consist of the follicle-stimulating hormone (FSH) and luteinizing hormone (LH). These hormones stimulate the gonads to secrete hormones. In females, these hormones regulate the ovaries. In males, FSH and LH cause the testes to secrete androgen and to produce sperm cells.

Hormones Secreted by the Posterior Lobe.   The antidiuretic hormone (ADH) secreted by the posterior lobe of the pituitary regulates water balance in the body. It does this by affecting the water permeability of the cell membranes in the collecting ducts in the kidney. Oxytocin stimulates muscle-like cells in the breasts to eject milk and it stimulates muscle tissue in the uterus to contract. This is a good example of a reflex involving both the nervous and endocrine systems. Sucking stimulates receptors in the nipples. These receptors send impulses to a portion of the brain that in turn stimulates the posterior lobe of the pituitary to release oxytocin into the general circulation which carries it to the breasts to bring about secretion of milk. The release of oxytocin can be inhibited by the central nervous system as a result of pain, fear, embarrassment or anxiety. Thus, in particular, stress can cause a new mother not to be able to nurse her baby.

Adrenal Glands

The adrenal glands are similar to the pituitary in that they consist of two parts of very different origin, cell composition, and function, see figure 7.20b. They are paired, yellow masses that lie at the upper ends of the kidneys. The inner part, the medulla, develops from primitive cells of the ANS. The outer portion, the cortex, develops like other glands from primitive epithelial cells. The adrenal cortex secretes many different hormones, including small amounts of androgen and estrogen. The two major chemical types of hormones it secretes are the glucocorticoids and the mineral corticoids. The major representative of the former is cortisol and that of the latter is aldosterone. ACTH from the pituitary stimulates the adrenals to synthesize and secrete cortisol. Cortisol influences the conversion of proteins into fats and sugars which are stored as glycogen, a large sugar molecule. Aldosterone, on the other hand, influences mineral metabolism. Its primary effect is to regulate sodium and potassium levels in the blood via an active transport system in the kidney (in the tubule cells). Aldosterone secretion is controlled by substances produced in the kidney. Stimulation of the ANS causes the adrenal medulla to secrete adrenalin. This hormone causes the blood pressure to rise by increasing the heart rate and strength of contraction of the heart and by constricting the arterioles almost everywhere except in the skeletal muscles where adrenalin causes vasodilation thus providing a greater blood supply in preparation for vigorous activity. It causes glycogen stored in the muscles and in the liver to break down into simple sugars, which are released into the bloodstream. It also causes the bronchioles to dilate; this is why adrenalin is given to people suffering from a severe asthmatic attack. Stimulation of the sympathetic division of the ANS and secretion of adrenalin have similar effects on the body.

Thyroid Gland

The *thyroid gland* consists of two lobes on either side of the trachea connected by a thin isthmus over the front of the trachea, see figure 7.20c. This gland secretes at least three hormones, of which the most important is thyroxin. In order to form thyroxin, a certain amount of iodine is required in the diet. Lack of iodine can lead to a goiter, which is an enlarged thyroid gland. Thyroxin does not have any one particular target organ; it acts on almost all tissues throughout the body. Thyroxin stimulates increased breakdown of substances, particularly of glucose, into CO_2 and water. As a result, more thermal energy is produced. Thyroxin also increases absorption of glucose from the intestine and enhances the conversion of protein into glucose. Consequently, the blood sugar level increases affecting the activity of other endocrine glands such as the adrenal cortex and the pancreas. Finally, thyroxin increases nitrogen retention and protein synthesis and influences fat metabolism.

A number of diseases arise from more or less than adequate amounts of thyroxin. Thyroxin helps to bring about normal growth and development. Hypothyroidism, i.e., greatly decreased levels of thyroxin, produces cretinism in children, a condition which is characterized by dwarfism and mental deficiency. Hypothyroidism in adults is associated with muscle weakness, intolerance to cold, and mental deterioration ranging from lethargy and forgetfulness to psychosis. Hyperthyroidism, i.e., greatly increased levels of thyroxin, in adults is associated with nervousness and other emotional disturbances. Usually an adult afflicted with this is unable to tolerate heat, loses a great deal of weight, and shows an increased rate of breathing and of heartbeat.

Parathyroid Gland

In humans the parathyroid gland lies on the back surface of the thyroid. However, accessory parathyroid tissue is common and may be scattered in the neck region. The gland secretes parathormone, which, along with calcitonin, another hormone secreted by special cells in the thyroid gland, regulates the blood level of calcium and phosphate ions. These two hormones have opposing effects. Parathormone increases the level of calcium in the blood in four ways: (1) it stimulates bone tissue to break down (as a result, calcium phosphate deposited in the matrix of bone is released to the bloodstream); (2) it stimulates elimination of phosphate in the kidney (as a result, more calcium will be retained in the blood); (3) it increases calcium absorption from the intestinal tract; and (4) it makes bone cells more permeable to the calcium in the matrix around the cells. The increased calcium inside bone cells stimulates active transport of calcium into the blood. Calcitonin, on the other hand, inhibits bone tissue breakdown. Thus, more calcium phosphate will be deposited into the bone matrix with a corresponding decrease in calcium level in the blood.

The secretion of both parathormone and calcitonin is determined by the calcium level in the blood. A certain calcium level is necessary for the proper functioning of the body cells, particularly nerve and muscle cells. Removal of all parathyroid tissue will very quickly lead to calcium imbalance and death.

Pancreas

The pancreas, in addition to secreting digestive juices into the duodenum, also functions as an endocrine gland. Scattered throughout the gland are patches of cells called islets of Langerhans which secrete the hormones glucagon and insulin into the blood. Insulin first enters the portal vein which runs from the pancreas to the liver and stimulates the liver cells to convert glucose into glycogen, the storage form of glucose. Insulin also has a potent effect on cells throughout the body. It stimulates muscles to convert glucose into glycogen, which then is stored in these cells. Insulin enables most cells to actively take up glucose which then is used as an energy source by the cells. Thus, the net effect of insulin is to decrease the concentration of glucose in the blood. Insulin also promotes the synthesis of certain types of fats and inhibits the breakdown of fat tissue. Finally, it stimulates the transport of amino acids (subunits making up proteins) into cells and then stimulates the synthesis of proteins from these amino acids.

Glucagon also is secreted into the portal vein and is first taken to the liver where it stimulates liver cells to convert glycogen into glucose. Secondarily, it causes amino acids to be converted into glucose. It has a number of other actions, but the physiological significance of these is still under debate.

The most common endocrine disorder, hyposecretion of insulin, results in a number of clinical symptoms referred to as diabetes mellitus. In this disease, there is a decreased secretion of insulin particularly when there is an increased intake of sugar. As a result, the blood sugar level rises, and not enough insulin is secreted to counteract this rise which is called hyperglycemia. Prolonged hyperglycemia while producing tremendous thirst results in dehydration and consequently rapid weight loss. While the liver secretes more glucose into the bloodstream, many cells of the body are less able to utilize glucose as a result of the decreased level of insulin. In effect, these cells starve in the midst of plenty. One result is a tremendous increase in appetite. The basic defect is the inability of the pancreas to respond to the stress of increased blood sugar level. The underlying mechanisms for this are still unknown.

Gonads

Ovaries in females secrete estrogen and progesterone in a cyclic manner which results in the menstrual cycle. Testes in males secrete androgen at a more or less constant rate; at least the cyclic variation of androgen is not nearly as dramatic as that of the female sex hormones. After puberty, a maintained level of estrogen in females and androgen in males sustains the female and male sex characteristics.

Human Neuroendocrine System

The ANS and endocrine systems cooperate to regulate many properties of the circulating blood. The circulating blood, in turn, tends to equilibrate with the internal milieu. Thus, regulation of properties of the blood indirectly regulates properties of the internal milieu. For example, glucose in the blood tends to diffuse into the tissue fluid which is the internal milieu. Therefore, maintaining a particular level of glucose in the blood indirectly maintains the same level of glucose in the tissue fluid.

The interactions of the nervous and endocrine systems are so extensive that these two systems are often regarded as a single system called the neuroendocrine system. The two major places where these two systems interact are (1) the ANS and adrenal medulla and (2) the hypothalamus and the pituitary gland. As described in the previous section, stimulation of the sympathetic division of the ANS results in generalized, physiological arousal. This generalized arousal is enhanced by motor neurons stimulating the adrenal medulla to secrete adrenalin. Adrenalin brings about many of the same responses brought about by sympathetic motor neurons such as increased blood pressure, dilation of bronchioles, and increased depth and rate of breathing. At the other site of interaction between the nervous and the endocrine systems, the hypothalamus regulates secretion of hormones by the pituitary gland. This in turn, regulates hormone secretion by many other endocrine glands.

Hypothalamus-Pituitary Axis

The relationship between the hypothalamus and the pituitary gland is called the hypothalamus-pituitary axis. This axis is the major site of interaction between the nervous and endocrine systems.

Anatomical Connections.   There are two types of connections between the hypothalamus and the pituitary glands corresponding to the two lobes of the pituitary gland. Type 1 connection consists of a neural interaction between the hypothalamus and the posterior lobe of the pituitary gland. Neuron cell bodies in the hypothalamus synthesize the hormones ADH and oxytocin. Axons of these cell bodies carry these hormones to the posterior lobe, where nerve endings of these axons store hormones, see figure 7.21. Type 2 connection is a vascular connection, called a *portal system,* between the hypothalamus and the anterior lobe of the pituitary gland. This system is similar to the liver portal system. Blood in the capillary beds in the hypothalamus—instead of passing into veins—flows into portal vessels which lead to secondary capillary beds in the anterior lobe of the pituitary gland, see figure 7.21.

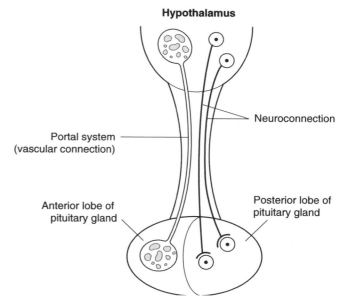

FIGURE 7.21

1. Nerve cells in the hypothalamus stimulate cells in the posterior lobe of the pituitary gland to secrete hormones.
2. Nerve cells in the hypothalamus secrete RH (or RF) into the portal system. The RH or RF stimulate cells in the anterior lobe to secrete hormones.

Functional Interactions.   The functional interaction associated with the type 1 connection simply involves cell bodies in the hypothalamus stimulating the release of hormones stored in the posterior lobe of the pituitary gland. The functional interaction associated with the type 2 connection involves the hypothalamus secreting releasing hormones (RH) and release-inhibiting hormones (IH). Cells in the hypothalamus secrete RHs or IHs into the hypothalamus-pituitary portal system which carries these hormones to the anterior lobe. RHs stimulate particular pituitary cells to secrete specific hormones, and IHs inhibit the secretion of specific hormones by particular pituitary cells.

Releasing Hormones (RH).   There are five releasing hormones: (1) corticotrophic-releasing factor (CRF) which stimulates production and release of ACTH; (2) thyroid-releasing hormone (TRH) which stimulates the production and release of TSH; (3) gonadotropin-releasing hormone (GnRH) which stimulates the production and release of FSH and LH; (4) growth hormone-releasing hormone (GHRH) which stimulates the production and release of GH and (5) melanocyte-stimulating hormone releasing factor (MRF) which stimulates the production and release of MSH. There also may be a sixth releasing hormone involved with stimulation of production and release of prolactin.

Inhibiting Hormones (IH).   There are three inhibiting hormones: (1) somatostatin (growth hormone release-inhibiting hormone, GIH) which inhibits the release of the growth hormone; (2) melanocyte-stimulating hormone release-inhibiting factor (MIF) which inhibits the release of MSH; (3) prolactin release-inhibiting factor (PIF) which inhibits the release of prolactin.

Negative Feedback Regulation of Endocrine Glands

Regulation of Growth Hormone.   The hypothalamus senses and evaluates the blood level of growth hormone. When this level is below a preset value, the hypothalamus secretes GHRH which stimulates the pituitary gland to increase secretion of growth hormone. When the growth hormone

Negative feedback regulation of cortisol

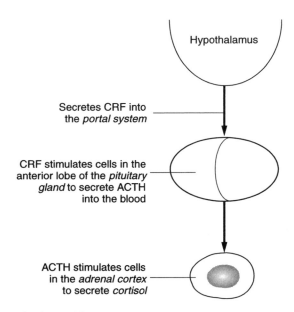

Hypothalamus

Secretes CRF into
the *portal system*

CRF stimulates cells in the
anterior lobe of the *pituitary
gland* to secrete ACTH
into the blood

ACTH stimulates cells
in the *adrenal cortex*
to secrete *cortisol*

FIGURE 7.22

level is above a preset value, the hypothalamus secretes GIH which inhibits and therefore reduces secretion of the growth hormone by the pituitary gland. Thus, the interplay of GHRH and GIH causes the blood level of the growth hormone to fluctuate near a preset value.

Regulation of Melanocyte-Stimulating Hormone (MSH). The regulation of MSH is similar to that of the growth hormone. The hypothalamus senses and evaluates the blood level of MSH. When this level is below a preset value, the hypothalamus secretes MRF which stimulates the pituitary to increase secretion of MSH. When the MSH is above a preset value, the hypothalamus secretes MIF which inhibits and therefore reduces secretion of MSH. Thus, the interplay of MRF and MIF causes the blood level of MSH to fluctuate near a preset value.

Regulation of Prolactin. The hypothalamus exerts a continuous inhibition of secretion of prolactin by the pituitary gland. During pregnancy and the time when mothers breast-feed their babies, this inhibition is removed. The details of prolactin regulation at these times is rather complex.

Regulation of Glucocorticoids, such as Cortisol, Secreted by the Adrenal Cortex. The pituitary gland is the intermediary for the nervous regulation of the adrenal cortex. This is the major mechanism though not the only mechanism for the regulation of glucocortcoids (cortisol). The hypothalamus senses and evaluates the blood level of cortisol. When cortisol is below a preset value, the hypothalamus secretes CRF which stimulates the pituitary gland to secrete ACTH. The resulting increased blood level of ACTH causes the adrenal cortex to increase secretion of cortisol. When cortisol is above a preset value, the cortisol inhibits and therefore reduces the secretion of CRF by the hypothalamus. Reduced secretion of CRF causes a reduced secretion of ACTH by the pituitary gland, and reduced secretion of ACTH causes a reduced secretion of cortisol by the adrenal cortex. Thus, the interplay of CRF, ACTH and cortisol causes the blood level of cortisol to fluctuate near a preset value, see figure 7.22.

1. If cortisol is above the preset value in the evaluation unit in the hypothalamus, then the hypothalamus decreases secretion of CRF; this leads to decreased secretion of cortisol.
2. If cortisol is below the preset value in the evaluation unit in the hypothalamus, then the hypothalamus increases secretion of CRF; this leads to increased secretion of cortisol.

Negative feedback regulation of thyroxin

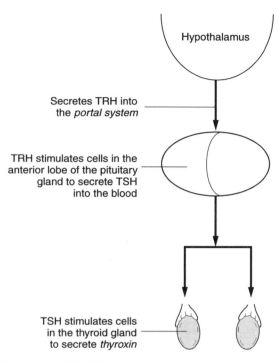

Secretes TRH into
the *portal system*

TRH stimulates cells in the
anterior lobe of the pituitary
gland to secrete TSH
into the blood

TSH stimulates cells
in the thyroid gland
to secrete *thyroxin*

FIGURE 7.23

Regulation of Thyroid Hormones Secreted by the Thyroid Gland.   A major mechanism for regulating thyroid hormones is similar to the regulation of glucocorticoids (e.g., cortisol). The hypothalamus senses and evaluates the blood level of the thyroid hormone (e.g., thyroxin). When thyroxin is below a preset value, the hypothalamus secretes TRH which stimulates the pituitary gland to secrete TSH. The resulting increased blood level of TSH causes the thyroid gland to increase secretion of thyroxin. When thyroxin is above a preset value, the thyroxin inhibits and therefore reduces the secretion of TRH by the hypothalamus. Reduced secretion of TRH causes a reduced secretion of thyroxin by the thyroid gland. Thus, the interplay of TRH, TSH, and thyroxin causes the blood level of thyroxin to fluctuate near a preset value, see figure 7.23.

1. If the thyroxin is above the preset value in the evaluation unit in the hypothalamus, then the hypothalamus decreases secretion of TRH; this leads to a decreased secretion of thyroxin.
2. If the thyroxin is below the preset value in the evaluation unit in the hypothalamus, then the hypothalamus increases the secretion of TRH; this leads to an increased secretion of thyroxin.

Tearout #1

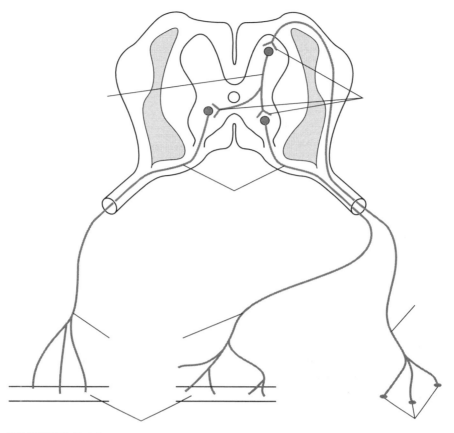

FIGURE 7.18a

Directions:

1. Using figure 7.18a as a reference, fill in all labels.

2. Using figure 7.21 as a reference, fill in all labels and write out the two statements that describe the interactions between the hypothalamus and the pituitary gland.

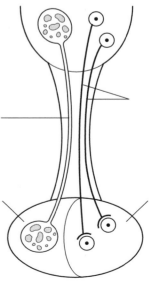

FIGURE 7.21

Tearout #2

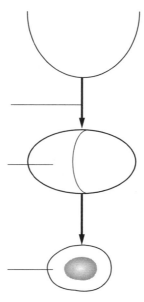

FIGURE 7.22

Directions:

1. Using figure 7.22 as a reference, fill in all labels and descriptions of the negative feedback process.

Neuroendocrine-Immune System

Overview of the Immune System

The immune system is one component of a network of three subsystems that collaborate to define and maintain an adapted homeostatic self. The endocrine system has been modified to be subordinated to and controlled by the autonomic nervous system. The resulting two-level system, called the neuroendocrine system, functions to define a functional homeostatic self of an individual animal. This "self" is the unity of the individual that separates it from other "selfs" and from the nonliving environment. All living, individual "selfs" consist of one or more cells, where each cell, in turn, is an individual, functional, homeostatic self. The neuroendocrine system produces a higher order homeostatic self for an individual animal by coordinating a set of negative feedback reflexes that maintain resting levels of steady state of the circulatory system, various organ systems, and the system of skeletal (voluntary) muscles. The pre-set value, see Chapter 4, that is, the steady state value, of many of these feedback mechanisms may be adjusted to overcome the stress of various situations. For example, the animal preparing to attack another animal needs more oxygen and glucose in the skeletal muscles to accomplish this task. Correspondingly, the steady state value of blood pressure, muscle tone, and a host of other variables are elevated to meet the demands of this stress situation. This physiological arousal in response to a stress situation defines a dynamic, functional, homeostatic self attempting to adapt to all situations ranging from resting state to various levels of stress.

Cell division in the immune system continually produces billions of cells that collaborate to maintain a molecular, cognitive homeostatic self. Protein molecules embedded in membranes of immune cells have a three-dimensional surface pattern that is complementary to a three dimensional surface pattern of some other molecule. These complementary patterns are analogous to the complementary, base pairing of nucleotides that hold together a single DNA molecule that consists of two strands of nucleotides. The molecular pattern in the membrane of an immune cell is like a lock and the complementary pattern of some other molecule is like a key. The key fitting into the appropriate lock is the first stage of molecular communication. If the key-lock interaction leads to an observable event, then the interaction is a molecular communication. For an immune cell the membrane "lock" interacting with a "key" molecule produces the observable event of eliminating in some manner the key molecule from the whole animal. The "key" molecule is thus a nonself entity. That is, if the elimination of an entity is initiated by a key-lock molecular communication, then: (1) whatever is eliminated is a nonself entity, (2) the "lock" molecule in an immune cell membrane is a receptor that is said to recognize the nonself entity, and (3) the key-lock communication is a stimulus communication that leads to an immune response. The nonself entity may be an individual molecule or a virus or bacteria or parasite that displays a molecular pattern complementary to a receptor on an immune cell. In effect, billions of immune cells collaborate to recognize and then remove nonself entities. In a roundabout way, this community of immune cells defines a "molecular, cognitive homeostatic self" as follows: all those cells or cell products not recognized by any immune cell are components of the homeostatic self.

Though the neuroendocrine and immune subsystems have different ways of defining homeostatic self, they express similar hierarchal functions. The parasympathetic division of the ANS consists of a network of homeostatic reflexes that respond to stress to local areas of the body. The sympathetic division of the ANS collaborates with the parasympathetic division to produce physiological arousal that adapts to stress to the animal as a whole. Via the primitive limbic system these sympathetic adaptive mechanisms collaborate with particular instincts to produce a more specific adaptation to stress to the whole animal. For example, the physiologically aroused animal will instinctively attack a "perceived" prey or "perceived" enemy, that is, an external nonself entity instinctively perceived as prey or as an enemy. In like manner, the immune system has a two-level hierarchal response to recognizing and removing nonself entities. The first level, which is called the innate immune response, consists of cells in tissues throughout the body that recognize general patterns on nonself entities. This first level immune response by individual cells is analogous to the response of the parasympathetic division of the ANS to local stress. Sometimes some of these immune cells secrete chemicals that initiate molecular communications that collaborate to produce what might be called an immune instinctive response. If the nonself entity is not quickly removed from the local tissue in which it resides, then some immune cell in this tissue initiates an adaptive immune response, which takes about 3 weeks to fully develop. In this more elaborate response the nonself entity is recognized in steps that are progressively more specific to the relatively unique pattern or several different patterns on the nonself entity. Some of the immune cells producing this more specific response communicate with molecules and receptors on cells involved with the innate immune response. Thus, the adaptive immune response, which is analogous to the response of the sympathetic division of the ANS, is a higher level response that includes aspects of the lower level innate immune response which is analogous to the parasympathetic division of the ANS.

The "instinctive, innate, immune response" called inflammation, incorporates functions of the neuroendocrine system. This interface between the neuroendocrine and immune systems is one basis for considering these two functionally distinct subsystems as components of one overarching homeostatic self system. Inflammation provides the interface between the dynamic, functional, homeostatic self with the higher level emotional self that associates physiological arousal with instinctive behavior. That is, inflammation may involve local, parasympathetic and general sympathetic responses that when coupled with instinctive, nonconscious behaviors produce an emotional behavior. For example, inflammation in the knee or lower back leads to the physiologically aroused instinctive response of sitting, standing, and walking in such a way to reduce physical stress on the inflamed area. Inflammation involves physical and chemical stimulation of sense receptors generating nerve impulses that are interpreted as pain. Remembered instances of pleasure/pain are aspects of animal consciousness called feelings; that is, feelings are conscious emotions. Inflammation provides an interface for incorporating the emotional self into the higher-level feeling self. By means of inflammation and some other molecular key-lock type communications that also produce pain, the higher level feeling self that includes the emotional self interacts with the neuroendocrine-immune system thus expressing a multilevel integral system.

The rhythm of the neuroendocrine system is like a musical theme. The theme maintained by the parasympathetic division of the ANS can be brought to relative chaos by stressful situations. The sympathetic division of the ANS brings the body back to the old or perhaps to a new individualized homeostatic theme. The rhythm of the immune system integrates with that of the neuroendocrine system to produce the more complex musical theme of the neuroendocrine-immune system. When one or the other components of this system becomes discordant, then maintaining the homeostatic self becomes less effective—the animal becomes less able to maintain its health. The discordant music of the neuroendocrine-immune system disturbs the harmony of the higher level feeling self. Likewise, disharmony of the feeling self will produce disharmony in the neuroendocrine-immune system resulting in diminished health.

Cellular Components of the Immune System

There are 9 types of immune cells all of which originate in the bone marrow. Six types develop from what are called *myeloid progenitor cells* and three types develop from what are called *lymphoid progenitor cells.* Myeloid progenitor cells generate (1) monocytes that in tissues differentiate into macrophages, (2) dendritic cells, (3) mast cells, and granulocytes that differentiate into (4) neutrophils, (5) eosinophils, and (6) basophils. Lymphoid progenitor cells generate (7) B-lymphocytes = B-cells, (8) T-lymphocytes = T-cells, and (9) natural killer cells, see figure 5.5.

Innate Immune Response

The neuroendocrine system maintaining the dynamic, functional, homeostatic self is what keeps the human body from beginning to die, but it also establishes the circulating blood and the tissue fluid throughout the body as ideal environments for small organisms to survive and multiply into large "foreign communities." Those potential foreign communities or aggregates of molecules that can bring local or systemic damage to the body are called pathogens. Pathogens include bacteria, parasites, viruses, and toxic substances. The skin and surface cells lining the tubes of the digestive, urinary, reproductive, and respiratory systems provide a barrier and associated mechanisms for preventing pathogens getting into the tissue fluids of the body. Likewise the integrity and thickness of the walls of the blood vessels and lymph vessels larger than capillaries prevent foreign pathogens from invading the blood and lymph fluid. The body is continuously exposed to billions of diverse pathogens just waiting for the opportunity to invade tissue fluid, lymph, or blood. This drive toward invasion is analogous to the "drive" of any system toward equilibrium. The neuroendocrine system reestablishes a desired steady-state in opposition to the body going toward some equilibrium, e.g., body temperature dropping to that of the environment. When there is a breach in the barriers to pathogens, these pathogens rush in and begin to damage the body. The immune system reestablishes "cognitive" homeostasis by removing these recognized "nonself" pathogens.

Simple Immune Phagocytic Reflexes

Individual immune cells express simple, immunological reflexes. Receptors embedded in the membrane of an immune cell have patterns that are complementary to general patterns expressed by various pathogens. The pathogen pattern is a stimulus to the receptor which is analogous to a sense receptor. The stimulus-receptor communication generates a biochemical change that sets off a sequence of biochemical events that lead to a response called phagocytosis. In this process the section of the cell membrane containing one or more (usually several) receptor-pathogen complexes indents, that is, moves inward, as neighboring membrane segments move outward and toward one another. The result of these membrane movements is that the pathogen is surrounded by membrane thus forming a membrane vesicle containing the pathogen inside, see figure 5.23. The pathogen containing vesicle moves inside the immune cell and then fuses with one or more internal vesicles called lysosomes which contain enzymes and other proteins that can degrade the pathogen, see figure 5.25. Thus, phagocytosis removes the pathogen from the site of its invasion into the tissue fluid, blood, or lymph of the body. A particular immune cell receptor is like a sense receptor that only communicates with a particular type of stimulus, such as, receptors in the eye only respond to light waves; receptors in the ear only respond to air waves, and so on. Since the simple patterns expressed by pathogens are never expressed by cells making up the body, these simple phagocytic reflexes are immunological. They indirectly acknowledge "self" by never being activated by self cells but are activated by nonself entities called pathogens.

Monocytes in the circulating blood continually migrate to spaces containing tissue fluid throughout the body and differentiate into macrophages. When a pathogen invades any site in the body and sets up an infection there, macrophages in the tissue are the first to recognize and begin to remove it by phagocytosis. Some receptors on the macrophage that recognize the pathogen also initiate what might be called an immunological instinct called inflammation. Analogous to a behavioral instinct, inflammation coordinates several different immunological reflexes and processes to produce an overall stereotypic, nonspecific immunological behavior. One aspect of inflammation is that it produces greater flow of blood to the site of infection thereby bringing neutrophils in the circulating blood to this site. Like macrophages, neutrophils phagocitize pathogens, but unlike macrophages they die after phagocitizing pathogens and thereby become a major component of the pus that forms in some sites of infections.

Inflammation

Inflammation is the instinctive, immunological behavior of vasodilation at the site of infection coupled with several other immunological reflexes and processes. The macrophages express several different nonphagocytic immune reflexes. Receptor recognition of pathogen leads to a sequence of biochemical reactions called intracellular signals that are analogous to nerve impulses in reflex arcs. These intracellular signals lead to the immunological, reflex response of producing a chemical that contributes to the process of inflammation. Each immunological reflex is specified by the chemical that it releases. Initially, signals in macrophages lead to enzymatic degradation of phospholipids embedded in their outer cell membranes to produce prostaglandins, leukotrienes and platelet-activating factor (PAF). Prostaglandins, which also are released by cells at sites of tissue damage, stimulate smooth muscles in arterioles to decrease their degree of contraction, thus bringing about vasodilation. This brings an increased quantity of blood flow to the area plus a decrease in velocity of blood flow. This is analogous to increasing the diameter of the nozzle on a hose which simultaneously results in a greater quantity of water flow out of the hose and a decrease in the velocity of water coming out of the hose. This increase of blood flow brings neutrophils to the site of infection. The platelet-activating factor (PAF) initiates a localized blood clotting process that seals punctured blood vessels and provides an internal barrier for pathogens to travel to neighboring areas of the site of infection.

Other intracellular signals in macrophages activate portions of DNA to direct the synthesis of several chemicals referred to as cytokines. Cytokines stimulate various cellular reflexes. The cells lining blood vessels are activated to express adhesion molecules that briefly combine to receptors on circulating leukocytes, which are white blood cells of which the major type is neutrophils. This results in these cells rolling along the surfaces of the local blood vessels. Then the leukocytes bind more firmly to other receptors on membranes of the lining cells of the blood vessels. This provides the occasions for these leukocytes to migrate into the site of infection. Other cytokine-driven reflexes facilitate this migration. Other cytokines activate the loosening of cell-to-cell binding of cells making up the lining of blood capillaries. This leads to fluid and proteins in the blood moving into and accumulating in the tissue. The accumulation of this fluid is called *edema*, see Chapter 7. Prostaglandins and the pressure of edema stimulate sense receptors in the inflamed tissues to generate increased nerve impulses that are interpreted as pain.

Tissue damage not coupled with infection also leads to the release of chemicals that stimulate cell reflexes that produce the coordinated responses that make up the inflammatory process. Inflammation promotes the repair of tissue damaged by infection or by nonpathogenic agents. This repair process in not an exclusive immunological response. Thus, inflammation overlaps the homeostatic activities of the neuroendocrine and the immune systems. The complement system dissolved in the fluid portion of blood, the plasma, also is caused to breakdown into a chemical (a peptide) called C5a that, in turn, stimulates several responses contributing to inflammation. C5a increases vascular permeability, induces the expression of some adhesion molecules, is a chemoattractant for neutrophils and monocytes, and activates phagocytes and mast cells located in tissues. Mast cells release histamine that locally brings about vasodilation. If large amounts of histamine are released, this chemical is transmitted throughout the body producing various degrees of systematic inflammation depending on the excess of histamine in the circulating blood.

The Complement System

The complement system consists of many different protein enzymes dissolved in the blood plasma in an inactive form called zymogen. These zymogens also are widely distributed throughout body fluids. The complement system is activated by a cascade of enzymatic reactions. An initiating stimulus breaks down a zymogen on the surface of a pathogen into a large and a small fragment. The large fragment remains bound to the surface of the pathogen while the small fragment becomes dissolved in the fluid near the pathogen. The large fragment then brings about the breakdown of the second zymogen in a sequence of zymogen cleavages. Each time a large fragment stays bound to the pathogen surface while a small fragment dissolves in the surrounding fluid. The sequence of several zymogen cleavages leads to the production of a molecule called C_3 convertase bound to the pathogen surface. This convertase activates several C_3 zymogen molecules in fluids around the pathogen to break down into a large fragment and a small fragment. Phagocytes in the infected tissue have membrane receptors that recognize this large fragment of C_3 bonded to the pathogen. As a result of this recognition, the phagocyte engulfs and thus removes the pathogen. The small fragment of C_3 as well as other released small fragments bring about aspects of inflammation including migration of other phagocytes to the site of infection. The large C_3 fragment also combines with C_3 convertase to produce C_5 convertase that, in turn, breaks down into a large and a small fragment. This large fragment initiates a sequence of reactions that terminate in the formation of a membrane-attack complex that creates a pore in the cell membrane of some pathogens, for example, bacteria. The pore allows water to rush into the pathogenic cell thus killing it.

The complement system is initiated by three different events that occur on the surface of a pathogen. Each different initiating event leads to three different pathways: the classical pathway, the MB-lectin pathway, and the alternative pathway. These three pathways converge in producing C_3 convertase. Each of these pathways exhibits positive feedback as described in Chapter 4. Each activated zymogen molecule catalyzes the breakdown of several other zymogen molecules representing the next step in the cascade of zymogen breakdowns. Thus, the number of zymogen molecules broken down in any step of the complement cascade leads to a much greater number of zymogen breakdowns in the next step. As a result of this positive feedback mechanism, pathogen activation of the complement system very quickly produces important aspects of an innate immune response. Each type of soluble molecule that initiates respectively each of the three pathways of the complement system interacts with a portion of a protein on the surface of a pathogen. This activating molecule is analogous to a lock and the portion of the pathogen protein is analogous to a key. Thus, this lock-key interaction is an immunological communication where the soluble initiating molecule is a soluble, i.e., nonmembrane-bound, receptor molecule that recognizes a pattern on the surface of the pathogen. The recognition process is operationally defined by the fact that the lock-key communication leads to the various innate immune responses carried out by the complement system.

Acute Phase of Innate Immune Response

Certain cytokines produced by macrophages (e.g., TNF-α, IL-1 and IL-6) initiate immune reflexes carried out by organ systems throughout the body. The liver decreases the production of some proteins and increases the production of others. In particular, liver cells are induced by signals from the cytokine-receptor communication in these cells to synthesize and secrete proteins that activate two of the pathways of the complement system and activate fibrinogen which is necessary for blood clotting. Likewise cytokine-receptor interactions in bone marrow cells (endothelium) produce more neutrophils. In cells of the hypothesis cytokine-receptor communication leads to increased body temperature thus producing fever. This slows the growth of most pathogens while increasing the intensity of an adaptive immune response. Cytokine-receptor communication in muscle and fat cells mobilizes protein and energy thereby contributing to elevating the body temperature. Cytokine-receptor communication in dendritic cells (see later section) leads to these cells migrating to lymph nodes where they mature and initiate an adaptive immune response. Many different proteins synthesized during this phase also promote macrophages, neutrophils, and monocytes to phagocitize pathogens. Thus, within a day or two after infection, the several different proteins produced during this phase collaborate to remove the pathogen and when necessary to initiate the adaptive immune response.

Interferons and Natural Killer Cells

The virus-receptor communication on a cell not only leads to a portion of the virus entering the cell, it also triggers the production of proteins called interferons. The name of these proteins represents their function which is to interfere with viral replication within cells. Interferons indirectly activate natural killer cells (NK cells) to kill virus-infected cells. NK cells have activating receptors that bind to stimulatory patterns present on most cells. This cell-to-cell communication leads to two reflex immune related responses. The NK cells release granules on the surface of the target cell which, in turn, leads to effector proteins in NK cells to penetrate the cell membrane and induce a program in which the target cell "commits suicide." NK cells also have "inhibitory receptors" which bind to stimulatory sites on MHC-I proteins (see next section). This cell-to-cell communication inhibits the killer action of the NK cells. Interferon provides a 20-fold intensification of the activating communication with NK cells. At the same time sometimes virus-infected cells have less MHC-I on the cell surface due to the action of interferons, and therefore, these cells produce less inhibitory messages when interacting with NK cells. The activating messages drown out the less number of inhibitory messages resulting in these killer cells "doing their thing—kill cells."

Individual Animal Unique Cellular Self

Cells transplanted from one human into another stimulate a massive, adaptive immune response by the human's immune system that received the transplant. This is true even in genetically closely related humans, for example, mother and son. By contrast cells transplanted from one genetically identical twin to another do not elicit an immunological rejection. Thus, an individual's unique genetic makeup imparts to most cells of her body a unique "immunological self" which I am referring to as a "unique, cognitive, homeostatic self." This uniqueness is due to most cells synthesizing two types of proteins (class of proteins). As will be described next, these proteins are presenting molecules type I and type II. Immunologists designate them as MHC class I and MHC class II molecules which I hereafter will represent as MHC-I and MHC-II. These MHC molecules when embedded in a cell membrane have a cleft running down the middle of the molecule on the surface facing outward, away from the cell, see figure 8.1. There are mechanisms inside most cells that degrade proteins to linear strands of amino acids. The chemical bond between any two amino acids is called a peptide bond; therefore, the whole strand or any segment of a protein strand is called a peptide. Cellular mechanisms bring peptide segments into the cleft of MHC molecules inside the cell. Via a key-lock interaction at one or both ends of the cleft a peptide segment binds to the MHC molecule. This binding stabilizes the MHC molecule, which then is transported to and incorporated into the outer cell membrane. The incorporated MHC-peptide complex now displays the peptide segment to all cells passing by. In effect, these MHC molecules present peptide segments to the outside environment of a cell.

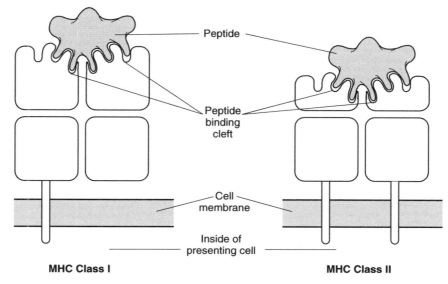

FIGURE 8.1

Several types of pathogens, for example, bacteria and single cell parasites are "eaten" by phagocytes. As a result, these pathogens become located inside membrane vesicles inside the cells that ate them, see figure 5.23 illustrating endocytosis. Sometimes the pathogen can survive and reproduce within these vesicles. Macrophages activated during inflammation are able to degrade these pathogens, cut them into peptide segments and then incorporate each segment into the cleft of a MHC-II molecule which then is sent to the cell surface. In a noninfected cell MHC molecules also bind and display peptide fragments from "self proteins." Because these fragments are from "self proteins," they will not be recognized by lymphocytes to initiate an adaptive immune response. The cell is in constant flux; so after awhile the MHC-peptide complexes are taken back into the cell, degraded, and new MHC-II molecules are formed. The number of MHC-II molecules varies from one type of cell to another. Macrophages make a large number of these presenting molecules so that when the macrophage does ingest a pathogen, there are many empty MHC-II molecules available to present peptide segments of the pathogen.

After invading a cell, a virus moves directly into the endoplasmic reticulum—the internal membrane system that compartmentalizes a cell, see Chapter 5. While in this membrane system, the virus cannot be detected by immune cells in the tissues. Moreover, unlike pathogens taken into the cytoplasm of the cell, the virus cannot be degraded, fragmented into peptide segments and then presented to tissue immune cells by MHC-II molecules. However, there is another completely different cellular mechanism associated with producing MHC-I molecules. Viruses take over the cell's biosynthetic mechanisms to make their own proteins; this occurs in the cytosol (cytoplasm) of the cell. As is true of normal cells, virus-infected cells continually degrade proteins in the cytosol and replace them with newly synthesized proteins. There is a mechanism that takes peptide fragments of degraded virus proteins or proteins from the viral membrane or viral secretions, for example, glycoproteins of viral envelopes, and loads them on to MHC-I molecules. These loaded presenter molecules then move to the cell surface where the viral peptide segment may be recognized by an appropriate immune cell.

Some viruses interfere with the cellular process of synthesizing MHC-I molecules. As a result, infected cells avoid detection because of the paucity or total lack of MHC-I molecules showing the presence of virus. However, as discussed in the previous section, MHC-I molecules also have specific sites that are complementary to receptors on natural killer cells. The NK cell-MHC-I communications generate signals that inhibit the killing activity of NK cells. The lack of MHC-I results in a lack of inhibition to NK cells which then proceed to kill the viral infected cell.

Each gene of a group of closely related genes guides the synthesis of a particular protein in the MHC complex. The several different proteins produced by this group of genes combine to form the complex MHC molecule. Each of the genes in this group is highly variable. That is, in two genetically, closely related individuals, for example, a son or daughter of the same parents, a gene associated with synthesizing MHC in one individual is likely to be slightly different than it is in the other individual. All the genes in the group of genes that contribute to synthesizing the complex MHC molecule are variable. The result of this gene variability is that, except for identical twins, MHC I & II molecules in the cells of any one individual will be to various degrees different from those types of molecules in cells of any other individual. This is why transplants not aided by suppressing the immune system are rejected. The receiver of the transplant (directly or indirectly) has T-cells that recognize all the donor cells with their somewhat different MHC molecules as nonself cells. As a result, the receiver mounts an adaptive immune response that kills all the donor cells.

Development and Creative Learning of Lymphocytes

Creating Diverse Antigen Receptors in Lymphocytes

B lymphocytes (B-cells) originate in bone marrow and remain there to differentiate and learn to recognize nonself entities by creating receptor complexes that move to the surface of the cell. The young B-cells have many diverse gene segments associated with guiding the synthesis of proteins that make up a receptor complex. There is a cellular mechanism that guides the rearrangement of these gene segments to produce a particular sequence of nucleotides in a portion of a DNA molecule. There are hundreds of thousands of possible rearrangements. Each developing B-cell of the millions of B-cells that are generated in the bone marrow expresses one of these possible arrangements, but once having created a particular arrangement, the mature B-cell keeps that arrangement and faithfully reproduces it when the cell divides and replicates its DNA. Each portion of gene segment arrangement produces a complex protein with a particular three-dimensional structure. If the complex protein is a stable variation of the overall structure of a B-cell receptor for antigen, then it becomes embedded in the cell membrane and the developing B-cell becomes an immature B-cell. The developmental goal is for the complex immunoglobin to have a structure, as shown in figure 8.2, that can bind to some complementary

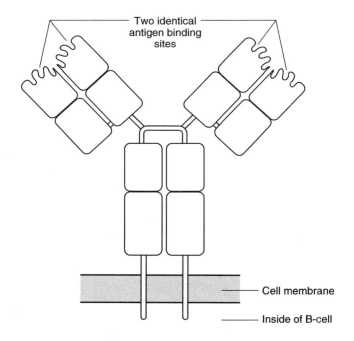

FIGURE 8.2

pattern on a nonself entity. The nonself entity is a possible antigen and the immunoglobin that binds to it is a possible antibody. If the possible antigen-possible antibody interaction can signal an immune response, then this interaction is said to be an antigen-antibody communication. The terms antigen and antibody only have meaning in relation to the whole process of a communication that initiates an adaptive immune response. As will be described in greater detail later in this chapter, activation of B-cells requires two signals: (1) possible antibody of a B-cell interacting with possible antigen, and (2) Helper T-cell interacting with a peptide segment derived from the same antigen and displayed by the B-cell. This second signal stimulates the Helper T-cell to secrete a cytokine that in conjunction with the antibody-antigen interaction activates the B-cell to begin to divide many times and differentiate to effector cells and memory cells, see figure 8.3.

Immature B-cells migrate from the bone marrow to peripheral lymphoid organs such as the spleen and lymph glands. There they undergo "creative learning" by discovering whether their receptor complex binds to patterns in self-cells, for example, patterns on self MHC molecules. B-cells that strongly bind to self-molecules send signals to themselves to start programs that produce cell death (a process called apoptosis). Sometimes before cell death occurs a portion of the receptor complex is changed so that the receptor no longer strongly binds to self-molecules. B-cells that weakly bind to antigens such as small soluble proteins enter a permanent state of inactivity; i.e., the receptor-antigen interaction does not signal them to further develop. Some become ignorant in that the signal generated by the receptor-antigen interaction is too weak to stimulate further development. Alternatively, these B-cells with receptors that could bind to self antigens are located in tissues where they do not encounter these self-molecules and thus are "ignorant" for that reason. The B-cells that do not commit suicide or are not ignorant survive and graduate to become mature B-cells which at this stage also are called naïve B-cells. They then begin to circulate in lymph vessels through other lymph nodes to the blood then to tissue spaces and back again to lymph vessels.

The T-lymphocytes (T-cells) migrate from the bone marrow to the thymus gland, see figure 7.20, and there undergo a more rigorous education than B-cells. As with B-cells, primitive T-cells undergo gene segment rearrangement to produce hundreds of thousands of proto-T-cell receptors. These immature T-cells undergo positive and negative selections. The T-cells with receptors that strongly bind to self-molecules die as a result of signals that stimulate cell death programs (apoptosis). Approximately 95% of the primitive T-cells die in this manner. Some T-cells become natural killer cells (NK cells) with a very limited diversity and specificity of receptors. As described earlier, these cells are activated against cells infected with virus during the late phase of the innate immune response. As a result of further developmental-evolution, T-cells in the thymus are negatively selected at various stages of their education. The surviving T-cells have differentiated into two types of cells. One type when activated becomes a cytotoxic effecter cell that kills virus-infected cells in the

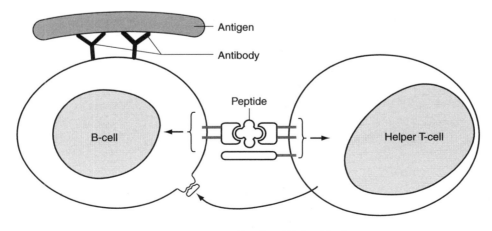

Dialogue between B-cell and Helper T-cell

FIGURE 8.3

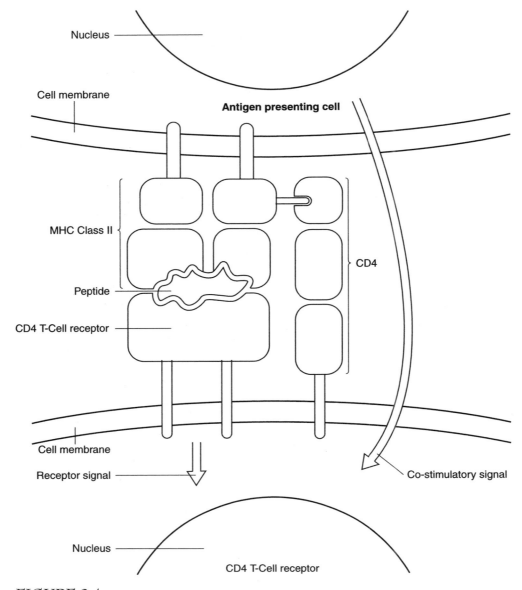

Nucleus

Cell membrane

Antigen presenting cell

MHC Class II

Peptide

CD4

CD4 T-Cell receptor

Cell membrane

Receptor signal

Co-stimulatory signal

Nucleus

CD4 T-Cell receptor

FIGURE 8.4

same way as do natural killer cells. The other type when activated becomes a cytokine-secreting cell. As a result of secreting a diverse set of cytokines, these T-cells participate in immunological processes. A major function of some of these T-cells is to stimulate a B-cell that has recognized a non-self antigen to begin to divide and differentiate. Thus, these T-cells are called helper T-cells.

The T-cells that can become cytokine-secreting cells have two defining characteristics. The T-cell receptor can loosely bind to complementary sites on self-MHC-II molecules and bind to the nonself peptide in the cleft of these MHC-II molecules. Secondly, these T-cells express membrane receptors that are complementary to stimulatory sites on MHC-II molecules, see next section. These other receptors are referred to as CD4 receptors. Separate signals from the MHC-II T-cell receptor and the CD4 receptor jointly stimulate the T-cell to become a cytokine-secreting cell. Therefore, the CD4 receptors are called co-stimulatory receptors and the T-cells possessing them are called CD4 T-cells. The T-cells that become cytotoxic effector cells also have two defining characteristics, figure 8.4. The T-cell receptor can loosely bind to complementary sites on self-MHC-I molecules and bind to the nonself peptide in the cleft on these MHC-I molecules. Secondly, these T-cells express membrane receptors that are complementary to sites on the MHC-I molecules.

These other receptors in the T-cell are referred to as CD8 co-stimulatory receptors and the T-cells possessing them are called CD8 T-cells, see figure 8.4. Thus, the two types of T-cells exhibit MHC restriction. The TD4 T-cells usually only bind to self-MHC-II molecules containing nonself peptides; whereas the TD8 T-cells usually only bind to self-MHC-I containing nonself peptide. Just as fully developed B-cells are called mature, naïve B-cells, fully mature T-cells are called mature, naïve CD4 T-cells or mature, naïve CD8 T-cells.

Antigen Presenting Cells (APCs)

General Description

An antigen presenting cell (APC) is an immune cell that can develop the capability of displaying on its cell membrane MHC molecules with peptide segments in their central cleft *and* stimulatory molecules that are complementary to co-stimulatory receptors on mature, naïve T-cells. The three types of these cells are: (1) dendritic cells, (2) certain macrophages, and (3) mature naïve B-cells. After a pathogen enters the endoplasmic reticulum of one of these types of cells, for example, viral invasion or into the cytosol (cytoplasm) of the cell by phagocytosis, these immature antigen presenting cells in tissues where infections occur migrate to lymph nodes and further develop into mature cells. Mature APCs lose their ability to engulf pathogens. Most of these mature cells become active APCs in that they synthesize and then display MHC molecules with nonself peptide fragments and stimulatory molecules complementary to co-stimulatory receptors on mature, naïve T-cells. An activated APC simultaneously communicates with a naïve T-cell through its antigen receptor and its co-stimulatory receptor. These two communications acting together stimulate the T-cell to divide and secrete cytokines which guide its further maturation. A weak or a total lack of co-stimulatory communication drastically effects the T-cell that recognizes the nonself peptide. Such a cell not only does not initiate an adaptive immune response, it thereafter is unable to be activated by an antigen presenting cell even when co-stimulatory communication also is present. After successful activation of a naïve T-cell and after hundreds of cell divisions, the single cell has become a clone of about 1000 T-cells. These differentiate into two populations: effector cells and memory cells.

Effector T-cells can participate in processes which kill and/or remove a pathogen or a cell infected with a pathogen, for example, virus or parasite. This effector activity no longer requires a co-stimulatory communication. The memory T-cells circulate around the body or remain dormant in some tissues. They are called memory cells because the same antigen that stimulated naïve T-cells to differentiate into them at a later date quickly can communicate with them so as to stimulate them to begin to divide and differentiate into new effector cells and more memory cells. Overall antigen presenting cells in tissues where an infection occurs stimulate an adaptive immune response of T-cells in lymphoid tissues, for example, lymph nodes and spleen. This adaptive immune response consists of three phases: (1) recognition of nonself antigen simultaneously supplemented by co-stimulatory communication, (2) clonal expansion in which a T-cell divides many times to produce about 1000 T-cells that have differentiated into effector T-cells or memory T-cells, and (3) the effector T-cells remove the nonself antigen.

Dendritic Cells

Dendritic cells arise from cells in the bone marrow, migrate via the circulating blood to tissues throughout the body and contribute to the innate immune response by phagocytizing nonself antigens. When the nonself antigen is not immediately removed, the tissue is said to be infected. Dendritic cells that have phagocytized an invading pathogen migrate to lymph nodes down stream from the site of infection. In the lymph nodes these cells lose their ability to phagocytize pathogens, but begin synthesizing and transferring to their cell membrane MHC molecules with nonself peptide segments and with stimulatory sites that are complementary to co-stimulatory receptors on naïve T-cells. Dendritic cells infected by virus will present peptide fragments of the viral proteins via

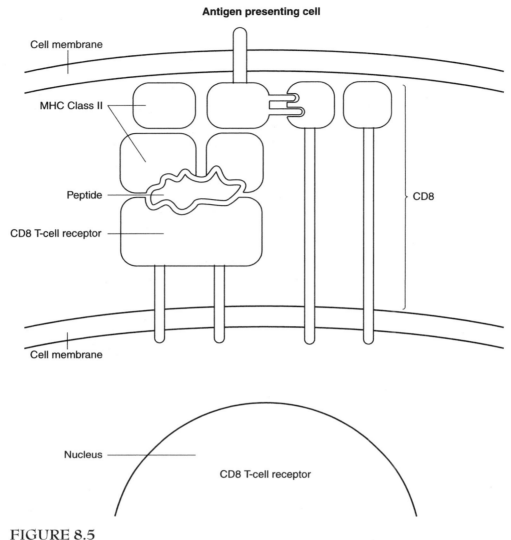

Antigen presenting cell

Cell membrane

MHC Class II

Peptide

CD8 T-cell receptor

Cell membrane

CD8

Nucleus

CD8 T-cell receptor

FIGURE 8.5

MHC-II molecules and thus activate CD4 T-cells that recognize the nonself peptide to undergo clonal expansion, see figure 8.4. Dendritic cells that have taken in a pathogen, for example, a virus particle or bacteria, into the cytosol will present nonself peptides derived from the pathogen via MHC-I molecules and thus activate the CD8 T-cells that recognize the nonself peptide to undergo clonal expansion, see figure 8.5.

Macrophages

As part of the innate immune response, macrophages phagocytize, destroy, and thus remove an invading pathogen. If the pathogen inside membrane vesicles in the cytosol of the macrophage is not destroyed, the macrophage is stimulated to become an antigen presenting cell. It presents the nonself peptide bound to MHC-II molecules that also have sites that can communicate with the co-stimulatory receptors on naïve T-cells. Naive CD4 T-cells that recognize the nonself peptide are stimulated to undergo clonal expansion and differentiation into effector and memory cells.

B-cells

Naïve B-cells have a membrane receptor, which is an antibody, that is specific for a dissolved antigen. The antigen-antibody communication leads the B-cell to internalize the antigen into the cytosol, degrade it and display nonself peptide fragments of it via MHC-II molecules. As will be described in the next section, "passive," no longer naïve B-cells now may communicate with an activated effector CD4 T-cell that has recognized peptides from the same pathogen. This appropriate effector T-cell is a helper T-cell whose receptor is complementary to the nonself peptide segment displayed by the B-cell; co-stimulation is not required, see figure 8.3. Communication between B-cell and helper T-cell leads to clonal expansion of the B-cell that includes differentiation into effector B-cells and memory B-cells. If various microbial constituents induce these passive B-cells to express stimulatory sites on the MHC-II molecules complementary to co-stimulatory receptors on T-cells, then the resulting "activated B-cells" can communicate with naïve T-cells just as dendritic cells can. The resulting dual communications lead to clonal expansion of the T-cells.

The Adaptive Immune Response

The Neuroeffector Analogy

The immune system exhibits the same pattern as the neuroeffector system. The simplest representation of this pattern is the reflex arc which consists of: (1) sensory neurons, (2) interneurons that process incoming information to make appropriate commands to adapt to the environment, (3) motor neurons that transmit commands to effectors, and (4) effectors that carry out the adaptation behavior, see figure 7.18. In like manner the immune system consists of sensory, information processing/commanding, motor, and effector components. The sensory component of the immune system consists of the millions of lymphocytes that originate from primordial cells, stem cells, in the bone marrow. Some early descendents of stem cells migrate to the thymus during fetal development and childhood to become primordial T-cells. Other descendents of stem cells remain in the bone marrow to become primordial B-cells. These two types of primordial lymphocytes become mature cells in two stages. In the first stage extensive gene segment rearrangement processes produce hundreds of thousands of different antigen receptors embedded in the cell membrane of lymphocytes which thereby become immature lymphocytes. In the second stage positive selection promotes the continued survival of lymphocytes with "appropriate antigen receptors," and negative selection eliminates those lymphocytes with "inappropriate antigen receptors." Some immature B-cells that weakly bind to self-antigen are allowed to survive to become ignorant mature B-cells. B-cells that strongly bind to self-antigen are eliminated. The remaining B-cells continue to develop to become mature B-cells. Immature T-cells go through a more rigorous education in the thymus. Those T-cells that loosely bind to self-MHC molecules as well as to self-peptide fragments receive signals for continued development. Those that strongly bind to self-MHC-peptide complexes, receive signals that lead to their death and removal. As a result of these positive and negative selections, each individual T-cell antigen receptor is specific for a particular combination of self-MHC molecule and nonself peptide segment. Thus, antigen receptors are said to be MHC-restricted for antigen recognition.

T-cell receptors also are selected in relation to co-stimulating sites bound to MHC molecules. The T-cells then diverge during development to become, when activated, a cytotoxic effector cell that is positively selected if it has a CD8 receptor that can communicate with a complementary site on the MHC-I molecules on antigen presenting cells. The T-cell is negatively selected if it does not have this CD8 receptor. In like manner, positive and negative selections lead to T-cells with the potential to become cytokine-secreting cells, to have CD4 receptors that can communicate with complementary sites on MHC-II molecules on presenting cells. Early stages of generation of T-cells produce two mutually exclusive classes of T-cell receptors. Each of these two classes differentiate into CD8 and CD4 T-cells. One class (γ-δ T-cells) is generated in the fetus and early childhood, but in adults these T-cells comprise only 5% of the T-cell population. The other class (α-β cells) makes up 95% of the T-cell population in adults.

The end result of generating hundreds of thousands of antigen receptors on billions of lymphocytes is an immunological sensory system for antigens in blood and tissue fluid, i.e., an internal sensory system for recognizing self and nonself cells. B-cells, for the most part, only recognize nonself entities; whereas T-cells recognize self-cells by loosely binding to them and thereby not generating an immune response to them. T-cells also recognize nonself cells by strongly binding to them and then generating an adaptive immune response to them. This immunological sensory system consists of three subsystems: population of B-cells, population of (α-β) T-cells, and population of (γ-δ) T-cells. Since the third subsystem consists of only 5% of the T-cells, let us focus on the first two subsystems. Each consists of hundreds of thousands of clones of cells where all cells in a particular clone have the same antigen-specific receptor and will reproduce that receptor during cell division. Thus, a clone is like a sense receptor in the neuroeffector system. A sense receptor for light will respond to light waves but not to sound waves, and a sense receptor for sound will respond to sound waves but not to light waves. In like manner each clone will initiate an adaptive immune response only to a particular pattern on a nonself entity. In this analogy the B-cell subsystem and the T-cell subsystem have hundreds of thousands of immunological sense receptors; that is, any particular clone is an immunological sense receptor.

Just as stimulated sense receptors in a neuroeffector system generate nerve impulses that travel to the central nervous system; so also recognition of antigens that without further intervention could cause disease/death, that is, pathogens generate signaling in B-cells and T-cells that produce the first stage of a response. This first stage is diverse cell-to-cell communications via released cytokines or cell source receptor communicating with cell receiver receptor. The central nervous system of the neuroeffector system processes incoming sensory information into a range of interpretations that are integrated to produce appropriate commands to muscle and gland cells which are the effectors. In like manner, the immune system processes recognition of nonself entities into interpretations that lead to appropriate commands for removal of the nonself entities. The first level interpretation generates the commands activating the innate immune response or incorporates this innate response into an adaptive immune response. Information processing of the adaptive immune response generates interpretations of the mode or modes of invasiveness of pathogens, the anatomical and physiological context of infection, for example, skin surface or lung mucosal surface, mechanisms of tissue damage, and the intensity of the infection. These interpretations lead to mobilization and integration of effector cells, antibodies and other substances, for example, the complement system in blood, that provide the most effective strategy to eliminate the invading pathogen. Immune effector cells collaborate with effectors of the innate immune system and of the neuroendocrine system to remove the pathogen.

Phases of the Adaptive Immune Response

Recognition of Antigen.   Pathogens not removed by the innate immune response persist in tissues. Nonspecific receptors on dendritic cells bind to and then engulf the pathogen by phagocytosis or macropinocytosis which is similar to phagocytosis but can take in large particles plus some fluid surrounding them, see figure 5.22b. The dendritic receptor site binding to the pathogen either directly sends signals that stimulate the tissue dendritic cells to migrate via lymph vessels to a lymph node down stream from the site of infection or signals secretion of cytokines that stimulate this migration to lymph nodes. If the pathogen is taken in by macropinocytosis in which there is no receptor-pathogen communication, the pathogen is degraded and one of its products is an internal receptor, i.e., an intracellular receptor, that triggers dendritic cells to migrate to lymph nodes. Dendritic cells infected with viruses produce double stranded RNA molecules that are recognized by intracellular receptors that signal activation of these cells to migrate to lymph nodes. Infected dendritic cells also may take up viral particles via macropinocytosis. In the lymph nodes the activated dendritic cells synthesize MHC molecules with stimulator sites complementary to the co-stimulatory receptors on T-cells. These dendritic cells then display MHC-I & MHC-II molecules each with appropriate stimulator sites for co-stimulatory receptors. Thus, these dendritic cells can present viral peptide segments to and consequently activate CD4 T-cells and CD8 T-cells. Dendritic cells activated by taking in bacteria, fungi or soluble nonself antigens will mature in the lymph node and present MHC-II stimulatory complexes which, in turn, will recognize and activate CD4 T-cells.

Tissue dendritic cells account for the recognition of pathogens in most infections and subsequent activation of the adaptive immune response. However, tissue macrophages via similar mechanisms also migrate to lymph nodes, mature and display MHC-II complexes. As before, CD4 T-cells recognize the MHC-II nonself peptide complex and its site for a co-stimulatory receptor. This two-fold communication leads to the T-cell component of the next phase of the adaptive immune response. Ordinarily B-cells while making an abundance of MHC-II molecules, do not make stimulatory molecules complementary to co-stimulatory receptors on T-cells. After a B-cell receptor, which is an antibody, binds to a soluble nonself antigen, the antigen is engulfed by the B-cell and presented as a nonself peptide fragment bound to a MHC-II molecule. This is the first step of B-cell recognition of a nonself antigen. Various microbial constituents can induce B-cells to express stimulatory sites on MHC molecules that are complementary to co-stimulatory sites on T-cells. When this happens, B-cells also can activate naive CD4 T-cells in the same manner as dendritic and macrophage cells do. The more usual situation is that the B-cell that presents a nonself peptide segment on MHC-II molecules communicates with an activated CD4 T-cell which has become a helper T-cell. The helper T-cell receptor communicating with the B-cell MHC-II-peptide complex activates the B-cell to begin to divide and later to differentiate into an effector cell. This communication, then, is the second and final step of B-cell recognition of a nonself antigen. B-cell recognition of an antigen that does not lead to the next phase of an adaptive immune response is not an immunological recognition. Thus, B-cells may communicate with soluble *self*-antigens, but for three reasons this communication by itself will not produce an immune response. One, this communication on its own does not stimulate the B-cell to proceed to the next phase. Two, there are no stimulatory molecules to communicate with co-stimulatory receptors on T-cells; so B-cells cannot activate naïve T-cells. Three, Helper T-cells do not recognize the B-cell MHC-II-self-peptide complex.

Immunological Matchmaking.   A key problem for lymphocytes recognizing nonself entities is: how do naïve T-cells find antigen-presenting cells, for example, dendritic and macrophages cells or B-cells, that have the specific antigen that is complementary to their receptors? The structure of lymph glands and the spleen solves this problem. Naïve T-cells circulate from lymph glands and spleen to lymph vessels, then to the blood which during inflammation leads to these lymphocytes moving into tissue spaces that drain into lymph vessels and back to lymph glands. Circulating blood contains lymphocytes that circulate through the spleen. Circulating T-cells and antigen-presenting cells that via lymph vessels migrate to lymph nodes are trapped in special places in the lymph gland. As a result of this trapping, naïve T-cells and antigen-presenting cells, especially dendritic cells, come into close proximity to one another. In various ways T-cells and dendritic cells loosely bind to one another. This allows the T-cell to survey the MHC-nonself peptide complexes on the antigen-presenting cells to "see" whether there is a match—a possible complementary interaction. The T-cells and dendritic cells dance with one another at the lymphoid "pickup bar." If there is a match, they stay together for awhile and then the "no longer naïve," activated T-cell breaks away and begins to proliferate. B-cells that present nonself peptides are trapped in areas where helper T-cells flow in and out. The B-cell and helper T-cell begin to dance at the lymphoid pickup bar. If there is a match, they stay together for awhile but eventually separate. The now activated B-cell stays in the lymph node and begins to proliferate.

Clonal Expansion and Differentiation into Effector Cells.   Naïve T-cells that find a match with a nonself peptide on a dendritic cell or macrophage in a lymph node stay there and begin to divide—about 2 cell divisions per day. Within 5 days or so there is a clone containing identical cells with respect to recognition of nonself antigen. These cells then begin to differentiate into effector cells and memory cells. Activated CD8 T-cells differentiate into cytotoxic effector cells. Activated CD4 T-cells differentiate into two classes of effector cells: T_H1 and T_H2 cells. T_H1 cells communicate with receptor sites on macrophages via secretion of cytokines. Signals from this communication lead to macrophages more vigorously attacking and engulfing pathogen. T_H2 cells are the helper T-cells that communicate with B-cells that display the same nonself peptide as that which led to the production of the helper T-cell. Communication with these B-cells leads to these cells differentiating into plasma cells that secrete nonself antigen-specific antibodies. Thus, cytotoxic effector cells (from CD8 T-cells) and T_H1 cells (from CD4 T-cells) provide what is called a cell-mediated attack and removal of pathogen. Communication between B-cells and T_H2 cells (from CD4 T-cells) eventually produces plasma cells secreting antibodies that provide what is called a humoral-mediated attack and removal of pathogen.

Activated B-cells secrete antibodies that specifically bind to soluble nonself antigen or nonself antigen on a pathogenic cell. By various mechanisms, this specific antibody-antigen binding leads to the neutralization and/or removal of the pathogen. However, activated B-cells migrate to the B-cell germinal center in a lymph node where they continue to rapidly divide as well as secrete antibodies. B-cells in these centers undergo somatic hypermutation which involves point mutations in DNA that sometimes produce antibodies that have greater specificity and therefore stronger binding to the nonself antigen. There is a mechanism for positively selecting those cells with a mutation that imparts a greater specificity for binding to a nonself antigen. The mutated cells divide and undergo more somatic hypermutation leading to selected cells with even greater specificity for nonself antigen, and so on. The somatic hypermutation coupled with positive selection leads to progressive evolution of B-cells that secrete antibodies that are highly specific for the nonself antigen. Eventually this evolutionary process stops, the B-cells stop dividing and instead differentiate into plasma cells and begin to secrete large quantities of antibodies.

Immunological Information Processing.   How intracellular signaling, cell-to-cell communications, and cytokines coordinate immunological information processing is one "cutting edge" of research in molecular biology. Valid new theoretical models will provide better control and/or cure for disease processes, allergies, autoimmune diseases and tissue/organ transplants. The validated theories present a very complex picture that will not be described in this short summary of functions of the immune system, but we will give an overview of the results of this information processing. Activated B-cells can be induced to secrete many different variations of an antibody specific for a particular nonself antigen. What is the most appropriate variation of antibody depends on many factors, such as the properties of the invading pathogen or allergy producing antigen and what tissue it invades. Immunological information processing or human medical intervention selects the most appropriate type of antibodies to be secreted by plasma cells. This same process also selects between primarily a cell-mediated response, a humoral-mediated response or one among several possible collaborations of cell-mediated responses with humoral-mediated responses.

Termination of Infection.   When an infection is effectively repelled by an adaptive immune response, three things happen. First of all, further adaptive immune responses to the pathogen are inhibited. Second, most of the pathogen-specific effector cells die and are rapidly removed by macrophages. Third, some effector cells remain dormant in tissues. During clonal expansion and differentiation, some B-cells become memory B-cells, and likewise some T-cells become memory T-cells. These cells represent immunological memory because they recognize the re-infection of an invading pathogen more quickly and within a couple of days rather than three weeks can mount an adaptive immune response to the pathogen. This response is referred to as a secondary immune response and is the basis for the effectiveness of vaccination against particular pathogens such as the polio virus. Most B-cells and T-cells die during their development into naïve mature cells. Likewise, all plasma cells die and most effector T-cells die after the termination of an infection. Therefore, the population of circulating lymphocytes is continually depleted. The thymus gland in an adult stops producing T-cells, but T-cells in lymphoid tissue continually divide to replenish the ones that are lost. Stem cells in the bone marrow continue to divide and replenish the macrophages, neutrophils, and B-cells that die and are removed from the body. There is an elaborate mechanism involving cytokines that maintains a steady state level for the various types of immune cells. This is immunological cell homeostasis analogous to physiological homeostasis maintained by the neuroendocrine system. These two types of homeostasis mutually interact with one another thus in another way pointing to the unity of the neuroendocrine-immune system.

AIDS

Acquired immune deficiency syndrome (AIDS) is expressed as a progressive decrease in CD4 T-cells. When the number of these immune cells drops below a critical level, the individual no longer can mount a CD4-controlled, cell-mediated adaptive immune response to diverse invading pathogens. CD4 H_1 T-cells secrete cytokines that bind to pathogens, such as bacteria, and target them for aggressive phagocytosis by macrophages. The decreasing number of these CD4 T-cells allows invading bacteria to flourish and eventually produce disease. Typically, oral Candida, and Mycobacterian tuberculosis begin to proliferate and produce thrush (oral candidiasis) and tuberculosis. Later, virus-infected cells are not removed which leads to shingles resulting from activation in cells of dormant viruses that can cause herpes zoster. Some virus-produced cancer also may appear such as EBV-induced B-cell lymphomas and Kaposi's sarcoma, a tumor of endothelial cells. Some fungi also become resistant to removal. In particular, Pneumocystis carinii causes pneumonia which often is fatal.

An individual *acquires* these immune deficiencies by infection with a virus known as human immunodeficiency virus (HIV) of which there are two types, HIV-1 and HIV-2. Populations in West Africa are infected by HIV-2, which now is spreading in India. The more virulent HIV-1 is the more usual cause of worldwide AIDS. The HIV virus binds to receptor sites on CD4 T-cells, dendritic cells and macrophages. The viral envelope fusses with the cell membrane of these cells thereby releasing the viral RNA into the interior. Inside the cell, the single stranded RNA uses cell machinery to produce a double stranded viral DNA molecule (via transcription, see Chapter 6). The viral DNA integrates into normal DNA in the nucleus and thereafter directs the duplication of viral DNA that leads to production of many new virus particles. Eventually T-cells infected with the virus become overwhelmed by the newly produced viral particles, die, and release the particles into the blood. These viruses then show up in other body fluids: semen, vaginal fluid or milk. Thereafter, the most common mode of spread of HIV infection is by genital, sexual interactions.

Macrophages and dendritic cells haboring duplicating viruses are not necessarily killed and therefore are an important reservoir of the infection and a means for spreading the virus to other tissues such as the brain. The viruses may remain latent and undetected in infected cells for several months to several years. Eventually they become active again, producing many more virus particles in the blood. During this active phase of the infection, mutation occurs that leaves some viruses resistant to any immune response or resistant to drug treatment. In particular, CD4 T-cells are killed by a combination of three processes: (1) direct viral killing of infected cells, (2) viruses induce infected cells to "commit suicide" (apoptosis), and (3) CD8 cytotoxic T-cells produced by an adaptive immune response kill infected CD4 T-cells. As a result of these three processes, the number of CD4 T-cells approaches zero and the infected patient comes down with several bacterial, fungal, or viral-cancer diseases. One or several of these diseases kills the individual.

Allergy

Some nontoxic substances, such as a component of dust mites, some drugs, e.g. penicillin and sulfa drugs, pollen (ragweed, birch), mold, various types of food, e.g., peanuts and shell fish, and cat and dog danders are called allergens. Allergens stimulate an adaptive immune response that produces symptoms characteristic of allergic reactions. Common features of these reactions are changes associated with inflammation, for example, edema and local accumulation of immune cells. The details of an allergic reaction depend on the route of entry of the allergen. Absorption of allergen into the blood by intravenous injection leads to systemic changes, e.g., generalized vasodilation and edema which may produce a life-threatening drop in blood pressure. Subcutaneous invasion of insect bites produces a local inflammation and corresponding local swelling. Inhalation of pollen or dust mites produces inflammation of nasal mucosal (allergic rhinitis) and inflammation of the head sinuses resulting in a "stuffy nose" or total inability to breathe through the nose. Inhalation of allergens may lead to asthma, which involves (1) constriction of bronchioles, (2) inflammation of mucosal of the bronchial tree, and (3) accumulation of immune cells in the tissue surrounding the bronchial passage ways. The end result is that the lungs can take in air but cannot exhale all of it. The air that remains in the lungs reduces lung capacity. This means that the lungs exchange less volume of air with the outside resulting in less oxygen going into the blood and less CO_2 eliminated from the blood. Food allergens taken in orally produce allergic reactions of the lining of the digestive tube that may produce vomiting and/or diarrhea.

The underlying mechanism for producing allergic reactions involves allergen-antibody communication. An antibody is an *immunoglobin*—a type of protein—that is represented as Ig. The Y-shaped structure of all antibodies, see figure 8.1, exhibits different sets of variations that have been grouped into different classes called isotypes. The major isotypes are IgM, IgD, various subclasses of IgG, IgA, and IgE. As discussed in the section, "Phases of the Adaptive Immune Response, immunological information processing," communication among immune cells leads to the production of the most appropriate type of antibody to be secreted by plasma cells (B-cells). The modes of invasion by pathogenic parasites are the same as the modes of invasion by allergens. The body protects against subsequent invasion by parasites by producing IgE antibodies. The first exposure to a parasite leads to an adaptive immune response that produces IgE antibodies that can specifically bind to the parasite. This antibody does not eliminate the parasite; rather IgE antibodies bind to mast cells located in tissues in the skin and in areas beneath the lining of the respiratory and digestive tubes. Then, when these tissues are invaded by the same parasites, they are recognized by IgE antibodies bound to the mast cells. Within seconds after recognition mast cells release granules that break down into chemicals that orchestrate inflammation. One of these chemicals, histamine, brings about local inflammation. Activated mast cells synthesize and secrete leukotrienes and prostaglandins that sustain chronic inflammation associated with influx of leukocytes. Later basophils and eosinophils are drawn to the inflammation area. These cells amplify the effect of mast cells and together bring about a great deal of tissue damage as well as destroying the invading parasites.

Populations in modern Western countries (or Westernized societies) are not exposed to many pathogenic parasites. For reasons not yet totally understood, various allergens elicit the same immune response involving IgE as do invading parasites. Thus, what has been described as an allergic reaction involving systemic or localized tissue damaging inflammation either eliminates an invading parasite or produces allergy. The allergy varies in intensity between discomfort, chronic disease, e.g. asthma, to death. The mechanism for producing IgE is as follows. A subtype of dendritic cell called myeloid dendritic cells (the other subtype is lymphoid dendritic cells) predominates in areas of invasion by parasites or allergens. These dendritic cells take up and process protein allergens very efficiently and in so doing become activated. They migrate to regional lymph nodes and differentiate into antigen-presenting cells. These cells as well as cells of the innate immune system in the regions of invasion by allergens are specialized to secrete cytokines that, in turn, stimulate potential CD4 T-cells to differentiate into CD4, T_H2 cells. The resulting T_H2 cells secrete cytokines and present co-stimulatory signals that stimulate B-cells to switch to producing IgE antibodies. (The co-stimulatory signal is a CD40 molecule on the surface of a B-cell communicating with a molecule with a section complementary to CD40—it is called a *CD40* ligand—that is, there is a key-lock interaction in which CD40 is the key and CD40 ligand is the lock.) Once B-cells are stimulated to switch to secrete IgE, this response is further amplified by activation by allergens of mast cells, basophils, and eosinophils. Eventually B-cells become plasma cells that only secrete IgE. The IgE, in turn, binds to mast cells which become ready to produce allergic reactions when re-exposed to allergens.

One mode of treatment is desensitization which shifts production of antibodies away from IgE to IgA. The predominating of IgA antibodies also can bind to allergens thus preventing them from binding to IgE bound to mast cells. This, of course, in effect blocks an allergic response. The other major mode of treatment is via prescription drugs that treat the symptoms of allergic disease. Antihistamines block the action of histamines released by mast cells. Inhaled bronchodilators act on receptors on smooth muscle of bronchioles, causing them to relax and thus increase the opening of these small, blind-ending tubes. As a result, trapped air can flow out into the larger tubes and then to the outside. Chronic inflammation tissue injury is treated with drugs related to cortisol, see Chapter 7, that suppress inflammation.

Autoimmune Diseases

In some people self-tissue substances are recognized by immune cells as if they were nonself antigens. This recognition may lead to an adaptive immune response that produces tissue damage. It usually is impossible for the immune system to eliminate the "misunderstood" self-antigens, and therefore, the immune system sustains an adaptive immune response that produces chronic inflammatory injury to tissues. As is true of severe allergic reactions, this may be lethal. Receptors with high affinity for self-substances are, for the most part, purged during the developmental-evolutionary education of B-cells and T-cells. However, there are some receptors that have a lowered affinity to self-antigens. Some nexus of events that may include stressors and psychic reaction to stressors, see Chapter 9, triggers an autoimmune response to self-antigens that produce an autoimmune disease. Also, genetic factors and other environmental factors may be involved.

Autoimmune diseases may be divided into organ-specific and systemic diseases. The organ-specific diseases include (1) type 1 diabetes mellitus, (2) Goodpasture's syndrome, which produces inflammation of a portion of the kidney leading to its dysfunction, i.e., glomerulonephritus, (3) multiple sclerosis, and (4) various diseases producing dysfunction of the thyroid gland, e.g., Graves disease. Systemic autoimmune diseases include (1) rheumatoid arthritis, (2) scleroderma, a disease of connective tissue resulting in forming scar tissue in the skin and in various organs, and (3) systemic lupus erythematosus, which produces inflammatory damage to various tissues and organs; the most common symptoms are extreme fatigue, painful or swollen joints, unexplained fever, skin rashes, and kidney problems.

<u>Tearout #1</u>

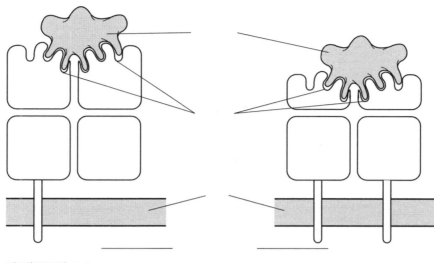

FIGURE 8.1

Directions:

1. Using figure 8.1 as a reference, fill in all labels.

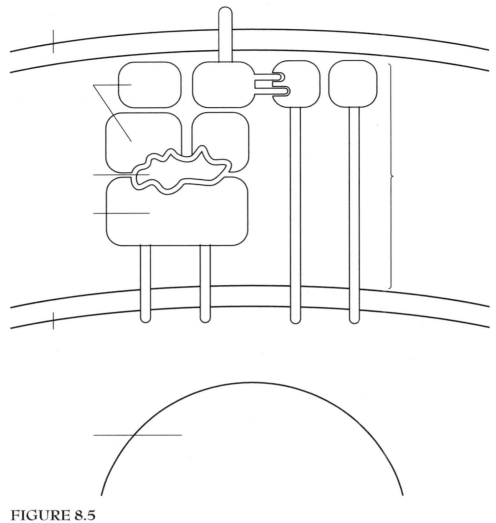

FIGURE 8.5

1. Using figure 8.5 as a reference, fill in all labels.
2. Summarize the function of CD8 T-cells that are activated to differentiate into effector cells.

Tearout #2

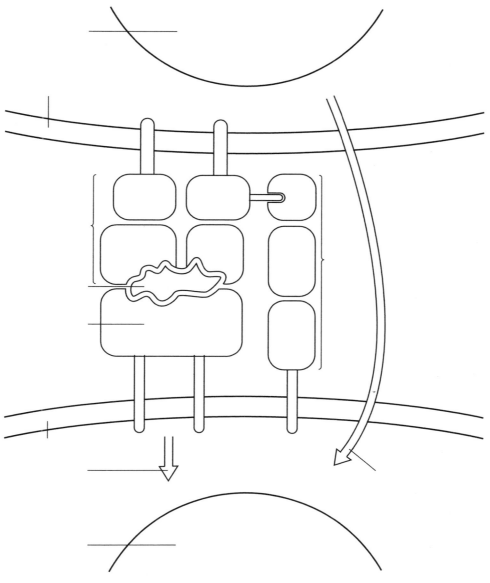

FIGURE 8.4

Directions:

1. Using figure 8.4 as a reference, fill in all labels.
2. Activated CD4 T-cells differentiate into T_H1 and T_H2 effector cells. Summarize the functions of each of these two types of effector cells.

Neuropsychic Homeostasis

Overview of the Human Nervous System

Subdivisions of the Nervous System

In the vertebrate nervous system most of the neurons become packed into a continuous mass of interconnected neurons enclosed and protected by the backbone and skull. This elaborate network of neurons consisting of the brain and spinal cord is called the central nervous system (CNS), see figures 9.1 and 9.2. The CNS receives information from sense receptors, "processes" this information (see following), and sends commands to all parts of the body. For the most part, the cell bodies of the neurons transmitting these commands are located in the CNS, while their axons are located outside the CNS. Axons or dendrites, which carry information to or away from the CNS at various places along it, are gathered into bundles surrounded by a connective tissue sheath. These cables, often containing thousands of axons and dendrites, are called nerve fibers. There are some neurons outside the CNS that operate relatively independent of it. They form networks called plexuses; for example, there are networks in the wall of the GI tract that control its motility. The cell bodies of neurons of a plexus or of neurons with sense receptors are packed into groups of neuron cell bodies. In animals with backbones, any group of nerve cell bodies located outside the CNS is called a ganglion (pl. ganglia). All parts of the nervous system such as nerve plexuses, ganglia, and nerve fibers not enclosed in the backbone and skull are components of the peripheral nervous system (PNS).

In any animal the nervous system should be understood in terms of the role it plays in the more comprehensive unit, the neuroeffector system. The reflex arc described in the previous section gives the overall pattern in any neuroeffector system, no matter how complex it is. The sensory neuron "translates" the information of some environmental pattern, the stimulus, into sensory information which then is sent to an interneuron which processes this information by sending it to a motor neuron. The motor neuron is said to have motor information because of this neuron's communication with an effector cell that expresses it in some organism activity, the response.

Within vertebrates, sensory information, represented by electrical activity generated by sense receptors, is carried to and motor information is carried away from the CNS by the PNS. Thus, the CNS translates sensory information into motor information. The spinal cord receives most of the

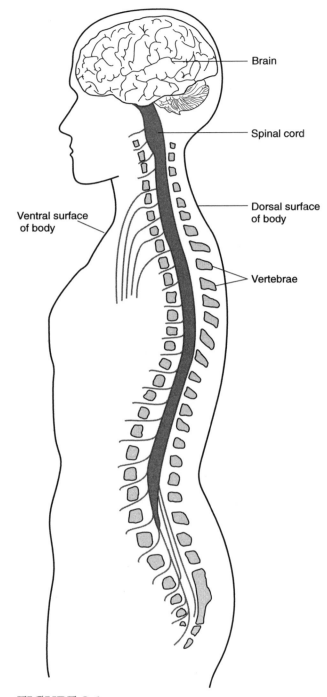

FIGURE 9.1

incoming sensory information and either translates it into simple patterns of motor information and/or carries the sensory information to the brain. The brain converts sensory signals into more complex signals which, in turn, are translated into complex patterns of motor information. For example, when a child first touches a hot stove, he withdraws his hand as a result of a spinal cord reflex. A short time later (less than a second), he experiences a painful sensation of "very hot" and begins to cry. He then may run to his mother for sympathy, and further, he may learn not to touch any object that is thought to be "very hot."

In the previous example, a simple environmental pattern, the temperature of the stove, is translated into many sensory and motor patterns of various degrees of complexity, e.g., withdrawal of

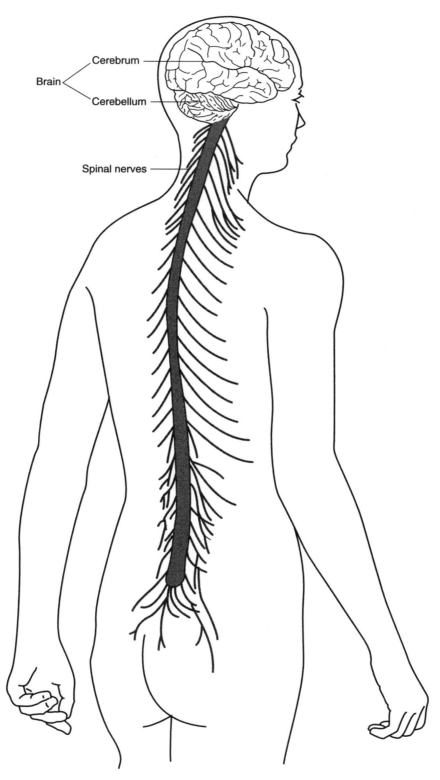

FIGURE 9.2

hand, sensation of hotness, crying, anger mixed with other feelings, running to mother, memory, learned behavior, and a new insight about the universe. This conversion of one piece of information into many more complex pieces of information is called information processing. The structure of the CNS is correlated to levels of complexity of information processing. The spinal cord performs the lower levels and the brain performs the higher levels of information processing.

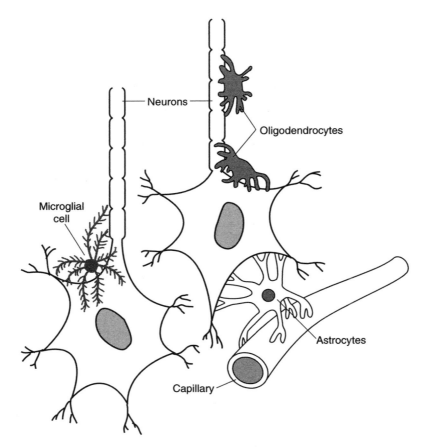

FIGURE 9.3

All axons in the PNS are associated with Schwann cells resulting in myelinated or nonmyelinated axons, see figures 7.14 and 7.15. In contrast, there are no Schwann cells in the CNS, but each neuron is closely associated with specialized cells called neuroglia (in humans there are 10 neuroglia per neuron). These cells help in the exchange of fluids and nutrients between the neuron and its surroundings. Many CNS neurons have a myelin sheath which is produced by specialized neuroglia which send out processes that spiral around axons, thus forming a myelin covering, see figure 9.3. Those portions of the CNS consisting of myelinated axons are called white matter because they appear milk white due to myelin. Bundles of myelinated axons in the CNS which carry sensory or motor information are called tracts. The portions of the CNS consisting primarily of neuron cell bodies and nonmyelinated axons are called gray matter because these areas appear gray; they lack myelin.

Developmental Anatomy

After the first few days of development of an embryo, ectoderm (see Chapter 12) forms a neural plate which produces two elevated ridges, the neural folds. Some cells of these folds proliferate downward and differentiate into the neural crest while other cells proliferate laterally and merge, forming a long cylinder, the neural tube, which develops into the CNS. The neural crest cells differentiate into various parts of the PNS including spinal cord nerves as well as ganglia and nerves of the autonomic nervous system (ANS), see figure 9.4. More rapid growth of the anterior portion of the neural tube divides it into a head region, the brain, and a longer tail region, the spinal cord. The brain becomes divided into three regions: the forebrain, the midbrain, and the hindbrain. The forebrain differentiates into two cerebral hemispheres and the interbrain; the hindbrain becomes divided into the pons, the medulla, and an outgrowth that develops into the cerebellum, see figure 9.5.

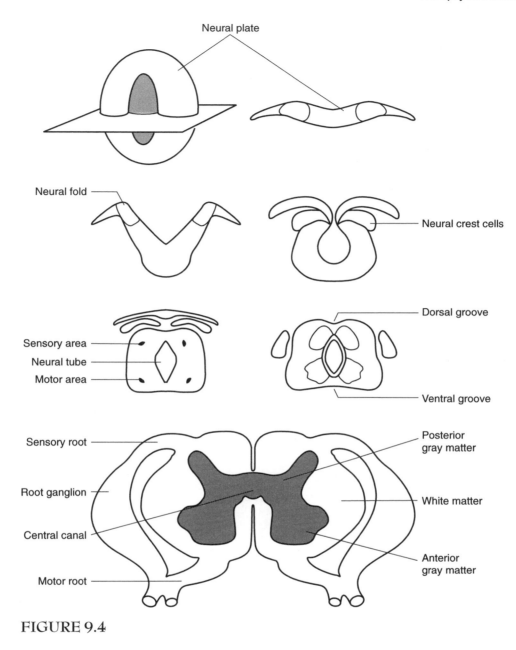

FIGURE 9.4

Figure 9.5 Development of the Human Brain.   A. The anterior portion of the neural tube has two constrictions which divide it into three sections, the forebrain, the midbrain, and the hindbrain. B. The forebrain has two lateral outgrowths which develop into the cerebral hemispheres on either side of the interbrain. C. and D. The midbrain does not undergo any further subdividing, but the hindbrain becomes divided into the pons, the medulla which is continuous with the spinal cord, and an outgrowth that develops into the cerebellum.

Adult Central Nervous System (CNS)

As a first approach to an overview of the human nervous system, it is useful to think of the CNS as having five major sections: (1) the cerebrum, (2) the interbrain, which consists of the thalamus, the hypothalamus, and the epithalamus, which includes the pineal gland, (3) the brain stem, which consists of the midbrain, the pons, and the medulla, (4) the cerebellum, which is connected to the brain stem, and (5) the spinal cord, which is continuous with the medulla of the brain stem, see figure 9.6.

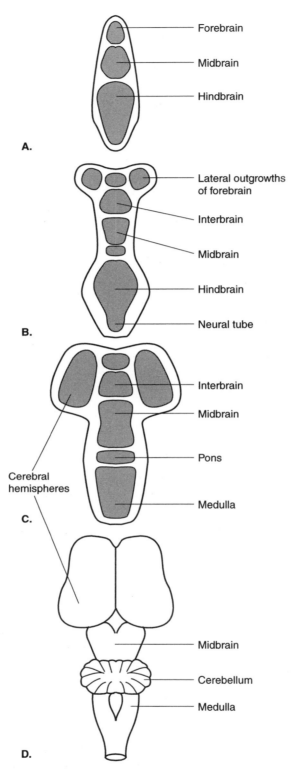

FIGURE 9.5

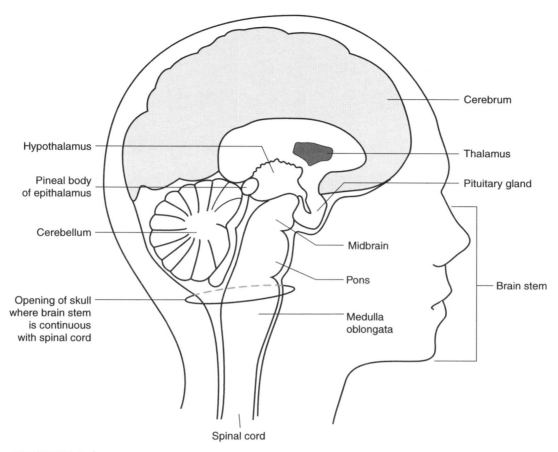

Cerebrum

Hypothalamus

Thalamus

Pineal body
of epithalamus

Pituitary gland

Cerebellum

Midbrain

Pons

Brain stem

Opening of skull
where brain stem
is continuous
with spinal cord

Medulla
oblongata

Spinal cord

FIGURE 9.6

The Spinal Cord

In all vertebrates the spinal cord is located near the dorsal surface of the animal, see figure 9.2, and consists of central gray matter surrounded by white matter, see figure 9.7. The spinal cord, protected by the bony vertebral column (the backbone), extends the length of the body. At successive levels of the vertebral column, 31 pairs of spinal nerves exit from the cord. Two portions, the cervical and lumbar enlargements, supply nerves to the arms and legs, respectively, see figure 9.8. Each spinal nerve emerges from the dorsal and ventral surfaces of the spinal cord via a dorsal and ventral root, respectively. Each dorsal root connects with a dorsal root ganglion which consists of the cell bodies of sensory neurons. The ventral root contains axons coming from motor neuron cell bodies in the ventral portion of the spinal cord. Thus, the dorsal root consists of afferent neuron processes bringing sensory information to the spinal cord, and the ventral root consists of efferent neuron processes sending commands to effectors, see figure 9.7.

The central gray matter consists of interneurons and cell bodies of motor neurons. The peripheral white matter contains two types of tracts: (1) descending (motor) spinal tracts and (2) ascending (sensory) spinal tracts.

Descending Spinal Tracts.   These tracts carry signals from higher brain centers to spinal cord motor neurons which participate in two types of reflexes. As components of somatic reflexes, some motor neurons send signals directly to voluntary muscles. Other motor neurons acting, as components of autonomic reflexes, send signals to secondary motor neurons located outside the CNS. These secondary motor neurons stimulate glands, or the heart, or smooth muscle in the walls of blood vessels and the GI tract. Thus, somatic reflexes terminate in the actions of voluntary muscles while autonomic reflexes terminate in the actions of effectors that are not voluntary muscles. Somatic

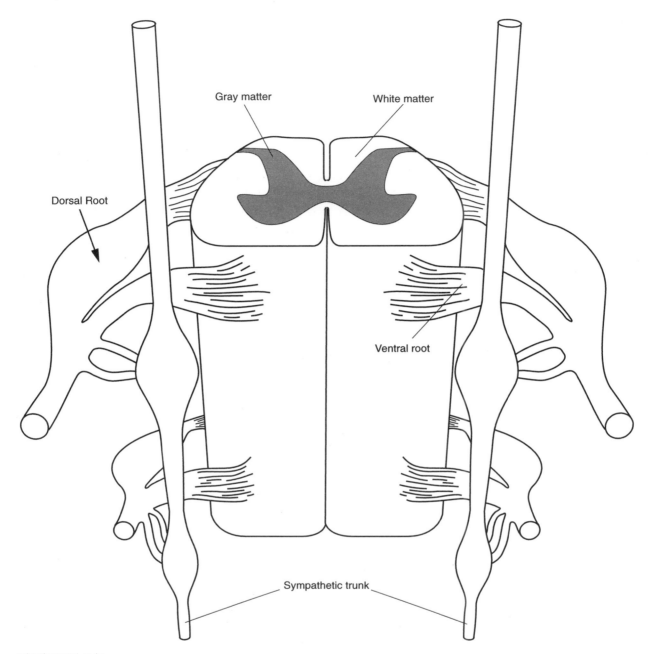

Gray matter

White matter

Dorsal Root

Ventral root

Sympathetic trunk

FIGURE 9.7

motor signals cross from one side of the CNS to the other. Thus, somatic motor signals from the right side of the CNS influence voluntary muscles on the left side of the body and vice versa.

Ascending Spinal Tracts. These tracts contain two broad categories of afferent fibers: (1) visceral afferent fibers which carry sensory signals involved in autonomic reflexes, and (2) nonvisceral afferent fibers which carry sensory signals to higher levels in the CNS. The nonvisceral afferent fibers participate in two successive synapses: (1) a synapse between a sensory neuron and an interneuron in the spinal cord or medulla, and (2) a synapse between an interneuron in the spinal cord or the medulla and a interneuron in the thalamus, a major sensory filter circuit in the CNS. Those sensory signals that leave the thalamus are interpreted by the cerebrum at the opposite side of the CNS from which they entered. Axons carrying sensory signals to the thalamus cross at the level of entry into the spinal cord, cross a few segments above entry level, or cross within the medulla. Thus, for exam-

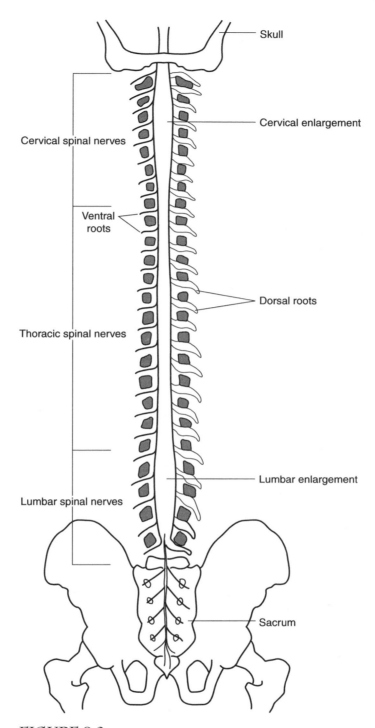

Skull

Cervical enlargement

Cervical spinal nerves

Ventral roots

Dorsal roots

Thoracic spinal nerves

Lumbar enlargement

Lumbar spinal nerves

Sacrum

FIGURE 9.8

ple, nonvisceral sensory information received by receptors on the right side of the body is interpreted in the left side of the cerebrum.

As suggested by its structure, the spinal cord carries out three functions: (1) it regulates local reflex activity; (2) it carries sensory information to higher areas in the CNS; and (3) it carries commands from higher areas in the CNS to motor neurons in the spinal cord which, in turn, cause responses in specific effectors of the body.

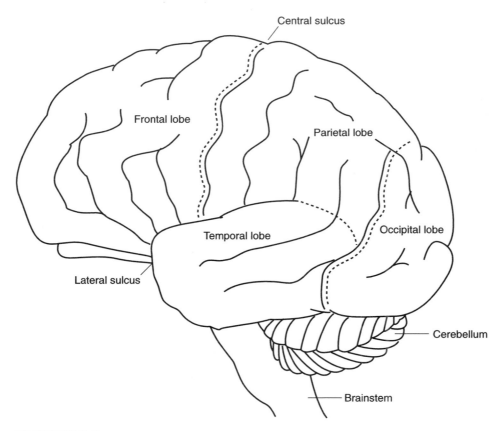

Central sulcus

Frontal lobe

Parietal lobe

Temporal lobe

Occipital lobe

Lateral sulcus

Cerebellum

Brainstem

FIGURE 9.9

The Forebrain

The forebrain consists of two cerebral hemispheres and the interbrain and, like the spinal cord, is composed of white and gray matter. However, unlike the spinal cord, some gray matter, called cerebral cortex, surrounds white matter in which are embedded masses of other gray matter called subcortical gray masses. The cerebral cortex contains neuron circuits which regulate and coordinate the activities of all other nerve circuits of the nervous system. Neurons with developmental origin, structure, and function similar to cortex neurons are present in all vertebrates. However, this system of command cells forms gray matter in a cerebral cortex only in mammals. With the great expansion of animal behavior in the evolution of mammals, a greater amount of cortex cells were needed relative to other portions of the nervous system. Therefore, to accommodate this disproportionately larger cortex system, evolution of the nervous system produced many folds, called *sulci*, in the cerebral cortex. Scientists estimate that when spread out flat, the human cortex would cover an area of 20 square feet. The deeper sulci are called fissures, and the cortex between sulci or fissures are called gyri.

The patterns of fissures, or sulci and gyri, differ somewhat from one human to another. However, there is enough similarity of pattern to use some of these sulci and fissures as landmarks for dividing each cerebral hemisphere into four lobes: frontal, parietal, occipital, and temporal lobes, see figure 9.9. Based on a number of clinical observations and experimental studies, neurobiologists have determined that certain areas of the cortex are related to specific functions. The major functional areas, as shown in figure 9.10, are (1) the primary motor area, (2) the premotor area, (3) the primary sensory area, (4) the special senses areas, and (5) the association areas.

The white matter consists of three types of tracts: projection tracts, association tracts, and commissural tracts. The projection tracts either carry motor signals from the cortex to lower neuron circuits or carry sensory signals from lower circuits to the cortex. The association tracts carry signals in fibers that connect various areas of the cerebral cortex within the same hemisphere. The commissural

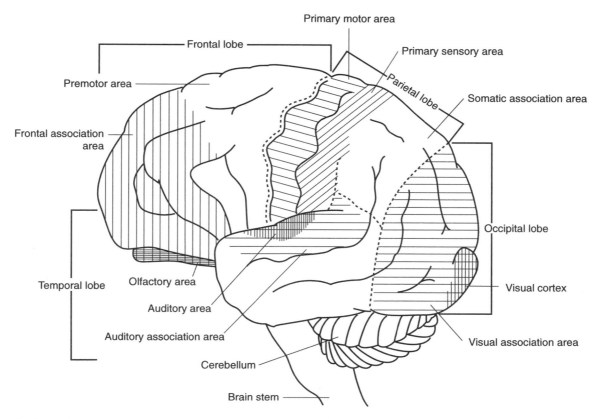

FIGURE 9.10

tracts carry signals in fibers that connect the right and left cerebral hemispheres. The two major commissural tracts are the anterior commissure and the much larger corpus callosum.

There are five masses of subcortical gray matter. The two major ones are the thalamus, and the basal ganglia, see figure 9.11. Note, the term *ganglia* is used loosely here, but this terminology is a firmly established carryover from a former way of naming nerve structures. (Ganglion usually refers to a collection of neuron cell bodies located outside the CNS). The smaller masses are the hypothalamus, the septal circuits, and the amygdala.

The Functional Subsystems

This section will list and briefly describe the overall activities of the functional subsystems of the nervous system. In general, all of these subsystems have components located in the CNS and the PNS.

The Thalamocortical System

The major CNS components of this system include the cerebral cortex, associated nerve circuits of the thalamus, and the sense receptors throughout the body. All sense organs in the body, except the olfactory sense organs for smell, send sensory impulses to the thalamus which routes some of these impulses to the cerebral cortex where they become sensations. The thalamus selects which sensory information is important enough to be further processed by the cerebral cortex. These unconscious decision-making circuits continually screen out all but a fraction of the nerve messages coming to them. It has been estimated that roughly one hundred million nerve impulses per second reach the CNS. We could not possibly attend to all these messages. All bits of information are not of equal value. A hierarchy among them has to be set up and only the most important ones can be allowed to get through to higher brain circuits. When we are put into circumstances where simultaneously many different stimuli seem important, we may throw up our hands in exasperation and not attend to any of them. Worse yet, we may try to process all these stimuli and work ourselves into a nervous frenzy.

There are five major activities and corresponding nerve tracts of the thalamocortical system: (1) the visual perception system, (2) the auditory perception (hearing) system, (3) the tactile perception (touch) system, (4) the somesthesis system, which involves sensory awareness of the body as a whole and sensory awareness of the surface and the interior of the body, including information interpreted as pain and temperature, and 5) the pyramidal system, which regulates fine control of voluntary movements.

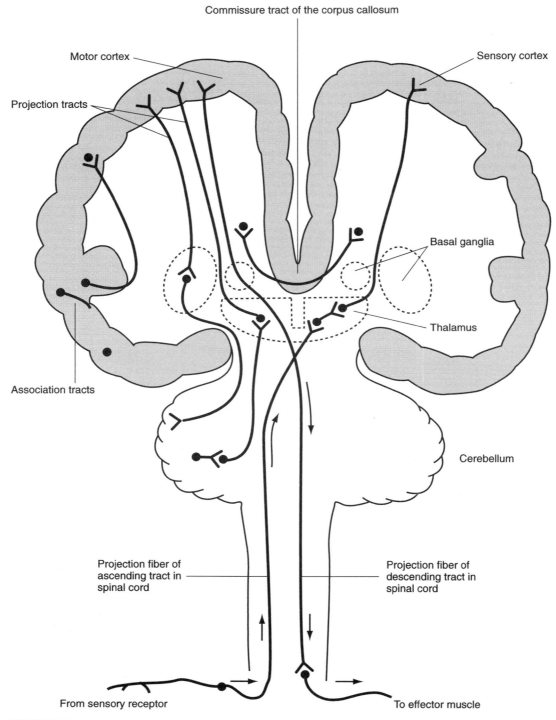

FIGURE 9.11

Lateralization

Lateralization refers to the fact that in humans each hemisphere is somewhat specialized in the type of activities it carries out. In most humans, the right hemisphere is involved with creative activities, holistic and circular thinking, and nonconceptual and analogical thinking. The left hemisphere is involved with a corresponding complementary set of activities: routine activities and habits; logical processes, for example, linear, sequential thinking; and conceptual thinking, which can be verbalized. The functions of the two cerebral hemispheres are complementary. For example, producing a work of art requires mastering routine techniques as well as having flashes of insight.

Ancient Subcortical Systems

The Limbic System.   A primitive limbic system found in all vertebrates consists of three principle structures: (1) the septal nuclei, (2) the amygdala, and (3) the hippocampus, which in mammals is a portion of the cerebral cortex in each hemisphere. In mammals the limbic system contains two major tracts: (1) the fornix, which connects the hippocampus to the hypothalamus, and (2) fibers (stria terminalis), which connect the amygdala to the septal nuclei and to the anterior portion of the hypothalamus, see figure 9.12. The key feature of the primitive limbic system is that it changes homeostasis in preparation for the animal to carry out some vigorous activity such as fighting, running away, or engaging in sexual activity. The primitive limbic system associates a particular instinctive behavior with a pattern of physiological arousal, which includes increased blood pressure, breathing rate, muscle tone, sweating, and CNS excitability. The anatomical basis for the association between instincts and physiological arousal is found in the connections among limbic structures and the hypothalamus.

An expanded limbic system found only in mammals (a similar version of this system is found in birds) is the basis for feelings, memory, and some types of learning, see figures 9.12 and 9.14.

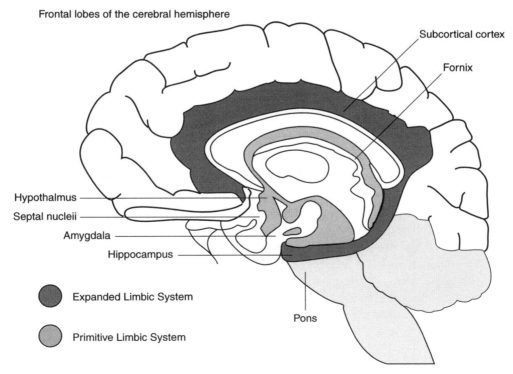

Animal communicative action

FIGURE 9.12

Medial Forebrain Bundle.   The medial forebrain bundle involves structures and functions associated with the primitive limbic system and with the autonomic nervous system (ANS).

Ancient Brainstem Mechanisms.   The ancient brain stem mechanisms include circuits regulating balance, swallowing, digestion, breathing, heart rate, and blood pressure. The core of the brain stem contains the reticular formation, which helps to regulate a human's sleep-awake cycles and overall arousal.

Animal Communicative Action

Concepts Related to Animal Social Structure

The trouble with the concept of ecological community is that it is impossible to define in any precise, nonambiguous and experimentally practical way. However, an animal community, even though temporary, can be specified by location and duration and characterized by social communications that achieve the *instrumental goals* of survival of the species and reproduction of the social community. Each individual fits into a niche in which its individual lifestyle simultaneously contributes to its survival and to the reproduction and possibly the long-term stability of the animal society by means of social communication. Any animal society presupposes social interactions defined as reciprocal social communication between two animals, i.e, social dialogue between two animals, leading to a joint behavior where each component behavior makes sense only in terms of the whole joint behavior. Accordingly, the component behaviors of a social interaction are social behaviors. Most animals and virtually all the more complex animals from arthropods on up to mammals perpetuate life via males producing sperm which fertilize eggs produced by females. These minimal sexual interactions are the most primitive forms of social behavior. The sexual behavior is a special case of what Habermas calls communicative action.

Three Types of Animal Communicative Action

I propose that Habermas's concept of communicative action is a complex, more differentiated version of instrumental communicative action that is operative in any stable ecological community. Instrumental communicative action is the network of communications between an individual organism and all aspects of its environment including other organisms that contributes to the survival of the organism. Instrumental communicative action in animal societies involves social dialogues of which there are three types: the nonconscious, instinctive communicative action and two types of conscious action, which are normative communicative action and expressive communicative action.

Animal Conscious Communicative Action

I take the view that animal consciousness is identical to conscious feelings which involve subjective orientations which, of course, cannot be observed. Those who believe in these subjective behaviors attribute feelings to other humans and to some nonhuman mammals via an analogy. The person may reason as follows: "I know what it is like to see a tree and upon realizing that you have eyes and a brain similar to mine, I confidently presume that you see this same tree in approximately the same way as I do. In like manner, I know what it is like to feel anger or fear, and I realize that I show these feelings via certain facial expressions, body gestures, and physiological arousal. Therefore, when I see another human or a nonhuman animal show analogous types of facial and body expressions and physiological arousal, I presume these other animals, human and nonhuman alike, also experience feelings analogous to mine." Joseph Le Doux, takes a similar view.

> . . . conscious feelings . . . are in one sense no different from other states of consciousness, such as [perception of roundness]. . . . States of consciousness occur when the system responsible for awareness becomes privy to the activity occurring in unconscious [I believe this term should be

non-conscious; unconscious is a type of consciousness not ordinarily available to ego consciousness.] processing systems. . . .There is but one mechanism of consciousness and it can be occupied by mundane facts or highly charged emotions.[1]

The preceding quote implies that feelings are conscious emotions where emotions may be defined in terms of behavior mechanisms for survival. Again according to LeDoux:

. . . the basic building blocks of emotions are neural systems that mediate behavior interactions with the environment, particularly behaviors that take care of fundamental problems of survival. And while all animals have some version of these survival systems in their brains [instincts], I believe that feelings can only occur when a survival system is present in a brain that also has the capacity for consciousness.[2]

Besides instinctive survival behaviors—instinctive communicative actions—instinctive adaptations also may include physiological arousal, but these only occur in vertebrates with a homeostatic body plan.

Homeostatic Body Plan (see Chapter 7)

In summary, the human body consists of trillions of cells each surrounded by tissue fluid which supports cellular life. Each cell continually pollutes the tissue fluid and depletes it of nutrients. Circulating blood continually reestablishes the life supporting properties of tissue fluid, but in doing this, the blood is polluted and depleted of nutrients. Four organ systems, (1) the integumentary system (skin), (2) the respiratory system, (3) the digestive system, and (4) the urinary system, carry out exchanges between the external environment and the blood resulting in the reestablishment of the blood's life supporting properties. The lymphatic system removes excess water and debris from the tissue fluid, and the immune system in conjunction with the lymphatic and circulatory systems removes foreign substances, for example, bacteria, viruses, and poisons, see figure 7.1.

The neuroendocrine system interacts with all other organ systems to establish homeostasis in which various properties of the body are kept at near constant values. A portion of the brain called the limbic system is associated with the experience and expression of feelings. As will be described later in this section, feeling is the connecting link between human consciousness and animal consciousness which is manifested through animal feelings. Animal feeling, in turn, is the connecting link between the animal psyche and the nonconscious body. Thus, human feeling is the connecting link between the human psyche and the nonconscious human body, see figure 7.2.

Emotions

Vertebrates, like the invertebrates, have nervous system circuits that coordinate the expression of instincts. Vertebrates also have circuits contained in a portion of the nervous system called the ANS (the autonomic nervous system) that coordinate homeostatic adjustments in the animal's body. Instincts are inherited behaviors that may be modified by learning or are themselves involved with programmed learning. Vertebrates also have neuron circuits which coordinate increased activity of the circulatory and homeostatic organ systems with some instinctual behaviors, especially drives (e.g., "sex drive") That is, situations that stimulate attack, escape or courtship and mating behaviors also stimulate the animal to prepare to carry out these behaviors. This uniquely vertebrate, instinctual behavior pattern is called emotion, which may be defined as the association of increased alertness and increased activity of the circulatory and homeostatic organ systems with a particular instinctual behavior.

---

[1]LeDoux, Joseph. (1996). The emotional brain, the mysterious underpinnings of emotional life. New York: Touchstone, p. 19.

[2]Ibid., p. 125.

Some neuron circuits, for example, the primitive limbic system, in conjunction with the ANS and portions of the brain stem bring about emotion via three steps: (1) a stimulus that ordinarily activates a particular instinctive behavior also stimulates the limbic system complex to inhibit the motor programs that carry out that behavior, (2) the limbic system then stimulates the ANS to bring on physiological arousal, for example, increased sensitivity to sensations, increased muscle tone, increased blood sugar, oxygen, and various hormones, increased blood pressure and blood flow to muscles, and many other changes, (3) after animal arousal, the motor programs are disinhibited thus leading to a particular instinctual behavior. For example, when a fish is stimulated to attack, it also becomes physiologically aroused so that it can vigorously carry out this attack behavior. The association of physiological arousal with instinctive attack behavior is "attack emotion." In mammals, this behavior is called rage. When rage overwhelms a human causing him to carry out acts of violence, the person is said to be temporarily insane. Fish, amphibians, and reptiles exhibit raw emotions from time to time whereas mammals and birds usually incorporate emotions into a behavioral innovation called feeling. However, in humans feeling may degenerate into a raw emotion.

Feelings = Animal Consciousness

As indicated previously, emotions may be defined in terms of overt behaviors and physiological changes that can be observed and studied via an experimental approach. Feelings that incorporate emotions also have objective aspects. However, feelings also involve subjective orientations which, of course, cannot be observed. As one might expect, this subjective aspect of feelings makes this topic controversial especially to those many scientists who desire to define all ideas totally objectively. Many thinkers use the terms *emotions* and *feelings* to refer to the same phenomenon. Some writers attempt to define emotion equivalent to feeling in totally objective terms; others, such as LeDoux, introduce subjective aspects. There is not the same kind of consensus about the meaning of emotion and feeling as there is about other behavioral concepts such as reflex, innate behavior, or programmed learning. The view presented here is that feelings are not equivalent to emotions but rather are emotions associated with subjective orientations or dispositions toward objects. All vertebrates have emotions, but only birds and mammals have emotions incorporated into feelings. Feelings have (1) perceptive, (2) experiential, and (3) expressive aspects. The perceptive aspect is an animal's imagination of his/her orientation (intention) to some aspect of the environment. One's understanding of the concepts of perception, imagination, and orientation must come from reflection on one's subjective experiences of various feelings that express different intentions. For example, anger expresses one type of orientation, fear another, curiosity still another.

The experiential aspect of feelings is that feeling orientations always are associated with direct or remembered experiences of pleasure or pain. Of course, the terms *pleasure* and *pain* also are subjective and not describable by mechanistic analysis. A scientist must project his own subjective experiences onto observed situations that mimic his experiences. This, however, can be done in a controlled experimental manner. In 1953, Dr. J. Olds designed an ingenious experiment that demonstrated areas in the brain related to the experience of pleasure. After anesthetizing rats, he inserted electrodes into various parts of their brains and cemented them into their skulls. After the operation, the rats moved around as if nothing had happened. Olds put these rats in cages where, if each rat pushed a lever, the implanted electrodes would stimulate the surrounding nervous tissue. When the electrodes were implanted in certain well-defined areas, the rats seemed to gain "pleasure" from pressing the lever. They ignored hunger and thirst in their pursuit of "self-excitement." Often the rats would continue pressing the lever until overcome with exhaustion. Other neurophysiologists also discovered other areas that when stimulated led to apparent rage, fright, and "pain." These areas were designated as punishment centers. Most of these so-called "pleasure" (reward) and "pain" (punishment) centers are located in the portion of the brain near the primitive limbic system (i.e., in the midbrain and hypothalamus).

Feelings are expressed through a change in the pattern of muscle tone, physiological arousal, and a chosen behavior. A change in pattern of muscle tone shows up in primates as a change in facial expression, body posture, and the way the animal moves. The physiological arousal is the same as found in emotions except that the arousal ranges in intensity from low to very high. The chosen behavior may consist of gestures along with some overt activity or suppression of any overt activity.

Triune Brain in Higher Mammals

Though invertebrates are capable of some degree of learning, they are dominated by instincts that may be modified to become learned but somewhat unalterable habits. Mammals, on the other hand, are dominated by complex learned behavior patterns which incorporate some instincts but repress others. For example, a dog may be taught not to attack any stranger that approaches his territory. These learned behaviors are quite flexible in the young though they tend to fossilize into unalterable habits as the animal grows older. The increased capacity of mammals to learn is correlated with an increase in size of the brain, especially of the cerebrum. In contrast to other vertebrates, mammals have a network of interconnecting nerve cells, called cerebral cortex, at the surface of the cerebrum, see figure 9.13. The cerebral cortex is the major landmark for locating the different animal behaviors of emotions, feelings, and complex learning.

The primitive limbic system, which coordinates the expression of emotions is located in the interior of the cerebrum in all vertebrates except mammals. In mammals, this system is located in that part of the surface of the cerebrum known as the primitive cerebral cortex. This primitive cortex is close to or on the side of each cerebral hemisphere which faces the brain stem upon which the hemispheres sit, see figure 9.14. The newly evolved nervous network involved with complex learning is called neocortex; *neo* means new. Neocortex is the lateral portion of each cerebral hemisphere. In primates, the neocortex makes up most of the cerebral cortex. A part of the cerebral cortex between the primitive cortex and the neocortex is called *mesocortex; meso* means in the middle. The mesocortex has been shown to be involved with the expression of feelings. Thus, all but the most primitive mammals may be thought of as having "three brains in one" referred to as the triune brain: (1) the emotional brain consisting of the primitive cerebral cortex in association with the brain stem and the spinal cord, (2) the feeling brain consisting of the mesocortex which has many connections with the emotional brain and the learning brain, and (3) the learning brain consisting of the neocortex, see figure 9.14. For the most part in all developmentally normal, healthy, nonhuman mammals, these three brains are integrated in the functioning of a unified triune brain. Feelings that incorporate emotions integrate with complex learning to make up unified behavior patterns.

The previous set of statements is a modified version of Paul MacLean's theory. Le Doux argues that the anatomical aspect of this theory no longer is tenable.

> Emotion is only a label, a convenient way of talking about aspects of the brain and its mind. . . . In a similar vein, the various classes of emotions are mediated by separate neural systems that have evolved for different reasons. The system we use to defend against danger is different from the one we use in procreation, and the feelings that result from activating these systems—fear and sexual pleasure—do not have a common origin. There is no such thing as the "emotion" faculty and there is no single brain system dedicated to this phantom function.[3]

I accept LeDoux's critique, but we still may use the idea of a triune brain even though there are no specific, recognizable brain structures corresponding to each brain. Rather there are separate neural systems scattered throughout the central nervous system that regulate the expression of emotions. We may refer to this set of separate neural systems as the emotion brain. The set of neural systems that associate awareness with emotions is the feeling brain. The set of neural systems that process information independent of emotions or feelings is the learning brain.

---

[3]Ibid., p. 16.

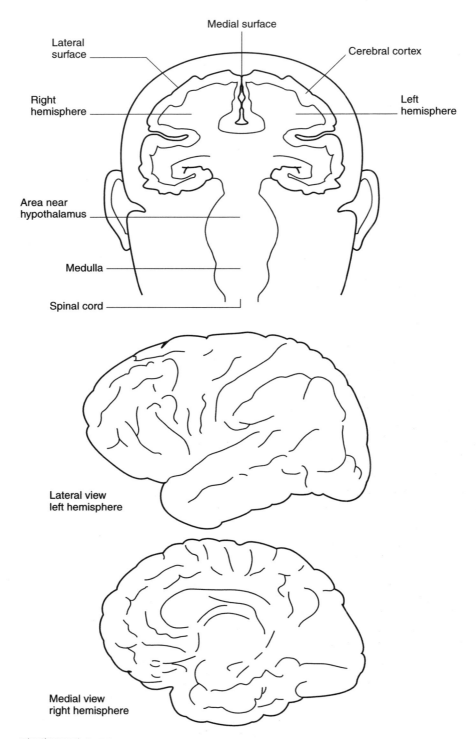

FIGURE 9.13

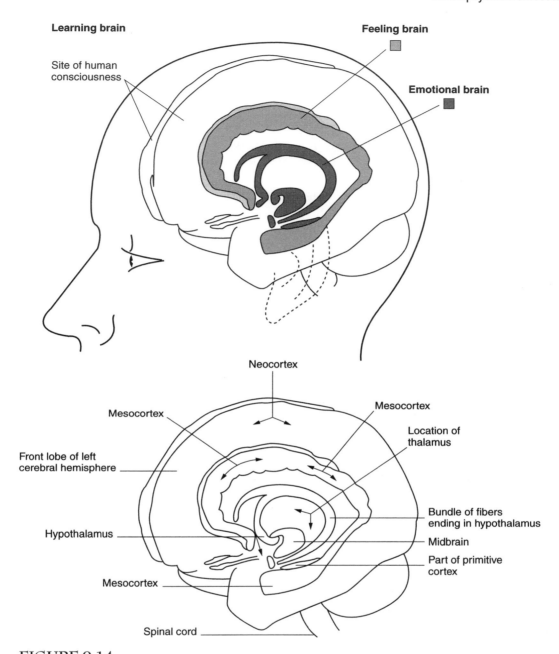

FIGURE 9.14

Homeostatic Reflexes of Physiological Arousal

Regulation of Blood Pressure

When we are at rest, all the blood passes through the body in about one minute. This rapid pass-through of blood is achieved by a high difference in pressure between blood coming out of the heart into the arteries and that flowing back into the heart via the veins. Thus, if the average pressure in the arteries drops still more, there will not be enough pressure difference to force all the blood around back to the heart. Blood will tend to accumulate in some body tissues rather than return to the heart.

The blood pressure must be maintained at a sufficient level at all times to ensure the circulation of all the blood. But if the blood pressure is always much higher than is required, the heart pumps harder than is normal and wears out faster. During times of activity ranging from walking to strenuous exercise, it is essential that the blood pressure be adjusted to meet the oxygen, nutrient, and waste removal demands of the body tissues. All this indicates how vitally important it is for the body to closely adjust the blood pressure to maintain the minimum value for circulation of all blood and to meet the changing needs of the body.

Since the blood pressure is near zero in the capillaries and veins, it is the pressure in the arterial vascular tree that must be regulated. This pressure is determined by how much blood is present. The more blood present, the greater the pressure. Blood enters the vascular tree from the heart and leaves through the arterioles. Increase in heart rate and strength of contraction of each beat will increase the rate at which blood enters the arterial vascular tree. Constriction of arterioles will decrease the rate at which blood leaves the vascular tree. Thus, an increase in heart rate, an increased strength of heartbeat, and constriction of arterioles will increase the amount of blood in the vascular tree and thereby increase blood pressure. Just the opposite set of events will decrease blood pressure. The regulation of blood pressure may be summarized in terms of the function of a sense receptor and a control center.

1. Sense receptor: nerve endings in the wall of the carotid sinus, see figure 9.15, which continually record the blood pressure in the carotid arteries. This "recording" represents the blood pressure in any large artery close to the heart.
2. Control center: nerve centers in the medulla and hypothalamus.
   a. These centers average the moment-to-moment blood pressure and compare it to some present value which can vary depending upon the level of activity of the individual.
   b. The center makes the appropriate decision.
      1) If the average blood pressure is below the preset value, motor neurons are stimulated to bring about:
         a) an increase in heart rate
         b) an increase in strength of each heartbeat
         c) a net constriction of arterioles throughout the body

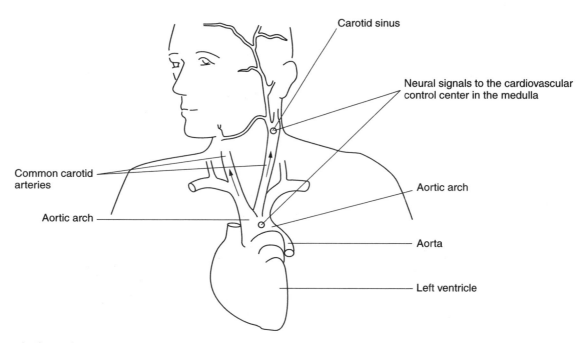

FIGURE 9.15

2) If the average blood pressure is above the preset value, motor neurons are stimulated to bring about:
   a) a decrease in heart rate
   b) a decrease in strength of heartbeat
   c) a net dilation of arterioles throughout body

All the other homeostatic mechanisms will be summarized in terms of the function of sense receptors and control centers.

Regulation of Body Temperature

Short Term Response

1. Sense receptor: cells in the hypothalamus which continually record the temperature of blood passing through it.
2. Control center: nerve center in the hypothalamus
   a. Compares blood temperature to the preset value of 37°C.
   b. The center makes the appropriate decision.
      1) If the temperature is below the preset value, motor neurons are activated to bring about:
         a) constriction of arterioles in the skin and dilation of arterioles in the deeper portions of the body.
         b) shivering. This rhythmic contraction of voluntary muscles produces more heat energy as a result of burning up more sugar.
      2) If the temperature is above the preset value, motor neurons are activated to bring about:
         a) dilation of arterioles in the skin and constriction of arterioles in the deeper portions of the body.
         b) sweating. The evaporation of water in sweat from the surface of the body decreases body temperature.

Long Term Response

The same sense receptors and control center as in short term response for temperature regulation are involved. If the body temperature tends to consistently fall below or rise above the preset value, the appropriate decision is made. If the temperature is below the preset value, the hypothalamus stimulates the pituitary gland which, in turn, stimulates the thyroid gland to increase secretion of thyroxin. This increases the rate of chemical reactions in all cells of the body thereby producing more heat energy which raises the body temperature. If the temperature is above the preset value, just the opposite set of events occurs.

Regulation of Carbon Dioxide Concentration of the Blood

Usually a high carbon dioxide concentration is associated with a low oxygen concentration, and conversely a low carbon dioxide concentration is associated with a high oxygen concentration. Therefore, regulation of carbon dioxide concentration indirectly regulates oxygen concentration as well.

1. Sense receptor: little bodies called carotid bodies and the aortic body suspended in the carotid sinus and the aortic arch respectively, see figure 7.1. These sense receptors respond to changes in carbon dioxide concentration.
2. Control center: medulla and other parts of the brain stem.
   a. Compares recorded carbon dioxide concentration to a preset value.

b. The center makes the appropriate decision.
   1) If carbon dioxide concentration is above the preset value, neurons are activated to bring about an increase in rate and depth of breathing.
   2) If carbon dioxide concentration is below the preset value, neurons are activated to bring about a decrease in rate and depth of breathing.

Regulation of Acidity in Blood

Carbon dioxide reacts with water to produce carbonic acid. Therefore, an increase in carbon dioxide concentration increases acidity and a decrease in carbon dioxide concentration decreases acidity. Consequently, by regulating carbon dioxide concentration the body also regulates the acidity of blood. The sense receptors and control center for regulating blood acidity are located in the hypothalamus.

Regulation of Concentration of Glucose in Blood

Glucose is a simple sugar molecule. Several thousand of these molecules may be joined to form a large molecule which in animals is called *glycogen*. All cells of the body continually take up and combine glucose with oxygen to produce carbon dioxide, water, and energy. Some of this energy is used to keep the cell alive. Since animal cells need a continuous supply of energy to stay alive, it is important that the concentration of glucose in the blood be carefully regulated. The regulation of glucose is quite complex; the following is an overview of its regulation in humans.

Primary Site of Regulation

The pancreas, which secretes *insulin*, is the primary site for regulating blood glucose concentration. Insulin stimulates many cells, particularly muscle cells and cells that store fat, to increase the rate at which they take up glucose from the blood. Insulin also increases the rate at which the liver converts glucose into glycogen. Thus, a change in the rate of secretion of insulin alters blood glucose concentration. Cells in the pancreas function as sense receptors and a control center by altering insulin secretion in response to changes in blood glucose concentration. If blood glucose concentration rises above a preset value, these cells increase the secretion of insulin. If blood glucose concentration falls below the preset value, these cells decrease the secretion of insulin.

Secondary Sites of Control

Brain cells are specifically adapted to rapidly take up glucose without requiring insulin. These cells are continuously active and begin to degenerate after only a few minutes of lower than normal supply of glucose. To insure survival of these sensitive brain cells, there are back up, secondary control centers that help keep blood glucose concentration sufficiently high.

1. Pancreas. Some cells in the pancreas secrete the hormone *glucagon*, which leads to an increase in blood glucose concentration. This hormone stimulates liver and muscle cells to break down glycogen into glucose and stimulates the liver to convert protein into glucose.
2. Hypothalamus-Sympathetic System. Activation of the sympathetic system leads to the secretion of *epinephrine* which is the hormone produced and secreted by the adrenal medulla. Epinephrine has the same effect as glucagon.
   1. Sense receptor: Cells in the hypothalamus which record blood glucose concentration.
   2. Control center: Nerve center in the hypothalamus.
      a. Compares glucose concentration to a preset value.

   b. The center makes the appropriate decision.
      1) If the glucose concentration falls below the preset value, the hypothalamus stimulates the sympathetic division to stimulate the adrenal gland to secrete epinephrine.
      2) If the glucose concentration is equal to or above the preset value, the adrenal gland is *not* stimulated to secrete epinephrine.
3. Hypothalamus-Pituitary Axis. The growth hormone secreted by the anterior lobe of the pituitary gland has the same effect as glucagon and epinephrine. It has the additional effect of decreasing glucose uptake by muscles and other tissues.
   1. Sense receptor: Cells in the hypothalamus which record blood glucose concentration.
   2. Control center: Nerve center in the hypothalamus.
      a. Compares glucose concentration to a preset value.
      b. The center makes the appropriate decision.
         1) If glucose concentration falls below the preset value, the hypothalamus, via the releasing hormone (GHRH) stimulates the anterior pituitary gland to secrete growth hormone.
         2) If glucose concentration rises above the preset value, the pituitary gland is not stimulated to secrete growth hormone.

Regulation of Water Concentration of Blood

1. Sense receptors: Cells in the hypothalamus which continually record water concentration of the blood.
2. Control center: Nerve center in the hypothalamus.
   a. Compares water concentration to the preset value.
   b. The center makes the appropriate decision.
      1) If water concentration falls below the preset value, the hypothalamus:
         a. stimulates the posterior pituitary to increase secretion of antidiuretic hormone (ADH). This increases the reabsorption of water by nephrons in the kidney resulting in lower amounts of water in the urine.
         b. stimulates the thirst center in the hypothalamus.
      2) If water concentration rises above the preset value, the hypothalamus:
         a) decreases stimulation to the posterior lobe of the pituitary gland resulting in decrease secretion of ADH and consequent loss of water in the urine.
         b) inhibits the thirst center.

Physical-Psychic Stress

General Adaptation Syndrome (GAS)

Stages of Response to a Stressor.   The most fundamental unit of life on earth is an organism-environment interaction. An organism's life is expressed in terms of continuous adaptations to its environment. A vertebrate carries out this continuous adaptation by maintaining homeostasis of its internal milieu. The homeostasis manifested in an animal's day-to-day survival may be thought of as an average level of resistance to stressors. A stressor-stress interaction is a departure from this average resistance. Usually the stressor either disappears as it is overcome. If it remains or continually reappears, the chronic stressor produces a sustained disease, what Selye called "disease of adaptation." The stages in the production of diseases of adaptation are (1) alarm reaction, (2) stage of resistance, and (3) stage of exhaustion.

The general adaptation syndrom (GAS) involves a specific pattern of (neuroendocrine) responses, but it is easier to describe the general nature of these stages of GAS in terms of a mechanistic medicinal example. Suppose a valve in the heart (the left atrioventricular valve which was described in Chapter 5) becomes leaky (perhaps due to rheumatic fever); this is the stressor. The alarm reaction occurs when the heart is unable to maintain sufficient blood pressure to sustain normal vigorous activity. The stage of resistance takes over with increased rate and strength of heartbeat and net constriction of arterioles. Now the individual can maintain normal blood pressure, but the heart is continually doing more work than in "normal" individuals. Loosely speaking, we say the heart is continually "under stress." After a period of time, years perhaps, the individual enters the stage of exhaustion and begins to show signs of heart failure. At this point he may adjust either to having heart disease or he may die.

Neuroendocrine Aspect of GAS.   Selye discovered that the pituitary secretion of the hormone called ACTH is a unifying feature of GAS. We now know that the hypothalamus stimulates the pituitary gland to secrete much higher than normal levels of the growth hormone and ACTH. By enhancing growth of body tissue in general, the growth hormone enhances growth of lymphatic tissues of the immune system and growth of connective tissue. High levels of ACTH stimulates the adrenal cortex, the outer portion of the adrenal gland, to secrete mineralcorticoids, a type of hormone, as well as another hormone called cortisol. By increasing sodium retention, the mineralcorticoids cause: increased water retention ⇒ increased blood volume ⇒ increased blood pressure. This sequence of events increases the body's potential to bring blood to a site of injury or site of local irritation. Overall, both the growth hormone and the mineralcorticoids enhance inflammatory and immune responses. Cortisol acts as a general tissue enhancer which supports the capacity of tissues to carry out peak rates of chemical reactions required to sustain homeostasis. In particular, cortisol enhances protein conversion into glucose, inhibits adrenal cortex secretion of mineralcorticoids, decreases local inflammatory processes, and decreases immune responses.

The growth hormone and mineralcorticoids dominate in the alarm stage followed in a few days by the stage of resistance in which cortisol dominates. In addition to facilitating tissue resistance to changes of homeostasis, cortisol decreases inflammatory and immune responses. Inflammation and immune responses have contradictory effects on the body's well being. For example, the increased blood flow that occurs at a site of inflammation increases the removal of toxins and dead cells from the area. However, inflammation leads to excess water (edema) in the area, and sometimes the harmful effects of this excess water (edema) override the benefits of local increase of blood flow. Initially an immune response counteracts the invasion by foreign substances. However, the prolonged effects of immune responses may cause local tissue damage. Thus, cortisol is the chemical equivalent to ice. Just as ice reduces excess water in tissues (edema) and the harmful effects stemming from it, cortisol reduces inflammation and immune responses and reduces the harmful affects stemming from them.

Repeated or sustained stressors produce chronic stress which leads to the stage of exhaustion. In this stage one may see kidney damage, high blood pressure or allergic reactions due to the repeated excesses of growth hormone and mineralcorticoids. One may also see many more infectious diseases due to inhibition of the immune system by cortisol.

Nervous System Aspect.   Via brain stem nerve centers and the ANS (the autonomic nervous system) the hypothalamus coordinates negative feedback mechanisms for maintaining homeostasis. In particular, the hypothalamus activates the sympathetic division of the ANS and stimulates the adrenal gland (its middle part) to secrete the hormone, adrenalin, as well as stimulating the pituitary gland to secrete ACTH. Thus, the GAS also includes the effects of sympathetic activation and adrenalin which are: increased heart output and blood pressure, increased rate and depth of breathing, breakdown of glycogen into glucose, diversion of blood flow from the digestive system to the skeletal muscles, desensitization to pain, and increased alertness. Overall, the nervous system aspect (ANS aspect) of GAS is the neuroendocrine system's alteration of homeostasis in an attempt to adapt to some stressor. A summary of the components of the ANS aspect of the GAS response follows.

The ANS is stimulated resulting in:

1. At first an increase and then a decrease of endorphins [this may lead to migraine and backaches].
2. Increase alertness—this may interfere with sleep and relaxation.
3. Increase in blood pressure.
4. SD overriding the PD control of the GI tract—this may lead to constipation or diarrhea.
5. Increase in muscle tone—this may lead to backaches and general body fatigue.
6. Increase air supply to lungs [due to bronchial dilation].
7. Increase sweating.
8. Increase secretion of adrenalin by the medulla of the adrenal gland—this reinforces the stimulation of the SD which brings about physiological arousal and local SD overriding the PD [SD and PD are divisions of the ANS].

Emotional Aspects of Stress in Humans

Nonhuman animals in the wild are well adapted to their environments. Behavioral adaptations involve fight, flight, or sexual activity in response to appropriate stimuli. The limbic system of vertebrates facilitates these behavioral adaptations in the following sequential manner: (1) the limbic system first inhibits the instinctual behavioral response to the animal's perception of some stressor; (2) by stimulating the hypothalamus, the limbic system then alters the animal's homeostasis in preparation for some vigorous activity, for example, flight, fight or sexual activity; (3) the limbic system then associates physiological arousal with a particular instinctual behavior. As described earlier, this last activity is called emotion, and when awareness is associated with it, the activity is conscious emotion = feeling. All vertebrates express emotions though only birds and mammals express feelings. Usually after a flight, fight, or sexual response, the animal's homeostasis returns to normal levels. In general, nonhuman animals in the wild do not experience chronic stress.

Civilized humans are misfits in nature. In civilized societies, humans are trained to repress many of their feelings which are conscious emotions. Most people would agree that the control of these conscious emotions is one factor that enabled humans to become the dominant species on earth. Nevertheless, humans had to pay for this adaptation, and the price is particularly high in modern societies. Many humans, on a day-to-day basis, encounter situations that start to activate flight, fight, or sexual responses, but the civilized human represses the behaviors associated with these responses. However, most humans cannot completely repress the associated emotional response. As a result, the civilized individual often may become intensely aroused in the course of each day. Eventually, this chronic physiological arousal associated with repressed feelings (conscious emotions) leads to the stage of exhaustion and thus produces some disease of adaptation. This is the mechanistic theory of stress-related diseases in humans.

Isolated Human Stresses

Modified Definition of Stress

As animals become more complex, their behavior becomes a more dominant feature which cannot be described totally in terms of physiological processes. In mammals and birds, complex learning becomes a major determinant of behavior, and this learning seems to emerge out of a kind of animal awareness which is manifested as intentions and motivated behaviors. Just as the body underlies function, so also the extensive learning which is the basis of an animal's individuality, called *psyche*, underlies awareness, intentions and motivated behaviors. The developed psyche and the body from which it emerged are separate but complementary aspects of adult mammals. Thus, any mammal, but especially a human being, may be viewed as a psyche-body unity. Stress, then, should be redefined as a departure from the psyche-body homeostasis. Body homeostasis is defined in terms of properties of the internal milieu, but what is psyche homeostasis? It is difficult to define psyche homeostasis, but we can readily discern obvious departures from it. In lower vertebrates, these departures are called emotions; in mammals, the analogous but much more diverse departures are called *feelings*. This is not to say that mammals, especially humans, do not have emotions. As described earlier, in mammals emotions are incorporated into the more complex behaviors called feelings.

Stressful Effects of Feelings

Passionate Stress.   Stress is like water; water is essential for life, but if one drinks too much water, it becomes a poison. The blood and then the tissue fluid becomes so diluted that cells swell and die. Likewise, stress, a type of adaptation, is one of the major themes of life, but too much stress brings illness. We see this in regard to passion which among other things is the expression of intense feelings. Suppose a person is consumed with excitement or curiosity in a chosen field of study, or exhilarated with personal performance as a speaker or artist, or wonderfully excited by being caught up in a whirlwind of activities, or passionately in love. Any one of these positive feelings impels one to say yes to life and be glad to be alive. However, these feelings involve physiological arousal and cost the body energy. This is a price we willingly pay, but if these highs are continually repeated for an extended period of time, the person enters the stage of exhaustion and becomes ill. Happy stress can produce disease as can any chronic passion.

Unresolved Feelings Produce Illness.   Modern society often puts people in situations where they dare not express how they feel. For example, a student may become enraged at a teacher, but in general it is not in the student's best interest to act out this anger. One may suppress overt behavior and hide behind a mask of composure, but the orientation underlying the anger remains and is expressed through some form of physiological arousal. The body becomes prepared to carry out some intense activity that never comes to pass. If a feeling activity such as screaming were to be carried out, *a* (not the) goal of the feeling orientation would be achieved and the physiological "energy" of the feeling would be released. The psyche-body unity would tend to return to normal homeostasis thus resolving the feeling. When feelings are not expressed in any overt activity, the psyche is frustrated and the body continues to be aroused; the feeling is unresolved. Resolved feelings accomplish some immediate goals with the minimum expenditure of energy; unresolved feelings accomplish no immediate goals while dissipating a great deal of energy.

Humans constantly choose immediate frustrations for the sake of long-term goals. These unresolved feelings fade after a while; they may cost us some energy, but usually no damage is done. Moreover, we usually experience other intense but resolved feelings that lead us back to psyche-body homeostasis. However, suppose an individual becomes consumed by an intense unresolved feeling. Often the unresolved feeling is guilt, anger or anxiety that may be provoked by one's school, home, or work situation. The nervous system has a "short circuit." Not only is there a continual drain of energy, but parts of the body are overloaded with physical activity. After a few months or years, the individual may develop a stress related illness.

Medical Model of Human Stress

Selye's original model for stress diseases may be represented as:

$$\text{Stressors} \Rightarrow \text{Stress} \Rightarrow \text{Illness}$$

His expanded model is:

$$\text{Stressor} \Rightarrow \text{Repressed Emotions} \Rightarrow \text{Chronic Expanded GAS} \Rightarrow \text{Illness}$$

Further research in the 1970s and 1980s indicated that stress in humans is much more complex than Selye's mechanistic model for it. More and more medical scientists have come to appreciate that humans have "psyches" which cannot be reduced to the physiology of the nervous system. At the same time, the human psyche (some call it the mind) has no meaning apart from the body; that is, the psyche has meaning in terms of the nervous system which has meaning only in terms of the whole body. Thus, psyche and body are complementary aspects of each human being.

The psyche-body complementary interaction implies that stress involves psychological as well as physiological aspects. The psychological aspects produce a diversity of responses to the same stressor. An objective situation or agent may produce a stressor-stress interaction in one individual but not in another. Current psychological-medical research indicates three factors that produce this diversity: coping mechanisms, appraisal, and mediating factors.

Humans have many coping mechanisms by which they reduce or terminate a stressor-stress interaction. For example, some students have developed techniques for relaxing, building self-confidence, and having a positive attitude toward taking exams which once produced great anxiety. A physical contest may produce anxiety, yet a seasoned athlete can stay calm or not be emotionally distracted during a key play or an important point in a game. Virtually all humans have some coping mechanisms for dealing with some stressors. One measure of maturity and success in life is the effective use of a number of coping mechanisms, but very few, if any, individuals can cope with all types of stressors. Also, humans differ relative to the kinds of coping mechanisms they possess. For example, some individuals cope well with physical dangers but experience anxiety from personal relationships. Others, however, can enjoy personal relationships but experience anxiety from physical risks. At any rate, the stress model for humans should be modified to:

$$\text{Stressor} \Rightarrow \text{Stress} \Rightarrow \text{Coping} \quad \nearrow \quad \text{Unsuccessful} \Rightarrow \text{Illness}$$
$$\searrow \quad \text{Successful} \Rightarrow \text{Health}$$

Many stressor-stress interactions are preceded by knowledge interactions, and according to an ecological perspective (and contrary to the notion of objective knowledge in the mechanistic perspective), knowledge always has a subjective aspect. Knowledge of the world to some extent is determined by one's attitudes and preconceived ideas. Knowledge (in the context of an ecological perspective) when applied to stress theory implies that many stressors are determined by one's appraisal of a situation. Going to the dentist for some people is "no big deal" while others feel the need to take tranquilizers. The student who does not care whether he passes a course or does not need high grades probably will not find taking exams very stressful, but the student of average ability who needs high grades in order to get into some professional graduate school may find exams very stressful. Therefore, the stress model should be modified to include the appraisal factor:

$$\text{Stressor} \Rightarrow \text{Appraisal} \Rightarrow \text{Stress} \Rightarrow \text{Coping} \quad \nearrow \quad \text{Unsuccessful} \Rightarrow \text{Illness}$$
$$\searrow \quad \text{Successful} \Rightarrow \text{Health}$$

Many psychologists maintain that still another idea should be added to the medical model of stress. Any social unit involves many different interactions among its members, and each of us is greatly influenced by the social units—society, family, set of friends, school, work, place of recreation—of which we are members. While some of these social interactions may be stressful, others may aid one in overcoming stress. Shared religious beliefs and rituals, for example, have helped many people through difficult times. The encouragement and love of family and close friends can give a student that extra "something" that enables her to cope with the stress of exams. These social factors which mediate between stressors and illness may be represented as follows:

$$\text{Stressor} \Rightarrow \text{Appraisal} \Rightarrow \text{Stress} \Rightarrow \text{Mediating factors} \Rightarrow \text{Coping} \nearrow \text{Unsuccessful} \Rightarrow \text{Illness} \searrow \text{Successful} \Rightarrow \text{Health}$$

Tearout #1

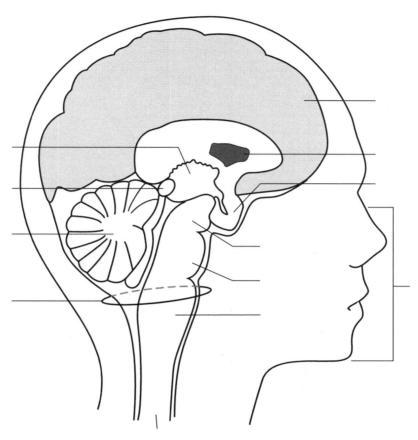

FIGURE 9.6

Directions:

1. Using figure 9.6 as a reference, fill in all labels.

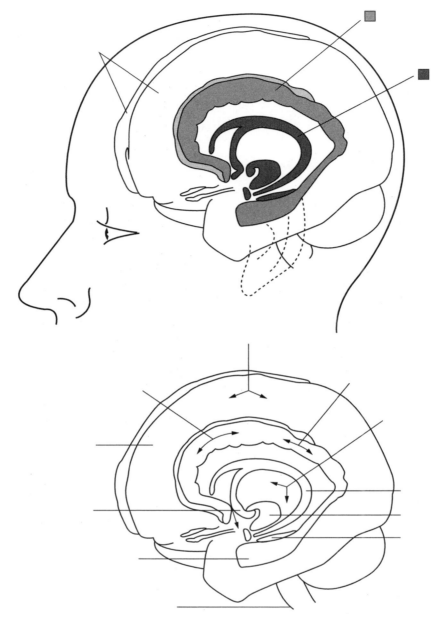

FIGURE 9.14

2. Using figure 9.14 as a reference, fill in all labels *and* indicate in this figure where each "brain" of the triune brain is.

Tearout #2

Directions:

1. Describe how unresolved feelings could lead to chronic high blood pressure.

Unsuccessful $\Rightarrow$ Illness

Stressor $\Rightarrow$ Appraisal $\Rightarrow$ Stress $\Rightarrow$ Mediating factors $\Rightarrow$ Coping

Successful $\Rightarrow$ Health

2. Explain the meaning of each term in the above schematic.

Ecosystem as a Homeostatic Machine

Ecology studies interactions among all things, living or nonliving, in the biosphere. Ecologists hope that this scientific knowledge will enable them to control the biosphere in order to enhance human life. However, the word *ecology* is used with two radically different meanings. As a division of biology, ecology is yet another discipline for objectively describing how things interact. Ecology also is used in contexts indicating that it has a meaning very different from scientific ecology. Social philosophers such as Robert Bellah and Jurgen Habermas point to ideas such as social ecology or moral ecology that bring about an integration ". . . between a public world of competitive striving and a private world supposed to provide the meaning and love that make competitive striving bearable."[1] This chapter focuses on scientific ecology but in such a way that it provides a foundation for formulating a systems perspective that includes the ideas of social and moral ecology.

Ecosystem as a Machine

Ecologists have worked out an ecosystem theory that is analogous to the cell theory. Just as the cell is considered the structural-functional unit of any individual life form, the ecosystem is considered the structural-functional unit of the biosphere. Thus, the overall structure, function, development and evolution of the biosphere may be understood in terms of interactions among the constituent ecosystems. Moreover, the biosphere is itself an ecosystem. Even if we cannot precisely define the boundaries and structure of constituent ecosystems, to some extent we can define these characteristics for the biosphere considered as a whole unit.

As an environmental unit of study, an ecosystem is a very complex pattern of interactions. One aspect of this pattern is that diverse organisms continually do the biological work of staying alive, reproducing, and maintaining certain ways of interacting with one another and with the physical-chemical environment. Thus, an ecosystem is like a machine. The sum of the biological tasks of all the component organisms is equal to the total biological work of the ecosystem. This work is accomplished as a result of energy flow through the components of the ecosystem and the cycling of energy couplers in the system.

---

[1]Robert Bellah, *Habits of the Heart* (New York: Perennial Library, 1985), p. 292.

Components of an Ecosystem

Any complete ecosystem consists of (1) abiotic factors, (2) producers, (3) consumers, and (4) decomposers. Abiotic factors include physical factors and the elements and the inorganic and organic compounds in the environment. Compounds exist in high, intermediate or low energy levels. For example, proteins, lipids, and sugars are substances at high energy levels—they contain a great deal of chemical energy that is released when they are broken down into small molecules. Ammonia (NH_3), amino acids, urea, uric acid, and glycerol are molecules with intermediate amounts of chemical energy. Carbon dioxide (CO_2), water carbonate (HCO_3), nitrate (NO_3), nitrite (NO_2), and phosphate (PO_4) have low amounts of chemical energy. Producers are organisms that are able to manufacture high energy compounds from intermediate or low energy substances (usually inorganic substances). All green plants, green algae, and some single-celled organisms produce energy-rich substances via photosynthesis; that is, they absorb energy from the sun to convert chemicals from low to high energy levels. Other single-celled organisms, for example, some bacteria, produce energy-rich compounds by a process called chemosynthesis. Consumers are organisms that ingest energy-rich organic compounds and/or other organisms. Consumers may be divided into herbivores which, are organisms that ingest plants or plant products, and carnivores, which are organisms that ingest animals. Decomposers are organisms such as fungi and bacteria that break down complex molecules of dead organisms, absorb some of the decomposition products, and release simple low-energy substances that can be utilized by producers. Decomposers may be divided into scavengers, which are animals that eat other dead animals and saprobes that include organisms like fungi and bacteria.

Energy Flow through Ecosystems

There are five major processes involved with energy flow through an ecosystem: (1) energy flow into ecosystems, (2) production, (3) consumption, (4) decomposition, and (5) heat energy loss.

Energy Flow into Ecosystems.   Energy from the sun is taken into the biosphere in three ways. Much of sunlight is converted into heat energy which is absorbed by the earth. Sunlight energy also causes water in the oceans, lakes, and rivers to evaporate. Finally, 1 to 1½% of sunlight is absorbed by pigments in photosynthetic organisms. Component ecosystems of the biosphere take in energy in a variety of ways. For example, rivers bring energy-rich organic substances, e.g., leaves, into lake ecosystems.

Production.   Photosynthetic organisms use absorbed sunlight to convert carbon dioxide and water and soil minerals into organic compounds. During this process, plants and green algae release oxygen to the atmosphere in contrast to many photosynthetic bacteria that do not produce oxygen as a by-product. Some bacteria produce energy-rich compounds via chemosynthesis. During this process, energy-producing chemical reactions (oxidations) enable bacteria to convert inorganic compounds into organic substances. In most ecosystems, bacteria are primarily decomposers rather than producers.

Consumption.   During consumption, consumers ingest and then chemically extract (oxidize) energy from organic compounds. These chemical reactions are coupled with energy-consuming reactions that underlie all life processes. Consumers consist of primary consumers, which are herbivores; secondary consumers, which are carnivores; and multilevel consumers, which are omnivores—they ingest producers and primary and secondary consumers.

Decomposition.   During decomposition, complex organic substances are broken down into simpler compounds that can be utilized by plants. All organisms carry out decomposition to some extent. For example, most cells break down (oxidize) organic compounds into carbon dioxide and water, which are used in photosynthesis. Besides this kind of decomposition, there are decomposers that obtain energy to sustain their life processes by breaking down organic compounds in dead organisms. This frees up essential nutrients to be used by other living organisms.

Heat Energy Loss.   In general, all energy that comes into an ecosystem passes out again usually as heat energy. In some ecosystems, such as a river, organic compounds bring energy into and carry energy away from the system. Sometimes incoming energy is used to make compounds such as petroleum and coal, which may keep chemical energy in the system indefinitely. All processes in the universe convert some energy into heat energy, and usually all the energy that accomplishes biological work eventually shows up as heat energy, which is radiated away from the ecosystem.

Trophic Structure of Ecosystems

Food Webs and Food Chains. The actual network of energy transfer between producers, consumers, and decomposers is called a food web, and any linear sequence of species in the food web is called a food chain. The various species within a food web may be put into groups according to how many transductions they are away from the light energy from the sun. These groups of organisms are known as trophic levels. The producers are the first trophic level; the herbivores are the primary consumers and represent the second trophic level. Very often ecosystems contain primary carnivore animals that eat herbivores; the carnivores are the secondary consumers and represent the third trophic level. Various bacteria and fungi decompose organic material from any or all of the other trophic levels. As a result they are grouped together as a single trophic level called the saprobes. The scavengers and saprobes collectively make up the decomposers, see figure 10.1.

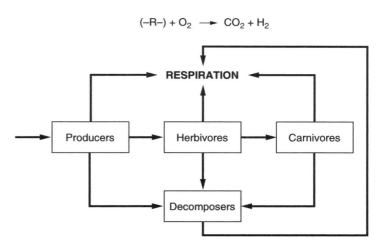

$$(-R-) + O_2 \longrightarrow CO_2 + H_2$$

FIGURE 10.1

With but a few exceptions, plants are exclusively only producers and therefore exclusively are in the first trophic level. Several species may occupy two or more trophic levels. For example, humans are omnivores; they eat plants, herbivores and meat (carnivores). Because humans usually are not prey to any other animal species, they exclusively occupy the top trophic level and thus simultaneously occupy three trophic levels. The same is true for many other animals. As a result, food webs tend to be quite complex.

Ecological Pyramids. Each organism is like a machine with regard to use of energy. Therefore, because no machine is 100%, some of the incoming energy goes off as heat energy even as it is being taken in. Much of the biological work of the organism, such as motion, is eventually converted into heat energy; so only a fraction of the incoming energy is stored in the organism as its organic constituents. Nevertheless, some energy always is stored in this way. The rate at which organisms at a particular trophic level produce organic material is called the productivity of that trophic level.

The rate at which energy comes into the first trophic level puts an upper limit on its productivity. The synthesis of high energy organic compounds cannot occur more rapidly than the inflow of energy into the producers. This upper limit is further reduced as a result of much of the incoming energy being immediately converted into heat energy or being used as maintenance energy that does the biological work of keeping the producer organisms alive. Thus, the maximum productivity of the producers is considerably less than the rate of energy inflow to this trophic level. If organisms of the second trophic level ate *all* of the organic compounds as they are produced in the first trophic level, then the producer productivity would equal the rate of inflow of energy into the second trophic level. Thus, similar to the energy inflow and productivity of the first trophic level, producer productivity puts an upper limit on the productivity of the second trophic level. However, this upper limit is further reduced as a result of much of the chemical energy coming into the second trophic level being immediately converted into heat energy or being used as maintenance energy. Thus, the maximum productivity of the second trophic level is considerably less than the productivity of the first trophic level.

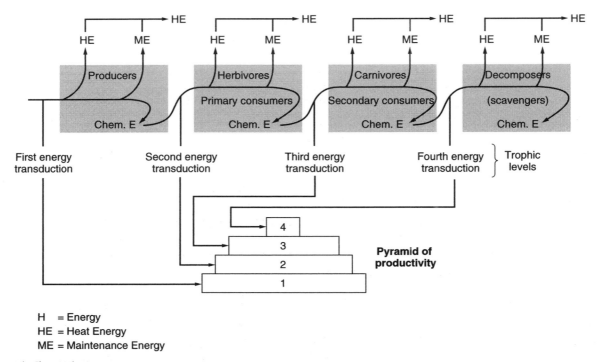

FIGURE 10.2

H   = Energy
HE = Heat Energy
ME = Maintenance Energy

The same ideas apply to all subsequent trophic levels. Therefore, when representing the amount of productivity by the width of a rectangle, the trophic structure of an ecosystem will look like a pyramid as shown in figure 10.2. In general, productivity always decreases in going from lower to higher trophic levels.

Biogeochemical Cycles

Various chemicals in an ecosystem cycle between high and low potential energy levels and in doing so focus incoming energy on doing the biological work of the ecosystem. These cycles, called biogeochemical cycles also constitute the cycles of energy couplers for the ecosystem. The pathways, rates, and degree of control of these cycles influence the complexity, diversity, and stability of ecosystems. Chemicals in the biotic portion of the cycle move through complex food webs. The same chemicals in the abiotic portion of the cycle may appear in the atmosphere, dissolved in water, in soil, and in sediments which travel back and forth between land and sea. The cyclic movements of elements involve a series of chemical and physical reactions, some of which occur in organisms and others in the physical environment. Consequently, biogeochemical cycles tend to be very complex. This section will describe the major constituents of five major cycles in the biosphere.

Water Cycle.   The water cycle nicely exemplifies the integration of energy flow through the biosphere with recycling of chemicals between low and high energy levels. In this cycle the chemical energy of water remains the same, but the water molecules cycle between high and low energy levels. As vapor, water in the atmosphere is at a high energy level, and as a liquid in the sea, it is at a low energy level. Heat energy from the sun causes liquid water to evaporate, and condensation returns water to the liquid state as precipitation. Evaporation exceeds precipitation over the oceans and precipitation exceeds evaporation over the continents. Therefore, there also is a cycling of water between the oceans and water reservoirs on land. The largest reservoir of land water is that stored beneath the surface. Much of this is so deep that very little of it actually circulates between land and ocean. The second largest amount of land water is stored in snow and ice, primarily glaciers. Most of this water also does not circulate.

As shown in figure 10.3, the general features of the water cycle are as follows. Water cycles between liquid and vapor over the oceans, but there is a net transport of some evaporated water in the air flow from oceans to continents. Water in clouds precipitates, thus raising the water level in mountain lakes. Lakes feed rivers which lead to a net discharge of water from continents into the oceans. Plants participate in the cycle by taking up water from the ground and loosing it to the atmosphere by evaporation from leaves (transpiration).

Carbon and Oxygen Cycles. The carbon and oxygen cycles are interrelated by the fact that one of the net results of photosynthesis is equivalent to the reversal of the breakdown of glucose.

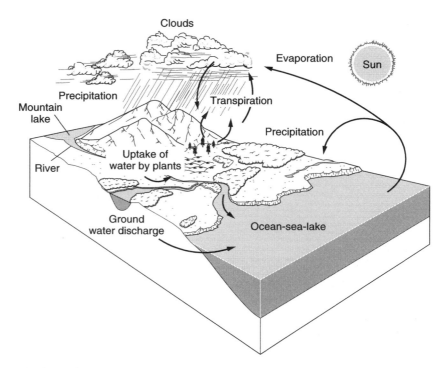

FIGURE 10.3

Photosynthesis

$$CO_2 + H_2O \Leftrightarrow Glucose + O_2$$

Oxidation

These cycles also nicely exemplify how energy flow through the biosphere is coupled with chemicals recycling between low energy states (CO_2 and H_2O) and higher energy states (glucose and O_2). The carbon in glucose and molecular oxygen have a high potential energy with respect to one another, analogous to a tennis ball in one's hand that has the potential to fall to the ground. Just as the ball-earth PE is expressed as kinetic energy when the ball falls to the ground, the carbon-oxygen potential is expressed as chemical energy when electrons in hydrogen atoms bonded to carbon in glucose transfer over to oxygen. As a result of this electron transfer, molecular oxygen becomes two water molecules [$O_2 \rightarrow 2\ H_2O$] and the carbon in glucose bonded to oxygen and hydrogen becomes exclusively bonded to oxygen as a result of loosing the hydrogens. [$-CH_2-$ or $-CHO- \rightarrow CO_2$]. Thus the carbon in glucose and the molecular oxygen go from a high chemical potential energy to a lower chemical potential energy as carbon in CO_2 and oxygen in H_2O, see figure 10.4.

There are two other processes that make these interrelated cycles more complex. (1) Carbon dioxide reacts with water to produce carbonic acid. This molecule, to some extent, dissociates into hydrogen and bicarbonate ions. Certain types of organisms such as sponges, corals, and clams convert dissolved bicarbonate ions into calcium carbonate ($CaCO_3$). Because this substance is the major component of these organisms' external skeleton, when they die, their external skeletons accumulate and eventually cement to form rocks. Thus, carbon is temporarily taken out of the cycle and stored as a compound at an intermediary energy level. (2) Sometimes plants and animals die in an environment where decomposers are ineffective in bringing about decay. Under these conditions over a long period of time (thousands of years) the organic material is converted into hydrocarbons, such as petroleum and coal, and oxygen is released during this process. In this case carbon is removed from the cycle and stored as a compound with high chemical potential energy. Using these hydrocarbons as a source of energy reintroduces carbon into the cycle in the form of carbon dioxide [Hydrocarbons + $O_2 \rightarrow CO_2 + H_2O$]. Figure 10.4 incorporates these processes in a simplified version of the carbon and oxygen cycles in the biosphere.

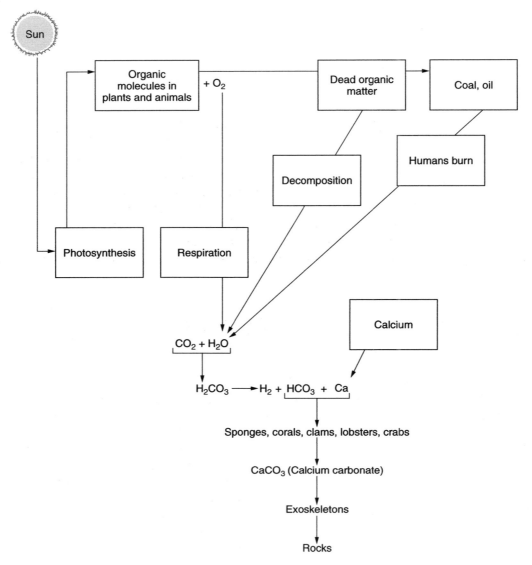

FIGURE 10.4 Carbon and oxygen cycles.

Nitrogen Cycle. Nitrogen cycles between organic material, primarily proteins, in organisms and nitrogen gas in the atmosphere. Neither plants nor animals can utilize nitrogen gas. Therefore, intermediary processes must occur to accomplish the recycling of nitrogen. Most plants can use nitrogen in the form of nitrate (NO_3) and some plants can use ammonia (NH_3). There are two processes in which nitrogen gas is converted into a form that plants can use. (1) Lightening causes atmospheric nitrogen to ionize and then eventually combine with oxygen to form nitrates (NO_3), and (2) certain types of organisms, called nitrogen fixers, convert nitrogen gas into ammonia (NH_3). Nitrogen fixers consist mainly of certain soil aerobic and anaerobic bacteria, but certain species of blue-green algae also accomplish this task. Some plants can directly take up and use ammonia. For the majority of plants that cannot do this, there are soil bacteria that convert ammonia into nitrite (NO_2). Other soil bacteria convert nitrites into nitrates (NO_3) that most plants can utilize. Plants incorporate nitrogen into proteins, and animals obtain nitrogen by eating plant proteins and/or other animals.

Nitrogen mainly circulates via inorganic nitrogen compounds in the environment and proteins in living organisms. Animals excrete large quantities of nitrogen in the form of urea and uric acid in their urine. Decomposer organisms convert these excreted substances as well as substances in dead organisms into ammonia. As already described above, the ammonia is directly converted into some plant proteins or is converted to nitrite, then nitrate, and finally incorporated into plant protein, thus completing the cycle.

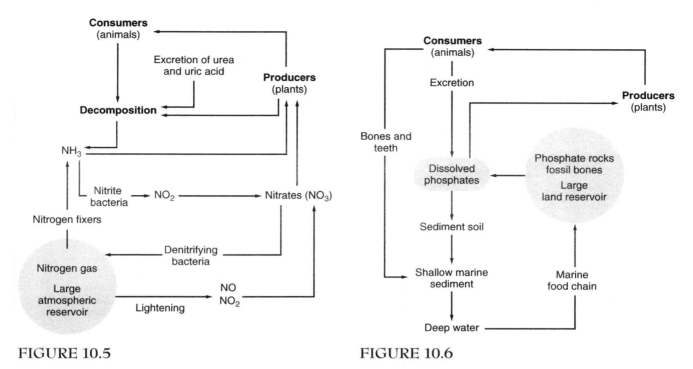

FIGURE 10.5   FIGURE 10.6

There are some organisms, the denitrifying bacteria, that convert nitrate (NO_3) into nitrogen gas in the atmosphere. This process completes the overall nitrogen cycle beginning with nitrogen gas. This and all the other processes described here are summarized in figure 10.5.

Phosphorous Cycle.   The phosphorous cycle is similar to the nitrogen cycle in having a large reservoir in the environment for the element that is circulated. The atmosphere is about 78 percent nitrogen by volume. This provides a gaseous reservoir for the nitrogen cycle. In contrast, the phosphate in rocks serves as a reservoir for the recycling of phosphorous. This element moves between phosphate dissolved in water and phosphate incorporated into organisms in the form of substances such as DNA, RNA, NAD, NADP, ADP/P_i, see Chapters 5 and 6, and phospholipids. The phosphate in rocks may become dissolved in water and transferred as a sediment to soil and to shallow and deep water. Plants take up and incorporate phosphate and pass it on to the animals that eat plants. Animals excrete phosphate which becomes dissolved in water, thus completing the phosphate cycle. Skeletons of some marine organisms contain phosphate. These may accumulate and form rocks, thus reintroducing phosphate into the reservoir of this cycle. A simplified version of the phosphate cycle is shown in figure 10.6.

Biosphere as a Machine

As illustrated in figure 10.7, the biosphere as a whole carries out three general functions that characterize any machine. Energy from the sun flows into the biosphere and directly or indirectly results in chemicals moving from low to high energy levels. As these chemicals return to low energy levels, work and heat interactions occur, which sooner or later lead to heat energy radiating away from the biosphere. Thus, energy from the sun flows through the biosphere. The cycling of chemicals between high and low energy levels results in coupling of some of the sun energy to accomplishing the biological work of diverse organisms staying alive, reproducing, and maintaining certain ways of interacting with one another and with the physical-chemical environment. Thus, energy couplers cycling between high and low energy levels connect incoming sun energy to biological work. The biological work of the biosphere results from the summation of billions of work interactions of the component living systems.

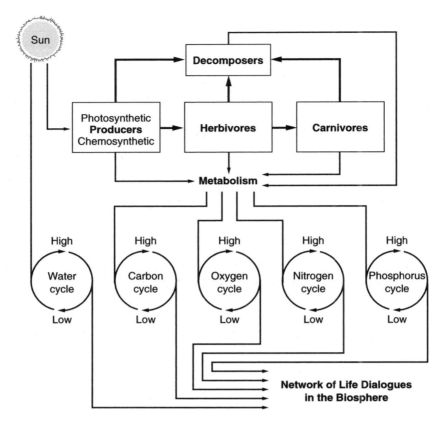

FIGURE 10.7

Ecosystem as a Homeostatic Machine

Homeostasis as an Ideal

Over short periods of time, e.g., from one to several hours, all cells exhibit negative feedback regulation of many of their properties such as volume, salt concentration, water concentration, some electrical properties associated with their membranes, and rates of chemical reactions that occur within the cell. Therefore, it is appropriate to think of the cell as a homeostatic machine so long as one realizes that this is a useful idealization. Many types of cells undergo development in which they express new properties. For example, primitive white blood cells differentiate into specialized cells such as immune cells (lymphocytes). The development that occurs in any cell leads to gradual changes in properties that result in cell division or cell death. All these developmental changes oppose a rigidly maintained steady state implied by the idea of homeostasis. Thus, in one context, a cell is a homeostatic machine, but in another context, the cell is in continual transition. However, throughout the transition some properties are maintained in steady state, and the various stages of the transition may exhibit homeostasis.

To some extent land plants and to a greater extent animals with backbones (for example, fishes and mammals) exert negative feedback regulation of properties of the whole organism. In humans, for example, blood properties such as temperature, concentration of glucose and water, and many other parameters are kept near preset values. However, in these animals with homeostatic body plans, developmental changes, especially aging, are much more overt than in individual cells. Some cells divide to produce new cells that temporarily better represent ideal homeostatic machines. In contrast, all vertebrate animals gradually depart from this ideal and eventually die. In effect, animals exhibit stability of a pattern of life in which many aspects are nicely described by the idea of homeostasis. Other aspects of this pattern require different explanatory principles. For example, each human being expresses the body pattern of the species, Homo sapiens. In the individual life span this pattern is maintained by stable structural and functional interactions among body parts and by homeostatic mechanisms. In time periods longer than a life span, this pattern is maintained on earth by the process of reproduction.

Ecosystems exhibit stable patterns in which some aspects are specific properties such as concentration of oxygen and nitrogen in the atmosphere, which are maintained by complex homeostatic mechanisms. Other aspects of these patterns involve networks of interactions among organisms and physical-chemical events that cannot be described by mechanistic models such as the idea of a homeostatic machine. A major difficulty here is that mechanistic models presuppose that the parts of a system are autonomous, whereas ecological networks presuppose that the components are not autonomous. Thus, homeostasis still is a useful explanatory idea for ecosystems but is less important than other explanatory principles that ecologists currently are creating. In fact, ecology is one source of a major transition in Western thinking from mechanistic to holistic models which incorporate mechanistic models as special cases.

Ecological Homeostatic Mechanisms

Constituent organisms of an ecosystem can alter their behavior, such as mating patterns and reproductive rates, and their physiological activity, such as productivity in plants or rate of decomposition by decomposers. A disturbance of an ecosystem leads to these alterations which in turn produces coordinated interactions among organisms that generate two major types of homeostatic mechanisms. These mechanisms bring the system back to the same or some new steady state.

One type of homeostatic mechanism involves negative feedback regulation of biogeochemical cycles. Well-regulated cycles show two features: (1) a large available abiotic pool of the substance being cycled and (2) the existence of many mechanisms for increasing and decreasing the flow of the cycled substance into and out of a large abiotic pool. A large available abiotic pool is the quantity of substance to which living organisms have free access. For example, the carbon and oxygen cycles involve the earth's atmosphere which contains large amounts of molecular oxygen (O_2) and carbon in the form of carbon dioxide (CO_2). Both of these gases are readily available to all organisms. The nitrogen cycle involves a large pool of nitrogen gas also present in the earth's atmosphere. Lightning fixation and nitrogen fixing bacteria (see previous section on biogeochemical cycles) make this gas available to plants and therefore indirectly available to consumers. Except in extremely arid regions, the water cycle involves a large abiotic pool in the form of water present in the soil of terrestrial ecosystems or water constituting the medium of aquatic ecosystems.

The nitrogen cycle exemplifies mechanisms for increasing and decreasing the flow of cycled substances into and out of a large abiotic pool. Nitrogen fixing bacteria may increase or decrease the flow of nitrogen out from the atmosphere and denitrofying bacteria may increase or decrease the flow of nitrogen into the atmosphere. Other forms of nitrogen, e.g., NH_3, NO_2, and NO_3, may be increased or decreased due to the activity of specific bacteria. As a result, a change in the steady state level of any of the molecular forms of nitrogen leads to a compensation by some bacteria that brings the cycle back to steady state.

The second type of homeostatic mechanism involves control of the numbers and biomass of organisms belonging to various trophic levels. For example, suppose there are increased numbers of herbivores in a terrestrial ecosystem. This would lead to increased grazing intensity, increased stress-producing interactions among herbivores of the same species, increased frequency of contacts with carnivores, and decreased selection pressure for high reproductive rates. As a result of these several interactions, the quality and quantity of plant food would decrease, the stressed herbivores would be less healthy, the carnivore population would increase as a result of eating more herbivores, and the rate of reproduction of herbivores would go down. All these factors, in turn, would lead to increased death rates and decreased birthrates of herbivores thus bringing the herbivore population back toward the original steady state. The decline in herbivores would lead to increased producer populations and decreased carnivore populations thus bringing these populations back toward the original steady state.

As one may surmise from this example, this type of homeostatic mechanism is quite complex and difficult to study. In fact, there are relatively few well controlled experiments that verify and suggest detailed models for this kind of mechanism. Yet, the need for these verified models is great. Outbreaks of pests, such as locusts, which may destroy thousands of acres of crops, come from failure of this kind of homeostasis. Also, such models would provide humans with limits of homeostasis with which humans must abide when exploiting resources of the biosphere.

Gaia Hypothesis

In 1969 at a meeting about the origins of life held at Princeton, N.J., J. E. Lovelock put forth his Gaia hypothesis which states:

> . . . the entire range of living matter on Earth, from whales to viruses, and from oaks to algae, could be regarded as constituting a single living entity, capable of manipulating the Earth's atmosphere to suit its overall needs and endowed with faculties and powers far beyond those of its constituent parts.[2]

Just like any "warm-blooded animal," for example, mammals and birds, the biosphere has maintained a climate within ranges compatible with life.

> The history of the Earth's climate is one of the more compelling arguments in favor of Gaia's existence. We know from the record of the sedimentary rocks that for the past three and a half aeons [billions of years] the climate has never been, even for a short period wholly unfavorable for life. Because of the unbroken record of life, we also know that the oceans can never have either frozen or boiled. Indeed, subtle evidence from the ratio of the different forms of oxygen atoms laid down in the rocks over the course of time strongly suggests that the climate has always been much as it is now, except during glacial periods or near the beginning of life when it was somewhat warmer. The glacieal cold spells [Ice Ages] . . . affected only those parts of the Earth outside latitudes 45° North and 45° South. We are inclined to overlook the fact that 70 percent of the Earth's surface lies between these latitudes. The so-called Ice Ages only affected the plant and animal life which had colonized the remaining 30 percent, which is often partially frozen even between glacial periods as it is now.[3]

One way of visualizing the GAIA hypothesis is in terms of an analogy with any warm-blooded animal such as the human body: (1) The human body keeps several properties of the blood, such as temperature, relatively constant, (2) The biosphere keeps several properties of the atmosphere such as temperature relatively constant. Therefore, the biosphere is similar to the human body and may be represented metaphorically as GAIA.

Lovelock's hypothesis seems to be a step toward a creative mindful ecology in two ways. Firstly, in calling the earth Gaia, he draws our attention to a time when Mother Earth was revered and humans participated in Her grand schemes rather than attempt to exploit Her treasures. This implies that all aspects of the biosphere are interrelated and contribute to the common good of the earth. Secondly, in drawing our attention to a worldwide order maintained since the beginning of life on earth, Lovelock sets a foundation for a worldwide ecological ethics. All ethics is based on some notion of the common good. With Lovelock's hypothesis the common good is taken from the parochial vision of any one society to the consensus of a technologically unified world community that acknowledges and studies an evolving steady state of the biosphere. This implies (1) just as modern medicine can modify the body's altered homeostasis to bring it back to health, humans can modify the biosphere's altered homeostasis to bring it back to its common good; that is, health, and (2) all human societies on earth should reach a consensus for a worldwide ecological ethics.

However, the GAIA hypothesis also sends a metaphorically message that is psychologically destructive to moderns struggling with current ecological problems. Christine Downing reminds us

---

[2]J. E. Lovelock, *Gaia: A New Look at Life on Earth* (New York: Oxford University Press, 1979), p. 9.
[3]Ibid., p. 19.

that Gaia is the Great Mother of Greek mythology and that all the goddess' "personalities" are highly developed and specialized forms of this primordial earth goddess.

> But there is in Greek mythology a "great" mother in the background—Gaia, grandmother to Demeter, Hera, and Hestia, great-grandmother to Athene and Artemis, and ancestress also of Aphrodite who is born of the severed genitals of Gaia's son-lover, Ouranos. Gaia is the mother of the beginning, the mother of infancy. She is the mother who is there before time. . . .[4]

It is true that for the Greeks, Gaia is first of all earth and all her goddess qualities follow from this. But in this context, earth is not understood but experienced as rebellious, alive, and eruptive.

> Gaia reminds us of all that cannot be brought under control. She is divine; she transcends the human. She is that very transcendence, but as an earthly, shaped, present, appearing reality. Yet to understand Gaia as *only* earth, to reduce her to a personification of an aspect of the natural world . . . is to miss the point. Gaia is earth made invisible, earth become metaphor, earth as the realm of soul. . . . souls live in her body. The Greeks understood that soul-making happens in earth, not in sky. Soul (unlike spirit) relates to the concrete imagination. . . .
> . . . she [Gaia] is nature moving toward emergence in personal form.[5]

Thus, Gaia is a personification of a psychological force.

> Freud called the force I had experienced *das Es* (usually and misleadingly translated as the id)—some primal "that" in us, sheer raw instinctual energy, nature alive within, nature just on the way toward being given human form. These psychic forces within that so clearly transcend the individual, that are an aspect of the unconscious that is inexhaustible and inassimilable but without which we could not live—these forces are Gaia.[6]

Lovelock's choice of Gaia as representing the biosphere is based on a profound misunderstanding: Gaia is not order, control, self-regulation; she is chaos from which orders emerge; she is uncontrollable; she transcends self-regulation.

> Gaia shows herself here as always for life and against any stifling order. Gaia cannot be subdued nor can her responses be predicted. She deceives and betrays. She is ever fertile: a drop of Ouranos's blood or of Hephaistos's semen impregnates her; but she is as likely to give birth to the monstrous as to the beautiful. . . . Gaia is for life but for ever-renewing life and so for life that encompasses death. Gaia rituals included animal sacrifice as well as offerings of cereal and fruit; in archaic Greece as in many vegetation cults, worship of her may have included human sacrifice.[7]

There probably are those who would forgive this ignorance of mythology, but I suggest that at the very least it is crass ignorance. Jacque Monod in contrast to Lovelock's crass ignorance has the courage and honesty to proclaim that modern science is a sacrilege.

> The fear is the fear of sacrilege: of outrage to values. A wholly justified fear. It is perfectly true that science outrages values. . . . it subverts every one of the mythical or philosophical ontogenies upon which the animist tradition . . . has made all ethics rest.[8]

Lovelock brings us back to a time when there was no distinction between sacred and the profane, but then, he reduces our transcendent, mysterious experience of nature to what is clearly profane and in fact, banal. He invites us to revere the biosphere as Gaia but then reduces her to a homeostatic machine. Who but an out-of-control addict to power can worship a self-regulating machine? Lovelock's hypothesis is a misguided return to the evolutionary force that produced our current life-denying fixation on patriarchal consciousness, which got its start afterall by objectifying woman as an object that can produce babies and that can be bartered as a result of this capacity.

---

[4]Christine Downing, *The Goddess: Mythological Images of the Feminine* (New York: The Crossroad Pub. Co., 1984) p. 135.

[5]Ibid., p. 147.

[6]Ibid., p. 147.

[7]Ibid., pp. 150–151.

[8]Jacques Monod, *Chance and Necessity* (New York: Vintage Books, 1972) p. 172.

Acknowledging and studying the evolution of biosphere homeostasis is analogous to the heliocentric revolutionary theory in astronomy. This ecological insight shocks us into a higher rational consciousness that separates us from the experience of mystery in the universe and in the core of our being. As necessary as cybernetics (which involves the study of positive and negative feedback mechanisms, see Chapter 3) may be for our continued control over nature, it does not lead to an ethic of care for nature or for one another. Furthermore, we dare not return to an ethics based on a rationally defined common good enforced by those who "know" what is objectively good for all of us. There is no turning back from the radical individualism of the modern commitment to science, technology, and democratic ideals. Somehow we must transcend the conflict to which this commitment leads, namely the apparent opposition between cultural diversity and a concern for the welfare of everyone in society. This is specified in greater detail in Chapter 7 which deals with ecological double binds.

Thus, the GAIA hypothesis appears to move us away from creative mindful ecology in two ways: (1) It implies that humans can rationally comprehend GAIA representing the earth and manage it rather than revere nature and participate in its mysterious fullness. (2) It implies that elite groups in the most powerful nations should establish the ecological, ethical rules for all peoples of the earth.

In 1978, Gregory Bateson proposed a "wider perspective" which he believed could be a foundation for a "trans-mechanistic perspective."

> The "Batesonian synthesis"—what might be termed the "cybernetic/biological metaphor" . . . is the extraction of the concept of Mind from its traditionally religious context, and the demonstration that it is an element inherent in the real world.[9]

As of 1981, Morris Berman believed that Bateson's perspective was the only one that could be a possible framework for a new metaphysics that "unites fact and value," that is, is a synthesis of control and participatory consciousness.

> As far as I can tell, his work represents the only fully articulated holistic science available today; one that is both scientific and based on unconscious knowing. Bateson's work is also much broader than that of Jung or Reich, in that it places a strong emphasis on the social and natural environment, in addition to the unconscious mind [which before the emergence of Ego consciousness was Gaia, the Great Mother]. It situates us in the world, whereas Jungian or Reichiian self-realization often becomes an attempt to avoid it.[10]

However, Berman cautions us that the Batesonian holism is grounded on systems science which according to Ludwig von Bertalanffy, one of its founders, is perhaps the ultimate technique to shape human society even more into a megamachine.

Thus, Lovelock's hypothesis as a special case of Batesonian cybernetics is an expansion from egocentric, ethnocentric, and human centered ethics toward a Geocentric ethics, but it does not take us beyond the horrors of what the poet Robert Bly calls "father consciousness."

> At the very point that the mechanical philosophy has played all its cards, and at which . . . in its attempt to know everything, it has ironically exhausted the very mode of knowing which it represents, the door to a whole new world and way of life is slowly swinging ajar. . . . We are witnessing the modification of this entity [father consciousness] by a reemergent "mother consciousness," [which according to Bly we cannot help but think about Her in terms of father consciousness; i.e., rational analysis]. . . . Bly credits the nonparticipatory consciousness of the Greeks and the Jews as producing cultures of "marvelous luminosity." . . . Bly's crucial point, however, is that the "marvelous luminosity" has reached its limits. It has become a hostile glare, . . . Its most creative outposts are now self-criticisms, . . . Regaining our health, and developing a more accurate epistemology, is not a matter of trying to destroy ego-consciousness [or rational analysis], but rather, as Bly suggests, a process that must involve a merger of mother and father consciousness. . . . It is for this reason that I regard contemporary attempts to create a holistic science as the great project, and the great drama, of the late twentieth century.[11]

[9]Morris Berman, *The Reenchantment of the World* (Ithaca: Cornell University Press, 1981) p. 196.

[10]Ibid., p. 196.

[11]Ibid., pp. 187–189.

The mother and father consciousness stand opposed to one another, so we cannot create an adequate holistic science by merely adding one to the other. Modern ecology stands between mother and father consciousness. Just as we can never return to blind submission to Gaia as we did in our childhood, we also must no longer attempt to dominate Her via a cybernetic model. Ecology must be transformed to include the participatory consciousness of our childhood and include rational analysis, the fruit of our adolescence. Process ecology is a step toward formulating a systems perspective that may achieve this integration of participatory and control consciousness.

Summary of the GAIA Hypothesis

1. GAIA hypothesis in terms of an analogy with any warm blooded animal, such as humans
   a. The human body keeps several properties of the blood, such as temperature, relatively constant.
   b. The biosphere keeps several properties of the atmosphere, such as temperature, relatively constant; thus, the atmosphere is analogous to blood.
   c. Therefore, the biosphere is similar to the human body and may be represented metaphorically as GAIA.
2. GAIA hypothesis as a step toward Creative Mindful Ecology
   a. The biosphere metaphorically represented as GAIA implies that all aspects of the biosphere are interrelated and contribute to its common good.
   b. In drawing our attention to a worldwide homeostasis in the biosphere, Lovelock sets the foundation for a worldwide ecological ethics. This leads to
      1) Just as modern medicine can modify the body's altered homeostasis to bring it back to health, humans can modify the biosphere's altered homeostasis to bring it back to health, i.e., greater harmony.
      2) All human societies on earth should reach a consensus for a worldwide ecological ethics.
3. GAIA hypothesis as a move away from Creative Mindful Ecology
   a. It implies that humans can rationally comprehend the biosphere by representing it as GAIA used as a metaphor of the biosphere being a homeostatic machine that can be managed in a way that represses humans' reverence for nature and participation in its mysterious fullness.
   b. It implies that elite groups in the most powerful nations should establish the ecological, ethical rules for all peoples of the earth.

Tearout #1

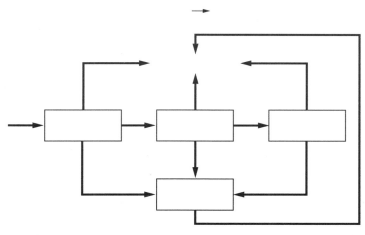

FIGURE 10.1

Directions:

1. Using figure 10.1 as a reference, fill in all labels *and* describe the meaning of each label including *respiration.*

Tearout #2

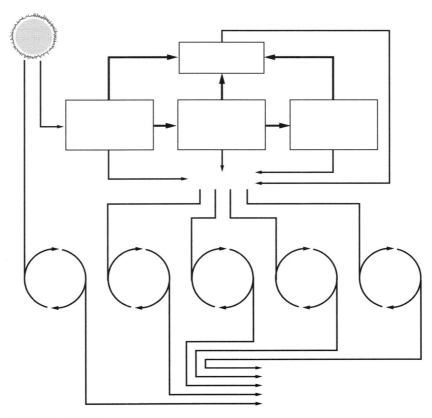

FIGURE 10.7

Directions:

1. Using figure 10.7 as a reference, fill in all labels.
2. Using the idea of an energy coupler, explain how figure 10.7 represents the biosphere as a machine.

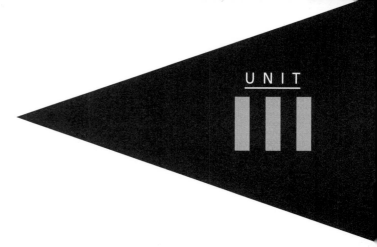

Individuation of the
Nonhuman Biosphere

Core Ideas of Genetics

It is common knowledge that in some families one son will resemble his father while another son will inherit many of his mother's features. Likewise, one daughter may resemble her mother, and another her father. In some cases these resemblances are very striking. At the other extreme a son or daughter may not look at all like either parent. For example, parents who may be considered average in appearance may have a beautiful daughter. Conversely, a handsome couple may have a son or daughter who is quite plain. How can we explain these observations? Many patterns of inheritance are quite complex in humans and they are further complicated by environmental effects. Because of this it would be virtually impossible to construct the elementary patterns of inheritance from a study of human populations.

The current theories of inheritance are founded on the rediscovered work of Gregor Mendel (born 1822). Mendel performed breeding experiments with pea plants and focused on the inheritance of a few simple traits such as stem length, seed color, and flower color. The results of these experiments enabled him to formulate a few simple rules that describe the inheritance of simple traits. These rules form the foundation of genetics, the science of biological inheritance. Mendel's ideas have been integrated with subsequent discoveries in cell biology. Biologists have used these rules to explain inheritance patterns in more complex organisms such as the fruit fly. In many cases, these rules had to be modified. Because of these extensions and modifications of Mendel's ideas, genetics has become sufficiently sophisticated to describe inheritance of many traits in humans.

Mendel's Insights

Shift to Process Perspective. Coitus is a type of sexual interaction that results in either male or female offspring. One cannot predict the sex of a new organism from any particular mating. Perhaps this all too obvious realization plus training in mathematics and physics led Gregor Mendel (between 1856 and 1868) to study breeding among plants in a radically different way. Breeding with respect to sex is a process which has two possible outcomes. Hence, any particular mating process is analogous to tossing a coin. Our usual experience with a coin is that the outcomes of many tosses will be about 50% heads and 50% tails. However, this need not be the case. Coins could be weighted so that among many tosses, 75% are heads and 25% are tails or vice versa, or some other unequal percent outcome like 60/40. The only way of determining whether the coin is not biased or of determining the bias of the coin is to perform the experiment of tossing the coin many times and counting the number of each type of outcome.

Suppose we toss a coin 100 times and heads occurs 60 times and tails occurs 40 times. We could summarize this by saying that the frequency of heads is 60/100 = 60% and the frequency of tails is 40/100 = 40%. The frequency of heads and tails may vary from one experiment to another, but the frequencies of heads and tails in any one experiment always will add up to 100%. If one tosses a biased coin 10 times, he may get a frequency of 70% heads (7 out of 10 tosses) but then get a frequency of 60% in 100 tosses. The frequency of heads after 1,000 tosses may be 61% and 60.5% after 10,000 tosses. If as the number of tosses becomes larger and larger (10 million, 100 million, etc.), the frequency of heads approaches a particular frequency, for example, 60% = 0.6, then that frequency represents the probability of getting heads whenever we toss the coin. Thus, *P(H)* = 0.6 means that the probability of getting heads for any toss of the coin is 60%. If one were to bet on the outcome of such a toss, it would be better to bet on heads while being aware that there still is a "good chance" of tails turning up, i.e., a 40% chance. This caution about the good chance that tails also could turn up follows from the understanding that the sum of the probabilities of all possible outcomes of a particular process is always equal to 1. With respect to the above example of tossing a coin, we have: P(H) + P(T) = 0.6 + 0.4 = 1. To be completely rigorous, we also should specify that each possible outcome is independent of any other possible outcome.

Mendel's First Theoretical Insight.   Sexual reproduction in humans is like tossing a coin in that one gets one of two possible outcomes, male and female. In general, sexual reproduction in any species is like tossing a coin. The random trial of tossing the coin is like the random coming together of a sperm and an egg. The outcome of such a trial is a "fertilized egg" called a zygote which is like a coin that may be heads or tails. Heads is equivalent to the zygote having messages that direct it to develop into a biological male, and tails is equivalent to the zygote having messages that direct it to develop into a biological female. Biological sex is a complex of many different traits, but the fact that sex usually is expressed in only two ways probably suggested to Mendel that there may be simple traits that are similar to tossing a coin. Just as tossing a coin results in heads or tails, the inheritance of some traits occurs only in one of two forms. For example, pea plants have flowers with a color that is either purple or white.

Mendel's Second Theoretical Insight.   Just as making something is the resultant of a set of directions, so also a complex trait is the expression of a set of messages in the individual received from each parent. A simple trait like flower color in pea plants must result from a "flower color message" that can occur in one of two forms: a form that directs the pea plant to make purple flowers and a form that directs the pea plant to make white flowers. As a result of his breeding experiments with pea plants, Mendel guessed that each parent has two messages for any particular trait; sometimes both messages are in the same form and sometimes each is in a different form. The offspring that develops from the zygote produced by the union of a sperm from one parent and an egg from the other parent also will have only two messages for any particular simple trait. Therefore, the sperm will carry only one of the two messages present in one parent, and likewise the egg will carry only one of the two messages present in the other parent.

Thus, genetics studies two processes. The first process is that of transferring messages from parents to offspring that may have different combinations of messages than either of their parents. For example, suppose one parent pea plant has two purple color messages and the other parent has two white color messages. All the offspring pea plants will have a purple color message from one parent and a white color message from the other parent. The second process gives rules for determining what trait any particular combination of messages will express. For example, it is clear that the parent pea plant with two purple color messages will have purple flowers, and the parent pea plant with two white color messages will have white flowers. But what color of flower will the offspring plant have if it receives a different color message from each parent?

Furthermore, genetics studies these two processs—message transfer leading to new combinations and message expression—in populations involving random interactions. On the one hand, genetics is the study of messages passed on to offspring via sperm and egg resulting in an expression of some trait. Each parent may have one or two forms of a message for a simple trait and only one of these messages will be randomly selected to be present in a gamete (sperm or egg); that is, random diversity of messages available to offspring. The different combinations of messages received from parents via gametes will lead to different expressions in an offspring. On the other hand, the traits that show up in an offspring result from random fertilization between populations of male and female parents. Thus, the production of gametes and the process of fertilization imply four types of random processes: (1) random diversity of messages: (a) any particular sperm may contain one of two different messages; (b) likewise any particular ovum may contain one of two different messages; and (c) the zygote will have two messages each received from one parent and these received messages may be the same or different; (2) random selection of sperm from the male population; (3) random selection of ova from the female population; and (4) random combination of sperm and ova to produce zygotes. Genetics exemplifies the process perspective in that the population rather than an individual is the unit of study and one must use statistical analysis based on the mathematical theory of probability to describe rules of inheritance.

Mendel's Experimental Insights. Mendel chose to study the inheritance process in pea plants because (1) some of their inherited traits are easy to identify, (2) these simple traits are expressed in only two forms, see table 11.1, (3) the generation time of pea plants is relatively short, and (4) one can easily study large populations of these organisms. Of course one can use statistical analysis only on homogeneous populations. Therefore, Mendel's first task was to artificially produce such homogeneous populations from the diversity ordinarily found in nature. Like anyone studying plant or animal breeding, Mendel observed that some pea plants "breed true" with respect to certain simple traits. For example, he was able to collect pea plants with purple flowers which produce offspring with purple flowers which cross pollinate to produce offspring with purple flowers, and so on. In like manner, he was able to collect true breeding pea plants with white flowers. Thus, true breeding populations are homogeneous with respect to the genetic messages they always pass on to offspring. We may operationally define a population of true breeding organisms as one in which the same message for a trait is passed on from one generation to another so long as individuals in the population with the trait do not crossbreed with individuals in some other population.

Experiments Leading to the Concept of Dominance. Sheer curiosity would lead almost anyone to wonder what would happen if true breeding purple plants were crossed with true breeding white plants. Mendel found that all the offspring plants were purple. Likewise, true breeding plants with

▶ Table 11.1 Simple Traits in Pea Plants

Simple Trait	Dominant Form	Recessive Form
Seed Coat	Round	Wrinkled
Seed Color	Yellow	Green
Flower Color	Purple	White
Pea Pods	Inflated	Constricted
Pod Color	Green	Yellow
Flower Position	Axial	Terminal
Stem Length	Long	Sort

yellow seeds crossed with true breeding plants with green seeds always produced offspring that had yellow seeds. The same pattern occurred for all of the other simple traits: a cross between the two true breeding forms of each trait produced offspring, all having only one form of the trait. The trait that appears after such a mating is called the dominant trait and the other trait that doesn't appear is called the recessive trait.

Using modern terminology, we say that P: parent generation: purple flower x purple flower where x means cross-pollinate (or more generally, crossbreed or mate). The offspring from the mating of two distinct populations is called the F_1 generation. Thus, true breeding purple plants crossed with true breeding white plants produce plants that always have purple flowers in the F_1 generation. Mendel generalized from these observations to two new operational definitions: dominant trait: the trait that all individuals of the F_1 generation exhibit when each parent population is a true breeding population which exhibits one of the two forms of a simple trait; recessive trait: the trait that no individual of the F_1 generation exhibits when each parent population is a true breeding population which exhibits one of the two forms of the trait.

At this point we may introduce three theoretical terms that will be used in formulating Mendel's theory of inheritance of simple traits. These theoretical terms must take their meaning from operationally defined terms. Gene is one of the two messages that determine the inheritance of a simple trait. Dominant gene is the form of a message that always results in the expression of a dominant trait. Recessive gene is the form of a message that results in the expression of a recessive trait only when the dominant gene for this particular trait is not present.

Experiments Leading to the Concept of Segregation.   Mendel's observation of dominance in inheritance must have puzzled him at first, but instead of being satisfied with some fanciful explanation, he devised a brillant experiment that could suggest a testable explanation. He crossed the purple flowered plants in the F_1 generation with one another and produced a new generation called the F_2 generation which contained individuals with purple flowers and other individuals with white flowers. The overall summary of these experiments is as follows. P: true breeding purple flowers × true breeding white flowers produce F_1 generation in which all offspring have purple flowers. Then F_1 × F_1 produces F_2 generation in which about 75% of the offspring have purple flowers and 25% have white flowers. The same pattern is exhibited for other simple traits.

As a result of interpreting these results Mendel invented the concept of segregation which will now be defined using modern terminology. An F_1 individual contains the dominant gene for purple flower trait (received from a "purple parent") and the recessive gene for the white flower trait (received from a "white parent"). During production of gametes, these genes segregate: some gametes only contain the gene for purple flower and other gametes only contain the gene for white flower.

Mendel's First Law

Mendel's explanation of the recurrence of recessive traits in the F_2 generation is the core idea of modern genetics. He knew that sexual reproduction involves a male gamete (pollen in plants) uniting with a female gamete, an ovum, to produce a fertilized egg (in a seed) that develops into an adult organism. Mendel made three interrelated assumptions:

1. Each simple trait is controlled by two genetic factors called genes.
2. A gene controlling a simple trait may exist in one of two forms: a) dominant gene which always is expressed as a dominant trait and b) recessive gene which is expressed as a recessive trait only when the dominant gene is not present.
3. When gametes are formed, the two genes controlling a particular simple trait segregate, that is, pass into separate gametes. (a) In the case where a parent has a dominant and a recessive gene for a particular simple trait, 50% of the gametes produced will get the dominant gene and the other 50% of gametes produced will get the recessive gene. (b) This segregation process is a random process analogous to tossing an unbiased coin. (c) The process of recombining genes by union of a sperm with an ovum also is a random process.

If both genes in an individual have the same form, the individual is said to be homozygous, that is, homo means same; either both genes are dominant or both genes are recessive. If the genes in an individual are different in form, the individual is said to be heterozygous, i.e., hetero means different wherein one gene is dominant and the other is recessive.

Statement of Mendel's First Law.   The expression of any simple trait is controlled by a pair of genes that segregate from each other during formation of gametes so that one half of the gametes carry one member of the gene pair and the other half of the gametes carry the other member of the gene pair.

Explanation of Results from the Cross between Two Heterozygous Parents.   True breeding plants are homozygous. By today's conventions, genes are represented by letters where a capital letter represents a dominant gene and a lower case letter represents a recessive gene. A homozygous purple pea plant represented as C/C produces gametes which always have a C gene, and likewise all gametes of a homozygous white plant, represented as c/c, have a c gene. The mating of a homozygous purple plant with a homozygous white plant, represented as C/C × c/c, produces offspring, C/c, which receive a C gene from one parent and c gene from the other parent. Because the C is dominant, all the offspring will have purple flowers. However, the plants of this F_1 generation produce two genetically different types of gametes; one contains a C gene and the other contains a c gene. The process of mating between any two F_1 organisms (i.e., C/c × C/c) has four possible outcomes: the C of a male parent could pair with a gene of the female parent which may be either C or c; thus we have (1) C/C or (2) C/c. Also, the c of a male parent could pair with C or c of the female parent; thus we have (3) c/C and (4) c/c. All these outcomes have an equal chance of occurring which is similar to the equal chance of getting heads or tails after tossing an unbiased coin. For coin tossing with two possible outcomes, $P(H) = 1/2$ and $P(T) = 1/2$. Likewise for F_1 mating with four possible outcomes, $P(C/C) = 1/4$, $P(C/c) = 1/4$, $P(c/C) = 1/4$, and $P(c/c) = 1/4$.

There are two ways F_1 mating can produce an organism containing both C and c genes: C from the male parent pairs with c from the female parent and c from the male parent pairs with C from the female parent. The pair c/C is equivalent to C/c and therefore, $P(C/c) + P(c/C) = P(C/c) + P(C/c) = 1/4 + 1/4 = 1/2$. Hence, for the F_1 mating, though there are four possible outcomes, there are only three different possible gene combinations, and their probabilities are: $P(C/C) = 1/4$, $P(C/c) = 1/2$, and $P(c/c) = 1/4$. According to Mendel's concept of dominance, any organism with at least one C gene will exhibit the dominant trait. The probability of an offspring receiving at least one C gene is $P(C/C) + P(C/c) = 1/2 + 1/4 = 3/4 = 75\%$. The probability of an offspring receiving no C genes is $P(c/c) = 1/4 = 25\%$. As stated earlier, Mendel observed that about 75% of pea plants in the F_2 generation had purple flowers and 25% had white flowers. He observed the same pattern in breeding experiments with respect to the other six simple traits of pea plants, see Table 11.1.

Cellular Basis of Mendel's Theory

Mendel's theory focuses on the transmission of simple genetic messages from a parent population to an offspring population which express these messages as simple traits. This process, called transmission genetics, is an aspect of life viewed as a developmental process. The central idea of this perspective is that all life on earth consists of diverse species each of which consists of a population of organisms that participate in a species life cycle which may be divided into three phases: (1) birth, (2) reproduction, and (3) development. All organisms of a species express *birth* which is the event when an organism establishes itself as an autonomous individual. Some but usually not all members of a species participate in *reproduction* which is the creation of the *potential* to express a species life pattern of traits. Development is the expression of this pattern by an organism interacting with a particular environment. For single-celled organisms, development begins after birth and culminates in the cell's death or division into two new cells. For multicellular organisms development usually begins before birth and always leads to aging which culminates in death.

The idea of a life cycle applies only to a population of organisms. The life of any individual organism is linear; it has a beginning, birth, and an end, death (or cell division for some single-celled organisms). However, a subpopulation of a sexually reproducing, multicellular species exhibits the

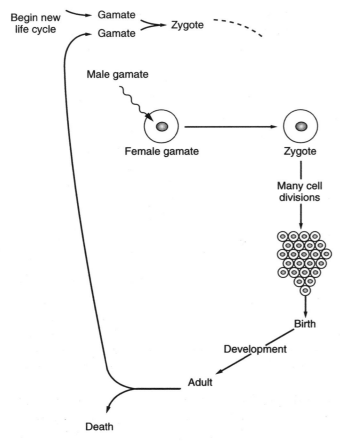

FIGURE 11.1

ideas of transmission genetics first described by Mendel. Each individual of this subpopulation participates in all three phases of the species life cycle. A gamete from each parent organism unite to form a single cell, the zygote, which begins to develop, is born, and continues to develop into a potential parent as a result of producing gametes. When one of its gametes unites with a gamete of another adult to produce a new zygote, a new life cycle begins, see figure 11.1. Twentieth century biology has studied how genetic messages are transferred to a zygote and how these messages lead to the expression of traits during development. The resulting mechanistic models of these processes enable us to picture Mendel's ideas.

Cellular Overview of Expression of Genetic Traits

Chromosome Theory of Gene Expression. As indicated earlier, each eukaryotic cell consists of pairs of "similar chromosomes," and when they can be seen under a microscope just before and during cell division, they are said to be similar because they have the same size and shape. Each member of a pair of similar chromosomes consists of two identical DNA molecules resulting from the DNA replication that occurred before the onset of cell division. A gene is a portion of a DNA molecule, and therefore, a chromosome is a linear sequence of genes. The sister chromatids of a chromosome are identical sequences of genes. The position of a gene in such a sequence is called a gene locus. A gene which may occur in two or more forms, e.g., dominant and recessive forms, is an allele. Each member of a pair of alleles occurs at the same gene locus on different members of a pair of similar chromosomes. Similar chromosomes which have the same size, shape, and sequence of genes are said to be homologous rather than the same. This is because pairs of alleles at any gene locus may be in different forms, e.g., there may be a dormant gene in one member of the pair of homologous chromosomes and the allele controlling the same trait and at the same gene locus on the other member of the homologous chromosomes may be a recessive gene.

Multicellular organisms consist of two types of cells: somatic cells and gametes. Somatic cells of a particular species result from cell division involving mitosis and have one or more pairs of homologous chromosomes. In contrast, gametes of the same species result from cell division involving meiosis and have half the number of chromosomes present in a somatic cell and no two of these gamete chromosomes are homologous. The somatic cell is said to be diploid, 2N, (di means two), and a gamete is said to be haploid, 1N, (ha means half).

A gene, which is a section of a DNA molecule, by itself has no more or less significance than any other chemical. A gene in the context of a test tube with requisite constituents (nucleotides, enzymes, etc.) can lead to the synthesis of a type of molecule called RNA which in turn is necessary for the synthesis of a protein. In this case we say the DNA has the information for directing synthesis of RNA. A gene in the context of an organism surviving in nature can lead to the synthesis of a protein. This gene product then may interact with other gene products and/or with other genes and with the cell environment to produce some trait. In this context the DNA has the information for directing the synthesis of a protein that leads to the expression of a trait.

The symbolic representation of genetic information, which under the right circumstances, is expressed as a trait of an individual is called genotype. The observable phenomena that a particular genotype produces in an organism is called phenotype. For example, the genotype, C/c, produces the phenotype, purple flower in pea plants; the genotype, c/c, produces the phenotype, white flower.

Types of Reproduction. Single celled organisms usually reproduce by a simple cell division which in eukaryotes always involves mitosis. Many plants and simple animals can reproduce by budding off portions of themselves that then develop into new adult organisms. Most multicellular organisms reproduce new individuals by one of two methods or, in some species, by an alternation of these two methods. In method 1, called *asexual reproduction,* a specialized cell called a germ cell is produced and, under the right conditions, enters into a series of mitotic cell divisions. The germ cell may start to divide while still in the parent organism or after having been released to the environment. The resulting mass of cells eventually develops into a new adult organism. Method 2, called sexual reproduction, involves individuals producing specialized cells which undergo cell division involving meiosis. The resulting gametes (usually from two individuals but in some organisms they may come from a single individual) unite to form a single cell, the zygote. The zygote initiates a series of cell divisions involving mitosis, and the resulting mass of cells develops into a new adult organism, see Chapter 12.

Cell Division Involving Mitosis

A cell ready to divide consists of chromosomes, each made up of two identical DNA molecules due to DNA replication. In a simplified overview, mitosis which now begins to occur, consists of six phases, see figures 11.2 and 11.3.

1. Chromosomes completely condense and become visable.
2. The nuclear envelope breaks down into membrane fragments; homologous chromosomes are not attracted to one another and therefore do not form pairs.
3. Chromosomes each consisting of a pair of chromatids line up on the cell's equitorial plane; i.e., the centromeres of chromosomes all lie in one plane midway between the two poles of the dividing cell.
4. The centromere of each chromosome divides and the identical chromatids begin to separate and move to opposite poles of the cell. Each former chromatid now is called a chromosome as a result of having its own centromere.
5. Each of the separating sets of chromosomes aggregates into a nucleus. The resulting two diploid nuclei are identical.
6. The cytoplasmic constituents are distributed to each pole and the dividing cell becomes two new cells called daughter cells.

Meiosis

The cellular basis for Mendel's first law is the process of gamete formation which involves meiosis. Gamete formation provides the cellular basis for each parent passing genetic information in gametes to an offspring. The first step in this process is DNA replication in specialized cells which then divide to produce gametes. A gamete from each parent unite to form a single cell, a zygote, which develops into the adult offspring. If cell divisions producing gametes were like mitosis, then the offspring would have twice as many chromosomes as each parent. However, offspring are always like each parent in having a diploid number of chromosomes. This only can happen if meiosis modifies the cell divisions to produce haploid gametes.

In order to produce haploid gametes, meiosis must involve a sequence of two cell divisions called first and second meiotic division, respectively. The first meiotic division results in separation of homologous chromosomes, thus converting one diploid cell into two haploid cells, see figure 11.4. However, each chromosome of the haploid cells consists of two sister chromatids. Each chromatid is equivalent to

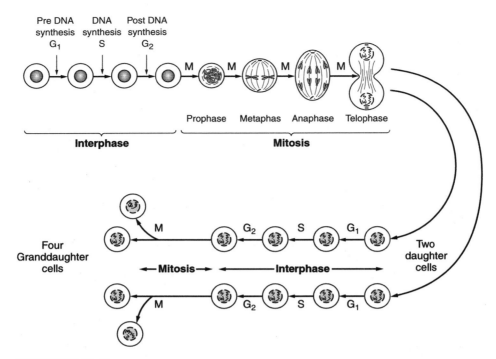

FIGURE 11.2

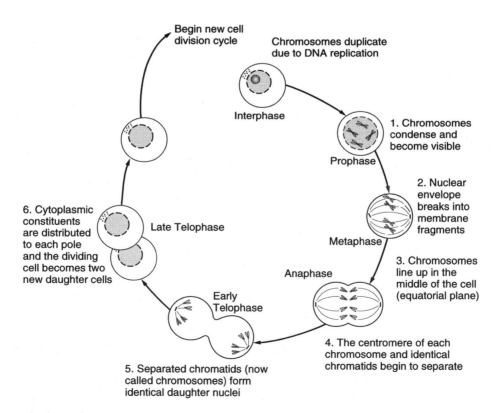

FIGURE 11.3

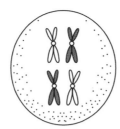

Diploid Cell

Two **Haploid Cells** at the end of the first Meiotic division

Each of the two chromosomes in each of these haploid cells consists
of two identical chromatids (DNA molecules bound to proteins)

Four **Haploid Cells** at the end of the second Meiotic division

Each cell has two chromosomes each consisting of a single DNA molecule

FIGURE 11.4

a chromosome except that a chromatid shares a centromere with another chromatid. Therefore, a second meiotic division is necessary to convert the cells with pairs of identical DNA molecules into gametes with a haploid number of chromosomes each consisting of only one DNA molecule, see figure 11.5.

First Meiotic Division.   From a genetic point of view there are four major events that occur during the first meiotic cell division. DNA replication prior to onset of cell division results in identical sister chromatids in each expanded but still invisible chromosome.

1. Chromosomes condense and become visible.
2. The pairs of homologous chromosomes attract one another and line up ("synapse") in the middle of the cell.
3. The centromere of each chromosome does not divide so identical chromatids remain together, but the members of each homologous chromosome pair separates forever.
4. The separated chromosomes, each consisting of a pair of identical chromatids begin to form two nuclei, and the cytoplasm divides. The resulting cells are haploid because they do not contain homologous chromosomes.

This division is said to be a reductive division because a diploid cell has divided into two haploid cells. Furthermore, the daughter cells are not identical because the homologous chromosomes that separated to form the new cells are not identical.

Second Meiotic Division.   Four major events occur during the second meiotic division.

1. The chromosomes, each already consisting of a pair of identical chromatids, line up in the middle of the cell.
2. The centromeres divide.
3. The identical chromatids separate.
4. The separated chromatids, now called chromosomes, regroup to form two nuclei, and the cytoplasm divides.

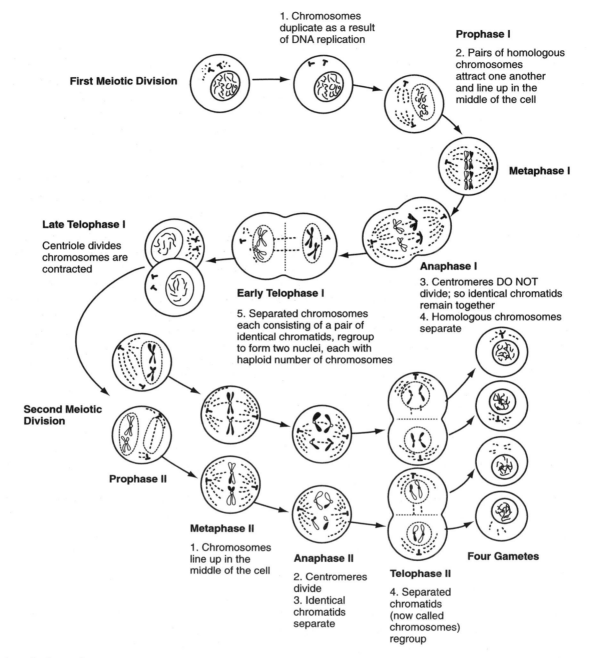

1. Chromosomes
duplicate as a result
of DNA replication

First Meiotic Division

Prophase I

2. Pairs of homologous
chromosomes
attract one another
and line up in the
middle of the cell

Metaphase I

Anaphase I

3. Centromeres DO NOT
divide; so identical chromatids
remain together
4. Homologous chromosomes
separate

Late Telophase I

Centriole divides
chromosomes are
contracted

Early Telophase I

5. Separated chromosomes
each consisting of a pair of
identical chromatids, regroup
to form two nuclei, each with
haploid number of chromosomes

**Second Meiotic
Division**

Prophase II

Metaphase II

1. Chromosomes
line up in the
middle of the cell

Anaphase II

2. Centromeres
divide
3. Identical
chromatids
separate

Telophase II

4. Separated
chromatids
(now called
chromosomes)
regroup

Four Gametes

FIGURE 11.5

The two daughter cells have identical chromosomes and are haploid. The cells resulting from the second meiotic division have the same number of chromosomes but half the number of DNA molecules as the cells resulting from the first meiotic division. That is, at the end of the first meiotic division each chromosome consists of a pair of chromatids. At the end of the second meiotic division each chromosome (formerly a chromatid) consists of a single centromere and a single DNA molecule.

Overall, meiosis produces two pairs of haploid cells. Each pair of cells, though coming from the same parent cell, has a different set of genes due to the separation of nonidentical homologous chromosomes during the first meiotic division. Crossing over, some random exchange of genes between homologous chromosomes always occurs during the first phase of the first meiotic division. This makes a somewhat complex process even more difficult to visualize. For the time being, let's ignore these crossing over events. Other than crossing over, figure 11.5 shows the genetically important events during meiosis.

> **Table 11.2   Cytological Differences Between Mitosis and Meiosis**

Mitosis	Meiosis
1. In prophase, homologous chromosomes are not attracted to one another and therefore do not form pairs.	1. In prophase of the first meiotic division, a chromatid from one chromosome forms an intimate point-for-point association with a chromatid of a homologous chromosome. Since each chromosome consists of a pair of identical chromatids, the resulting paired homologous chromosomes consist of four strands and is called a tetrad.
2. In metaphase, each duplicated chromosome lines up on the cell's equitorial plane.	2. In metaphase of the first meiotic division, pairs of homologous chromosomes line up on the cell's equatorial plane.
3. In anaphase, the centromers of each chromosome divides and identical chromatids begin to separate.	3. In anaphase of the first meiotic division, the centromere of each chromosome does not divide and therefore identical chromatids stay together. Instead, homologous pairs of chromosomes begin to separate.
4. Telophase produces two identical nuclei. Each chromosome is a single chromatin fiber that was called a chromatid in metaphase.	4. In telophase of the first meiotic division, paired homologous chromosomes separate and regroup to form two nonidentical haploid nuclei. Each chromosome consists of a pair of identical chromatids.
5. A second mitotic division again produces two identical diploid cells.	5. The second meiotic division produces two identical haploid cells. Each chromosome consists of a single strand because it came from the separation of identical chromatids produced in the first meiotic division.

Mitosis vs Meiosis.   Mitosis and meiosis differ in two ways that are important for understanding genetics. Mitosis produces somatic cells in contrast to meiosis which produces gametes. Mitosis produces identical, diploid cells in contrast to meiosis which produces nonidentical, haploid cells. These differences between mitosis and meiosis also characterize the major differences between mitosis and the first meiotic division. There are, of course, various cytological differences between these two types of division some of which are summarized in table 11.2. As shown in figures 11.2, 11.3, 11.4, and 11.5, cell division whether it is mitosis or meiosis, is divided into 4 phases: (1) Prophase, (2) Metaphase, (3) Anaphase, and (4) Telophase.

Mendel's First Law in Terms of Meiosis

The two genes that control any simple trait are located at the same gene locus in a pair of homologous chromosomes. Prior to the first meiotic division, there are two copies of each gene. During the first meiotic division these genes separate as a result of separation of homologous chromosomes, see figure 11.4. Each of the cells produced from this division has two copies of each gene that controls a simple trait. After the second meiotic division, each gamete has only one copy of one member of the original pair of genes each on different chromosomes of the pair of homologous chromosomes. If the original pair of alleles is heterozygous, then one-half of the gametes will have the dominant gene and the other half will have the recessive gene. The term genome refers to all the genes carried by a single gamete. A heterozygous individual produces two types of genomes with respect to one simple trait.

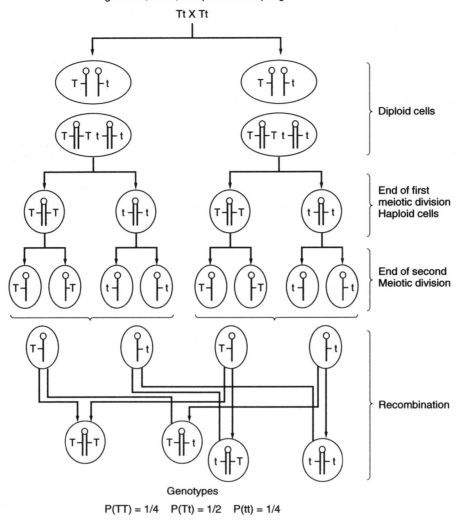

Mendel's 1st law plus the idea of genetic recombination in terms of
two heterozygous individuals for a simple trait who produce
gametes, mate, and produce offspring

FIGURE 11.6

Because these genomes occur in equal numbers, there is a probability of 1/2 of each type of genome participating in sexual reproduction, i.e., P(one type of genome) = 1/2 and P(second type of genome) = 1/2.

Let T/t represent a heterozygous genotype for some simple trait. If both parents are heterozygous, then there are four possible outcomes of a mating:

T/t
t/T
T/T
t/t

T/t = t/T so there are three types of combinations: t/t, T/T, and T/t.

The T/t = t/T = genotype of each parent, but T/T and t/t are different from either parent. These new combinations of genes that may occur in the offspring generation are called recombinants with respect to meiosis. (Note, however, that there are other ways of getting new gene combinations other than by meiosis.) Mendel's first law, then, is the basis of the core idea of genetics, which is that genes segregate during production of gametes and recombine in new ways in the offspring, see figure 11.6.

▶ Table 11.3   Core Ideas of Mendel's First Law

Concrete observable	Theoretical not observable
1. Simple traits expressed in only two ways: dominant trait and recessive trait.	1.
2. Diploid cells: have pairs of homologous (similar) chromosomes which are observable just before mitosis or meiosis begins to occur	2.a. A simple trait is controlled by a pair of genes each located in a different chromosome making up a pair of homologous chromosomes. 2.b. A gene occurs in one of two forms: (1) dominant; (2) recessive. 2.c. (1) When one or both genes are in dominant form, the pair of genes are expressed as a dominant trait. (2) When both genes are in recessive form, the pair of genes is expressed as a recessive trait.
3. Meiosis: each chromosome in a pair of homologous chromosomes separates from one another in the process of producing haploid gametes.	3.a. Each gene is a pair of genes controlling a simple trait separates from one another. 3.b. 50% of gametes contain one of the genes and 50% of gametes contain the other gene.
4. Sex: a sperm unites with an egg during fertilization.	4.a. Sperm are randomly selected from a population of sperm 4.b. An egg is randomly selected from a population of eggs. 4.c. The union of sperm and egg is a random process. 4.d. Genes separated during meiosis are randomly recombined to produce zygotes that develop into adults in the next generation.
5. The next generation can be observed to have different phenotypes each with a specific frequency.	5. The next generation can be represented as having different genotypes each with a specific frequency.

Table 11.3 represents the core ideas of Mendel's 1st law in relation to concrete observable facts/events vs. nonobservable facts/events.

Examples of Mendel's First Law of Inheritance

Dominant Traits

Woolly hair, see figure 11.7, is a well-documented dominant trait whose inheritance pattern follows Mendel's first law. There are many types of hair forms, such as woolly, wavy, curly, and straight. For simplicity of analysis, let us divide hair form into two categories: woolly and nonwoolly hair. Let W symbolize the dominant form of the allele that produces woolly hair and w symbolize the recessive form of the allele. Suppose two heterozygous individuals for this trait, i.e., Ww, mate and produce an offspring. What form of hair will the offspring have? According to Mendel's law, we can never know for sure. All we can do is calculate the chances of getting woolly hair and non-woolly hair.

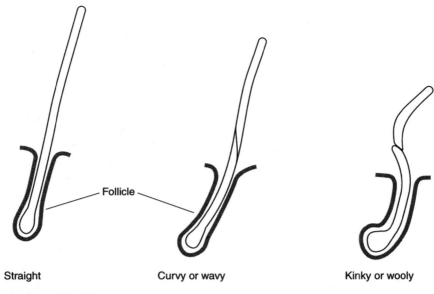

Straight   Curvy or wavy   Kinky or wooly

FIGURE 11.7

▶ Table 11.4

		Female Parent Ova	
Male Parent Sperm	**W**	WW Woolly Hair	Ww Woolly Hair
	w	Ww Woolly Hair	ww nonWoolly Hair

All problems dealing with the inheritance of a single trait obeying Mendel's first law are solved in the same way. First, one must determine the genotype of each parent. In this case each is Ww. Second, one determines the possible genomes resulting from the production of gametes by each parent. In this case meiosis in each parent produces two different genomes, gametes containing W and those containing w. Third, one determines the various different combinations of gametes from each parent. Table 11.4 represents one convenient way of representing these different combinations. As seen from this table, there are four different possible combinations: WW, Ww, wW, and ww. We say there is a one-in-four chance of obtaining each of these combinations. Since Ww and wW, though produced in different ways, are the same genotype, there are only three different possible genotypes. WW has a one-in-four chance of occurring, Ww has a two-in-four chance, and ww has a one-in-four chance. Alternatively, we say the probability of WW, represented as P(WW), is 1/4, P(Ww) = 2/4, and P(ww) = 1/4.

The phenotype, woolly hair, will occur whenever at least one W is present in the offspring. From table 11.4 we can see that P(woolly hair) = 3/4 and P(nonwoolly hair) = 1/4.

There are several normal dominant traits that obey Mendel's first law. Some of these are listed in table 11.5. Some diseases or developmental deformities are brought on by a single dominant allele. Some of these are listed in table 11.6.

▶ Table 11.5   Simple Dominant Traits

Dominant Form	Recessive Form
Hapsburg lip and chin	No trait
White forelock	No trait
Prominent and convex nose	Moderate and straight
High and narrow bridge of nose	Low and broad bridge
Straight eyes	Slanted eyes
Mongolian fold	Lidded European type eyes
Long eyelashes	Short eye lashes
Free earlobes	Affixed ear lobes
Thick lips	Thin lips
Straight chin	Receding chin
Chin cleft	Smooth chin
Dimples	Smooth cheek
Normal eyes	Deep set eyes

▶ Table 11.6   Simple Dominant Traits Producing Diseases

Dominant Form	Recessive Form
Brachydactyly (short fingeredness)	Normal
Polydactyly (extra finger)	Normal
Pseudo-achondroplastic dwarf (trunk and head similar to that of unaffected person but limbs are very short).	Normal
Dystonia muscularum deformans (disordered tonicity of muscles)	Normal
Congenital stationary night blindness (a type of defective twilight vision)	Normal
Huntington's Disease (Huntington's Chorea): A slow destruction of brain tissue; symptoms appear between 30–40 years of age; death follows about 10 years after onset of symptoms. Folksinger Woody Guthrie died of this disease.	Normal
Marfan's Syndrome (production of abnormal form of connective tissue)	Normal
Comptodactyly (permanently bent little finger)	Normal

Recessive Traits

Recessive traits are traits whose inheritance pattern follow Mendel's first law and are produced only by a homozygous pair of recessive alleles. The following are some examples of these traits.

Cystic Fibrosis. This is one of the most common genetic diseases among white children. It occurs in 1/2,000 births. The disease consists of abnormalities in the secretion of various glands, including the pancreas, the glands of the duodenum, sweat glands, and glands in the bronchial tree. The affected children are very susceptible to pneumonia as a result of the thick, viscous mucus produced by their bronchi. About 4% of the white population carries one of the recessive genes for this disease.

β-antitrypsin Deficiency. This also is a very common genetic disease. It occurs in 1/1,700 births. The disease leads to cirrhosis and emphysema. About 5% of the white population are carriers of a recessive allele for this disease.

Tay-Sach's Disease. This genetic disease is relatively common among Jewish people from central Europe. The rate of occurrence is 1/2,500. In the non-Jewish population, the disease occurs in 1/360,000 births.

Classical Albinism. In this defect skin cells fail to produce the pigment melanin which produces skin, eye, and hair color. This condition occurs in about 1/200 births among the Hopi Indians of Arizona and the Indians of New Mexico. It occurs in 1/20,000 births among Europeans.

Phenylketonuria (PKU). This is a metabolic abnormality that, when left untreated, leads to mental deficiency.

Myotonic Dystrophy. This is a type of muscular dystrophy. The muscles gradually degenerate so that they no longer function properly.

Others. (1) Total color blindness (rare disease in which individual sees the world in shades of gray), (2) inability to taste Phenylthiocarbamide [PTC], (3) alkaptonuria (a metabolic disease), and (4) hereditary deaf-mutism.

Epistemology of Chromosomal Theory of Inheritance

The chromosomal theory of inheritance is the synthesis of Mendelian genetics with cell, molecular, and developmental biology. Besides illustrating many of the characteristics of process science, this synthesis allows one to mechanistically picture the phenomena described by transmission genetics. Furthermore, the genetic theory illustrates a modified version of the scientific method associated with mechanistic science.

Summary of Genetic Terminology

A major distinguishing characteristic of experimental science which, however, often is neglected by textbooks, is that all theoretical terms are defined in relation to operationally defined terms, see table 11.3.

Operationally Defined Terms.

1. *Simple trait:* a trait that is expressed in only two recognizably distinct forms (analogous to tossing a coin and getting either heads or tails).
2. *True breeding plants:* population of plants in which a particular expression of a simple trait is passed on from one generation to another so long as individuals in the population expressing this trait do not cross-pollinate with individuals in some other population.
3. Symbols: P, F_1 and F_2
   a. P = parent generation
   b. F_1 generation = the first filial (of or relating to a son or daughter) generation of offspring from the parent generation
   c. F_2 generation = the offspring generation from random crossing (mating) the F_1 individuals among themselves

4. *Dominant trait:* the trait that *all* individuals of the F_1 generation exhibit when each parent population is a true breeding population (or pure line) which exhibits one of the two forms of a simple trait.

5. *Recessive trait:* the trait that no individuals of the F_1 generation exhibit when each parent population is a true breeding population which exhibits one of the two forms of a simple trait.

6. *Similar chromosomes:* two chromosomes seen under a light microscope to be of the same size and shape (as influenced by the position of the centromere).

Theoretical Terms

1. *Gene:* one of the two factors that determines the inheritance of a simple trait.
2. *Dominant gene:* when present this form of a gene always results in the expression of a dominant trait.
3. *Recessive gene:* when present this gene results in the expression of a recessive trait only when the dominant gene for this particular simple trait is not present.
4. *Homozygous individual* (with respect to a particular simple trait): all somatic cells have a pair of genes in the same form.
5. *Heterozygous individual* (with respect to a particular simple trait): all somatic cells have a pair of genes each of which is in a different form, e.g., one is dominant and the other is recessive.
6. *Genotype:* the form of the genes in the pair of genes that regulates a simple trait, e.g., both dominant or both recessive or one dominant and one recessive.
7. *Phenotype:* the expression of a particular genotype as a dominant or recessive trait.
8. *Chromosome:* a body consisting of a single centromere and containing one or more DNA molecules associated with special proteins.
9. *Sister chromatids:* the resultant of DNA replication which produces two DNA molecules and their associated proteins joined to a single centromere.
10. *Gene locus:* the position of a gene on a chromatid or chromosome.
11. *Allele:* any of the two or more forms of a gene controlling the expression of a simple trait.
12. *Homologous chromosomes:* a pair of similar chromosomes with the same sequence of alleles.
13. *Diploid cells:* cells that have one or more pairs of homologous chromosomes.
14. *Haploid cells:* cells that have no homologous chromosomes.
15. *Testcross:* the cross between a homozygous recessive parent with the parent expressing the dominant trait (note that such a parent could be homozygous dominant or heterozygous).
16. *Backcross:* the cross between the F_1 generation and one of the parent populations (NB: sometimes in the genetic literature, backcross is used synonymous with testcross).

Scientific Method of Mendel's Theory

As described in Chapter 3, the scientific method of mechanistic science has four characteristics: (1) operational terms, (2) empirical observation, (3) analytic-synthetic thinking, and (4) scientific public consensus of truth involving experimental prediction or (lack of) falsification.

The summary of genetic terms indicates that Mendel's theory uses operational terms. The events studied by Mendel's theory are empirical; that is, the inheritance of traits can be observed to occur many times. The model for the inheritance of a simple trait involves analytical-synthetic thinking: (1) the parts, a pair of genes, (2) are characterized as homozygous dominant or recessive or heterozygous, (3) interact to produce dominant or recessive traits, and (4) the sum of the probabilities of the different genotypes = 1 as well as P(dominant trait) + P(recessive trait) = 1.

Scientific public consensus of truth is illustrated by crossing individuals in an F_1 population of heterozygous individuals to obtain the F_2 generation, for example, $C/c \times C/c$ determining flower color in pea plants. According to Mendel's theory we would expect to get 25% C/C, 50% C/c and 25% cc. Of course, we cannot directly observe genotypes, but according to the theory, these genotypes would produce 3/4 purple flowers and 1/4 white flowers, which we can observe. If we observe color frequencies "close to" the ones the theory predicted, then the theory is proven to be true. Of course if the observed frequencies are "far from" the ones predicted, then the theory is proven to be false. Methods in modern statistical analysis provide precise definitions for "close to" and "far from." Thus, Mendel's theory nicely illustrates hypothesis testing which is a core idea of the scientific method.

Patriarchal Science Is Like Patriarchal, Institutional Religion

I. SAINTS/PROPHETS, e.g. Gregory Mendel: Shifted to a Process Perspective
II. FAITH & HOPE IN SCIENTIFIC POSITIVISM
III. REVELATION represented by metaphorical concepts organized into a scientific myth = scientific story.
    A. Metaphorical Concepts
        1. Sexual reproduction in humans or in flowering plants is *like* tossing a coin
        2. Traits of an offspring are *like* a system expressing itself in a particular way as a result of messages sent to it.
            a. Zygote = fertilized egg develops into an offspring that expresses traits as a result of receiving messages from each parent.
            b. Simple traits are expressed in one of two forms; the form that is expressed depends on the message the offspring receives in the sperm (or pollen grain) from the "father" and the message the offspring receives in the egg from the "mother."
    B. PEA PLANT STORY of how parent pea plants produce offspring with either purple or white flowers
        1. If both parents only have messages for expressing purple color flowers, then represent C for purple color and c for white color. Both parents will have genotype: CC and therefore only will produce sperm or eggs with one of these genes. The offspring will receive a C gene from each parent.
        2. Likewise, if both parents only have message for expressing white color flowers, then both parents will have genotype: cc. Therefore, the offspring will receive a c gene from each parent
        3. If each parent has a message for purple flowers, C, and a message for white flowers, c, then
            a. Genotype of offspring will be ¼ CC, ½ Cc, and ¼ cc.
            b. Phenotype of offspring will be ¾ purple flowers and ¼ white flowers.
IV. STORY CONVERTED INTO A RATIONAL MODEL: this involves
    A. Operationally Defined Terms
        1. True breeding purple flower pea plants
        2. True breeding white flower pea plants
        3. F_1 generation: cross between 2 types of true breeding pea plants: purple × white
            a. All offspring have purple flowers which is called dominant trait
            b. White flowers called the recessive trait
    B. Theoretical Terms
        1. Gene = message for expressing a trait.
        2. Dominant gene = always leads to expressing the dominant trait even in the presence of a recessive gene.
        3. Recessive gene = leads to expressing the recessive trait only if the dominant gene is not present.
    C. Testable Hypothesis
        1. F_1 generation: Cc
        2. Cc mate with Cc
        3. Predict: population of offspring
            a. CC, Cc = purple flowers 75%
            b. cc = white flowers 25%
        4. Carry out mating between plants taken from F_1 generation

5. Result: approximately 75% purple flowers and 25% white flowers
6. Hypothesis is valid.
7. Story from which hypothesis was created is valid.

V. FUNDAMENTALISM OF POSITIVISTIC PHILOSOPHY OF SCIENCE
   A. Validated Scientific Theories Produce Statements
      1. That are taken to be literally valid; i.e., it is valid to believe that the world is literally like the way the scientific theory proposes.
      2. Any other way of understanding the world is less valid or even totally invalid, e.g., any religious story about the world is considered to be invalid.
   B. Objectivity of Science
      1. Subjective insight leads to metaphors and a story that is the basis for formulating a rational model that can be validated by the scientific method.
      2. Once the scientific theory is validated via scientific, public consensus, it thereafter is considered to be objectively valid and its subjective basis is ignored or repressed.
      3. Then science is considered to be totally objective and literally valid even though it is based on subjectively held metaphorical insights.
      4. In fact, a scientific theory is never totally objective. Rather the creation of scientific theories depends on the mutuality of subjective and objective knowing. Positivistic scientists ignore this epistemological fact.
   C. The Positivistic, Philosophy of Science Is Like Any Fundamentalist, Creator-God Religion
      1. Like in any fundamentalist religion, the devotion to "science as religion" makes a leap of faith in:
         a. Metaphorical insights plus the scientific method will produce valid theories.
         b. No human can create an absolutely true description of any aspect of the world.
         c. No human can experience absolute Truth.
            a) One may acknowledge that such an experience can be described but never can be adequately represented by language.
            b) The believer in science as a religion claims
               1. that no one can have such an experience in the first place.
               2. any description of such an experience is self-delusion.
         d. Scientific type of knowing is the only valid way of knowing
            a) Other ways of knowing are to various degrees invalid.
            b) Some types of knowing, such as inspirations by various religious persons or philosophers, are totally invalid.
      2. Though positivistic scientists claim that their theories only are valid, in practice they treat these theories as if they were approximately true.
      3. Many positivistic scientists believe (make a subjective act of faith) that a scientific theory
         a. Is valid for all contexts
         b. Is valid independent of context
      4. The war between positivistic science and traditional values or religions or philosophies or the spiritual vision underlying American democracy is unavoidable and irresolvable.

Modifications of Mendel's Law

There are many examples in the history of science where a proposed theory is very successful in explaining certain phenomena, but then new observations are made that cannot be explained by the theory. Instead of discarding the old theory, scientists first attempt to modify it so that it explains these new observations. The old theory becomes a special case of a more all-inclusive theory. This is why science tends to become more and more abstract; it becomes more general in order to explain a variety of apparently different kinds of phenomena. This is what happened with Mendelian genetics. It became apparent that there are genetic patterns that cannot be explained by Mendel's first law. So Mendel's theory has been modified and extended to cover the new patterns.

Gene Expression

As indicated earlier, genes are genetic potentials that specify ways a developing individual can interact with a particular environment to produce a particular trait. The earlier stages of development are primarily determined by genes, but these stages can be modified, radically altered or completely repressed by various environmental stresses. As a result of such changes, sometimes a dominant gene will not be expressed in a particular individual. Eye color provides a notoriously embarrassing example of this. For years biology teachers following elementary textbooks have taught that dark eyes are dominant, resulting from the expression of a B gene, and blue eyes are recessive, produced only by b alleles. Theoretically, if both parents have blue eyes, then all their children must also inherit blue eyes. After carefully explaining this, many teachers have been confronted or even accosted by an irate brown-eyed student whose parents have blue eyes! Of course, infidelity is one possible explanation, but another plausible one is that the B gene in one heterozygous parent (Bb) did not express itself; so he or she developed blue eyes, but this parent passed on the B allele to one of his or her offspring, who, of course, developed brown eyes.

Alternatively, a particular gene or pair of genes may show degrees to which it is expressed; or the same allele or pair of alleles may express itself simultaneously in several different ways. Let us look at each of these modes of gene expression.

Penetrance.   When a particular genotype sometimes fails to express the appropriate phenotype in an individual, the pair of alleles is said to be incompletely penetrant. The particular genotype either is expressed or not expressed in a particular individual. Penetrance is the percentage of individuals of a particular genotype that show the expected phenotype under a specific set of environmental conditions. Many dominant, abnormal traits such as brachydactyly and Marfan's syndrome are incompletely penetrant. The dominant trait, comptodactyly (stiff little finger) has a 75% penetrance; that is, 25% of the individuals with the dominant allele for this trait do not express it.

Expressivity.   Some genes express themselves in some individuals but not in others (penetrance), but other genes may deviate by degrees from the expected phenotypic expression. Thus, expressivity is the degree to which a given allele manifests the expected phenotype. For example, myotonic dystrophy is a rare dominant trait that can be so debilitating that the person does not marry or survive very long. On the other hand, there are milder forms of this disease. Sometimes the expression of this trait is so mild that it goes unnoticed. This extremely mild form of the disease is equivalent to the trait being incompletely penetrant. Marfan's syndrome is another dominant trait that shows variable expressivity.

Pleiotropy.   Often an allele may produce several phenotypic expressions. This phenomenon is called pleiotropy. The basis of pleiotropy is that the protein whose synthesis is regulated by a particular allele influences or participates in many processes. Each process may produce a distinct phenotype. For example, the dominant trait, Marfan's syndrome, results from the production of an abnormal form of connective tissue. This, in turn, leads to several abnormal traits such as defects in the heart valves and large blood vessels because of weakness of the connective tissue, eye defects resulting from connective tissue not adequately holding the lens in place, and skeletal defects. Because the points of growth of bones do not close as early as they should, the long bones of the body tend to grow to extreme lengths. Abraham Lincoln may have had this defect. Also, the ligaments and tendons at the joints are weaker than normal allowing people with this syndrome to do unbelievable contortions. The allele producing this syndrome also exhibits variable expressivity. Some individuals, such as Abraham Lincoln, may have long limbs and loose jointedness. Others with the same allele suffer greater disability. In reality, most genes are probably pleiotropic though their secondary effects may be relatively insignificant.

Relativity of Dominance

The expressivity of a pair of alleles differs at the macroscopic and microscopic levels. A particular phenotype may result from the production of one or more enzymes in certain cells (or all the cells) in the body. The production of these enzymes is controlled by a pair of alleles. An allele is said to be dominant when the amount of enzyme produced by a single allele is sufficient to produce the expected or very near to the expected phenotype in the heterozygous individual. However, two doses of the allele (a homozygous individual) often produce about twice as much gene product (enzyme) as one dose. Thus, at the macroscopic level, it will not be possible to tell the difference between a homozygous and a heterozygous individual, but it may be possible to distinguish subtle differences at the biochemical level.

For example, the disease PKU is a recessive trait which, if left untreated in newborn babies, produces mental deficiency. Carriers of one of these recessive alleles, of course, will not have the disease. However, suppose a man and a woman want to marry but each has a history of PKU in his family. Each may be a carrier for the recessive allele for PKU. Knowing whether they both are carriers will help them decide whether to risk having children. There would be a 25% chance that the baby would inherit this disease. Because of the relativity of dominance, it is possible to biochemically determine whether an individual is a carrier of the recessive trait. Such a heterozygous individual would have one dominant allele that would result in the production of an enzyme that causes the breakdown of phenylalanine and thus prevent the PKU disease under normal conditions. However, this individual would have less of the enzyme than an individual who was homozygous. In such an individual under the stress of excess phenylalanine, not all of it will be broken down. The phenylalanine tolerance test measures the amount of phenylalanine in blood plasma after phenylalanine ingestion. The amount present for heterozygous individuals is about midway between that found for normal homozygotes and the disease-afflicted person. Thus, by giving this test, one can detect a heterozygote and give appropriate genetic counseling to prospective parents with a family history of PKU.

Codominance

Sickle-cell anemia is a genetic disease in which red blood cells take on a sickle shape when exposed to an environment with a low oxygen pressure. This routinely occurs in veins where the circulating blood has given up much of its oxygen while passing through the capillary beds. The sickled red cells tend to clump and block small venules in various tissues such as the bones, spleen, and lungs. Blood flow through these areas will cease and the local tissue will die. This is painful to the patient and eventually leads to death. The red cells return to their normal shape (biconcave discs) when the blood becomes reoxygenated. However, the continual stress of going from a normal shape to a sickle shape tends to shorten the life of these red cells. The number of red cells in the blood tends to be lower than normal so that the oxygen-carrying capacity of the blood is decreased. Any condition that leads to a lower than normal oxygen-carrying capacity is called anemia (hence sickle-cell anemia).

Red blood cells taking on a sickle shape is due to the presence of abnormal hemoglobin. Let the genotype for normal hemoglobin (HbN) be NN and that for the abnormal hemoglobin (HbS) in sickle-cell disease be SS; that is, the one pair of alleles controlling HbN or HbS is homozygous. However, neither the N nor the S allele is dominant, for the heterozygous genotype NS produces some HbN and some HbS hemoglobin. Thus, both N and S always express themselves so that a single pair of alleles has three possible genotypes (NN, NS, or SS) and correspondingly three phenotypes (HbN, HbN + HbS, and HbS). This genetic pattern where neither allele of a pair of alleles is dominant is called codominance.

People with the genotype NS are said to have the sickle-cell trait. Ordinarily, there are no clinical manifestations of red cell sickling; however, a sample of this blood can be caused to sickle in the laboratory. Furthermore, these individuals may show some manifestations of the trait under extreme environmental stress, such as high altitudes (therefore low oxygen pressure in the blood) or poor circulation of blood through some tissues. The sickle-cell disease occurs most frequently in Equatorial Africa and to a lesser extent in the Mediterranean area and in India. This disease (SS) is often fatal in early childhood. Among Afro-Americans 0.25% are born with this disease, whereas 8% have the sickle-cell trait.

Multiple Alleles

Mendel postulated that alleles can exist in one of two forms. However, we now appreciate that sometimes an allele may exist in more than two forms. A particular individual will have a single pair of alleles that controls a particular trait. However, in the population of individuals possessing this trait, the allele may be in three or more forms referred to as multiple alleles. As a result, there are more than just three genotypes and correspondingly there are several possible phenotypes. The ABO blood types and hair form exemplify phenotypes resulting from multiple alleles.

The ABO Blood Types.   It is possible to classify blood as type A, B, AB, and O according to the type of antigen(s) present on red blood cells and the type of antibodies present in the plasma. An antigen is a large molecule that when introduced as a foreign body stimulates the individual to produce antibodies that will combine with it thereby neutralizing it, see Chapter 8. Antigen-antibody reactions are the mechanisms underlying the body's defenses against foreign bodies or disease-causing organisms, the development of immunity and allergies, the rejection of tissue transplants, and the development of certain self-induced diseases. Type A blood has A antigens on red blood cells and anti-B antibodies (antibodies that can neutralize B antigens) in the plasma. Type B blood has B antigens and anti-A antibodies. Type AB blood has A and B antigens and no antibodies and type O blood has no antigens but has anti-A and anti-B antibodies, see figure 11.8.

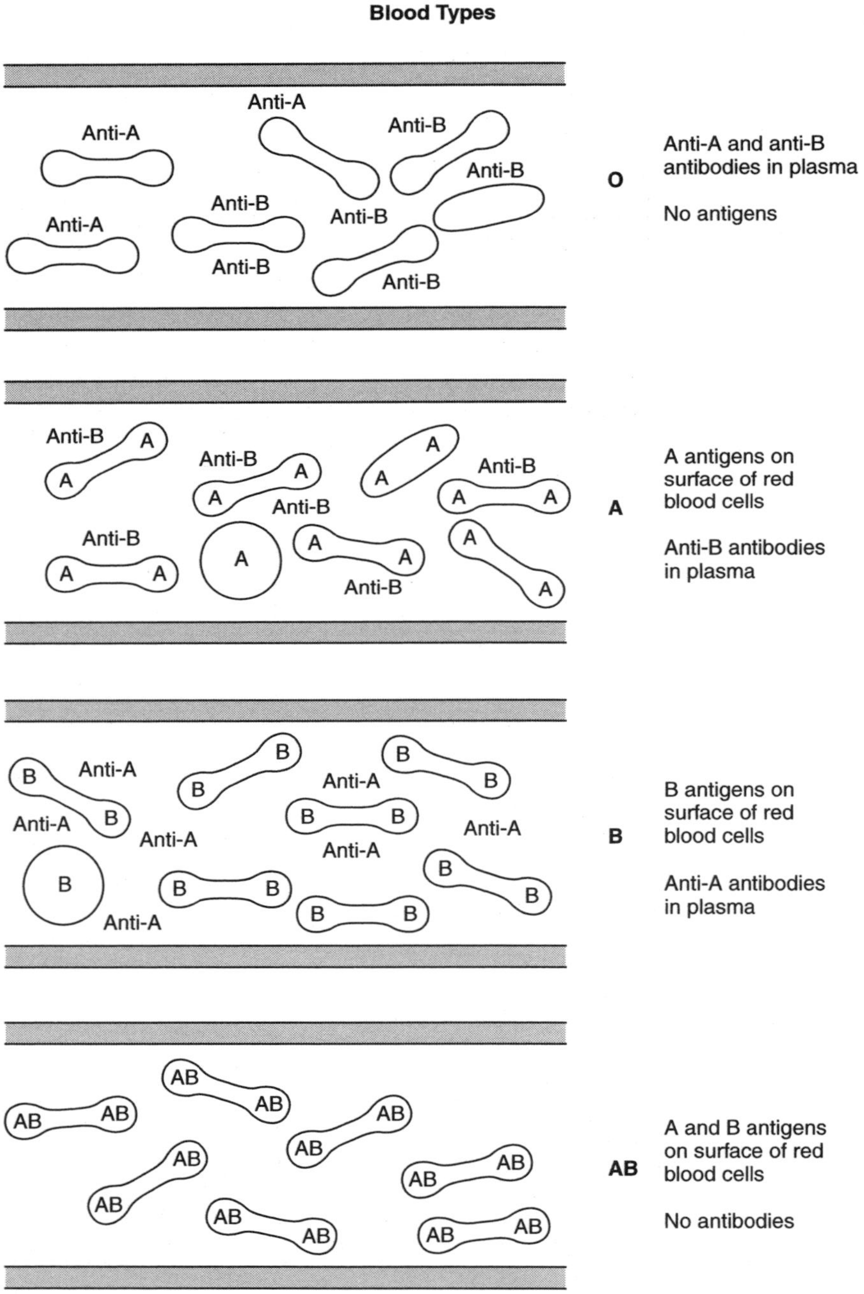

When blood transfusions are necessary, a person should only be given a blood type that is the same as his or her own. If blood types A and B are mixed, the B antibodies in type A blood will react with the B antigens on the red blood cells and cause these cells to clump (agglutinate). Likewise, the A antibodies in type B blood will react with the A antigen on red blood cells causing them to clump. Type O blood sometimes is designated as a *universal donor* because, provided the quantity is not too large, it can be given to individuals with type A, type B or type AB blood. Even though type O blood contains both anti-A and anti-B antibodies, when given in transfusion, these antibodies

Blood Types

Anti-A and anti-B antibodies in plasma

No antigens

O

A antigens on surface of red blood cells

Anti-B antibodies in plasma

A

B antigens on surface of red blood cells

Anti-A antibodies in plasma

B

A and B antigens on surface of red blood cells

No antibodies

AB

FIGURE 11.8

in the donor plasma quickly diffuse throughout the blood in the recipient. As a result, the foreign antibody level is too low to bring about red blood cell clumping. In contrast, red cells from a type A, a type B, or a type AB person would quickly clump when transfused into a type O individual. Type AB blood sometimes is designated as a universal recipient because, provided the quantity is not too large, types A, B, or O can be given to a person with type AB blood. This is because the antibodies in the plasma of the donor will be quickly diluted by the plasma of the recipient. As a result, the antibody concentration will be too low to bring about a reaction.

A single gene locus controls the inheritance of ABO blood types. Alleles at this locus may be in one of three forms: I^A, I^B, and I^O. The I^A is expressed as type A blood, I^B is expressed as type B blood, $I^A I^B$ is expressed as type AB blood, and $I^O I^O$ is expressed as type O blood. Thus, there are six genotypes and four phenotypes of the ABO system of blood types, as shown in table 11.7

Mendel's Second Law

Sometimes a complex trait such as one's "looks" may be considered to result from a number of simple traits. Mendel's second law describes the simultaneous inheritance of two or more simple traits where the inheritance of one trait has no effect whatsoever on the inheritance of the other trait(s). Such traits are said to be independent; the probability of one occurring is not influenced by the probability of the other trait(s) occurring.

Simultaneous Inheritance of Two Independent Traits

Let us consider the simultaneous inheritance of eyelash length and lip thickness. Let L represent the dominant form of the allele which determines long eyelashes. Then ll, the homozygous recessive alleles, determines short eyelashes. Let T represent the dominant form of the allele which determines thick lips. Then *tt*, the homozygous recessive alleles, determines thin lips.

Cellular Basis of Independent Traits. If traits are independent, then the gene loci regulating the inheritance of these traits must be on different pairs of homologous chromosomes. During the first meiotic division, homologous pairs of chromosomes line up in the middle of the cell. An imaginary plan called the equatorial plane, separates each member of a chromosome pair. Suppose the individual is a heterozygote, i.e., LlTt. At the beginning of the first meiotic division there are two equally possible arrangements that the chromosomes containing these alleles can take: (1) the dominant alleles, L and T, are on the same side of the equatorial plane; or (2) the dominant alleles are on opposite sides of the equatorial plane, see figure 11.9. The first meiotic division results in the separation of homologous chromosomes and production of haploid cells. The first arrangement just described produces two types of cells, those with dominant alleles, LT, and those with recessive alleles, lt. The second arrangement produces two types of cells, those with Lt and those with lT. A second meiotic division of

Table 11.7 Genotypes and Phenotypes for the ABO Blood Types

Genotype	Phenotype Type of Blood
$I^A I^A$ $I^A I^O$	A
$I^B I^B$ $I^B I^O$	B
$I^A I^B$	AB
$I^O I^O$	O

**Two equally possible arrangements of chromosomes
at the beginning of the first meiotic division**

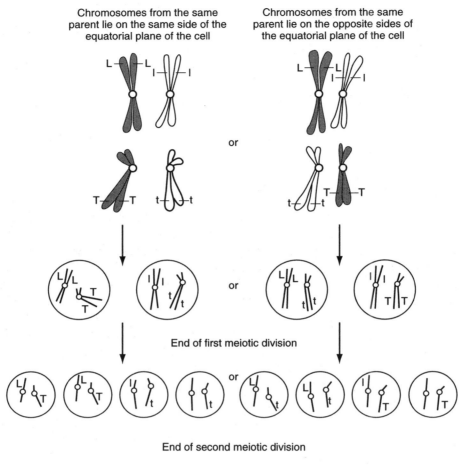

Chromosomes from the same
parent lie on the same side of the
equatorial plane of the cell

Chromosomes from the same
parent lie on the opposite sides of
the equatorial plane of the cell

or

End of first meiotic division

or

End of second meiotic division

Summary of possible gene combinations of the gametes

LT   lt   Lt   lT

FIGURE 11.9

each of these four cells produces eight gametes consisting of four different genomes: LT, lt, Lt, and lT. Each genome has a two in eight chance of occurring. That is, P(LT) = P(lt) = P(Lt) = P(lT) = 2/8 = 1/4, see figure 11.9

Figure 11.9 implies that the association between any two different loci is purely by chance. Thus, for example, L could be associated with T or with t. Therefore, if an offspring receives a T allele from one of its parents, one still cannot predict which form of the second gene it also will receive.

Punnet Square Method of Analysis.   All problems dealing with the simultaneous inheritance of two independent traits are solved in the same way as problems involving inheritance of one trait. First, we must know or deduce the genotype of each parent. Second, one determines for each parent the possible genomes resulting from the production of gametes. Third, one determines the various possible combinations of gametes from each parent. The punnet square method achieves this by drawing a square and putting the possible gametes produced by each parent on the top and the left side of the square. Table 11.8 represents the possible combinations of mating between two heterozygous parents (LlTt). Figure 11.9 shows that each parent produces four types of gametes: LT, Lt, lT, and lt. There are a total of 16 possible combinations, that is, elementary outcome events, but of these there are only 9 different genotypes (many of the genotypes are compound events). By inspecting

▶Table 11.8

Male Sperm	Female Ova			
	LT	**Lt**	**IT**	**It**
LT	LLTT long eye lashes thick lips	LLTt long eye lashes thick lips	LITT long eye lashes thick lips	LITt long eye lashes thick lips
Lt	LLTt long eye lashes thick lips	LLtt long eye lashes thin lips	LITt long eye lashes thick lips	LItt long eye lashes thin lips
IT	LITT long eye lashes thick lips	LITt long eye lashes thick lips	IITT short lashes thick lips	IITt short lashes thick lips
It	LITt long eye lashes thick lips	LItt long eye lashes thin lips	IITt short lashes thick lips	IItt short lashes thin lips

the punnet square one can determine how many times a particular genotype occurs out of the 16 possible combinations. Then, the probability of an offspring having that genotype will be the number of times it can occur divided by 16. Thus, for example, P(LlTt) = 4/16 or 2/8. In a similar way one can determine the probability of occurence of each of the four possible phenotypes. For example, P(long eyelashes and thick lips) = 9/16; P(long eyelashes, thin lips) = 3/16; P(short eyelashes, thick lips) = 3/16, and P(short eyelashes, thin lips) = 1/16.

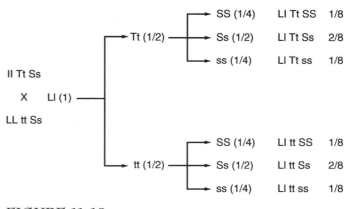

FIGURE 11.10

Simultaneous Inheritance of Three Traits

Consider the same two traits just described plus a third trait for chin shape. Let S represent the dominant form of the allele which determines a straight chin; then ss, the homozygous alleles, determines a receding chin.

Tree Diagram Analysis of Genotypes.   The tree diagram method provides a more direct approach for working out possible outcomes for simultaneous inheritance of two or more traits. First, one determines the genotype of each parent. Second, one considers for each locus the probability of possible combinations of alleles gotten from each parent. For example, suppose the genotype of the father is llTtSs and that of the mother is LLttSs. We work out or recall that ll × LL produces only one outcome whose probability is P(Ll) = 1. Tt × tt produces two possible outcomes whose probabilities are P(Tt) = 1/2 and P(tt) = 1/2. Ss × Ss produces three possible outcomes whose probabilities are P(SS) = 1/4, P(Ss) = 1/2, and P(ss) = 1/4. Third, one writes down in a tree diagram the various possible combinations and their associated probabilities of all the above listed pairs of alleles, see figure 11.10.

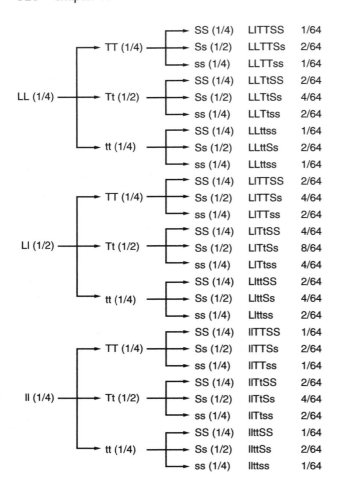

FIGURE 11.11

Finally, for each outcome genotype say for example, LlTtSs, one computes its probability by using the multiplication rule, i.e., since P(Ll) = 1 and P(Tt) = 1/2 and P(Ss) = 1/2, we have $1 \times 1/2 \times 1/2 = 1/4 = 2/8$. The genotype, LlTtSs, is a compound event that is said to occur only if its component events Ll, Tt, and Ss all occur simultaneously, which in this case means that all these pairs of genes are present in a single zygote resulting from the union of gametes from each parent. Fig. 11.11 represents the tree diagram analysis of the probabilities of possible genotypes resulting from the mating of two heterozygous parents.

Tree Diagram Analysis of Phenotypes. The tree diagram analysis for phenotypes is done in a way similar to that for genotype analysis. Let L- represent long eyelashes and ll short eyelashes; T- thick lips and tt thin lips; S- straight chin and ss receding chin. The dash in some of these pairs means that whatever allele, dominant or recessive, appears in that place is irrelevant to the phenotype because the other dominant allele in the pair determines what the phenotype will be. Consider again the possible outcomes from mating between two heterozyugous parents, LlTtSs × LlTtSs. The probability of the various individual traits are: P(L-) = P(LL) + P(Ll) = 1/4 + 1/2 = 3/4 and P(ll) = 1/4; P(T-) = 3/4 and P(tt) = 1/4; P(S-) = 3/4 and P(ss) = 1/4. One then constructs a tree diagram as shown in figure 11.12. Then one can calculate the probability of any particular phenotype as for example, P(long eyelashes, thin lips, receding chin) = P(L-ttss) = $3/4 \times 1/4 \times 1/4 = 3/64$.

Linked Traits

Two traits are said to be linked if they do not follow Mendel's second law; that is, they are not independent. What does this mean? Suppose traits #1 and #2 are linked. Mendelian inheritance implies that the inheritance of trait #1 is totally independent of trait #2. Suppose an individual has received alleles that produce trait #2. Linkage implies that the probability that the individual also will have received genes that produce trait #1 is greater than what one would have predicted using Mendel's second law. Linkage may be between two nonsexual traits or between a biological sex trait and a nonsexual trait.

Somatic cells of humans contain 22 pairs of autosomal chromosomes (chromosomes not

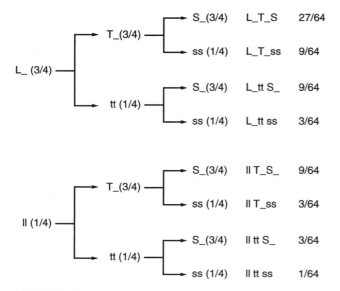

FIGURE 11.12

directly implicated in the determination of biological sex) and one pair of sex chromosomes. Traits determined by two gene loci on the same autosome are called autosomal-linked traits. Biological sex is a complex trait for which the inheritance pattern has not been determined. Geneticists still do not know the one or several pairs of alleles that regulate biological sex, but they do know the chromosomal mechanism for determining biological sex. This varies for different species. For humans (and other mammals) a pair of X chromosomes (usually) initiates development of a biological female, and an X chromosome paired with a Y chromosome (usually) initiates development of a biological male. Biological sex may be linked to non-sexual traits in two ways. If the alleles producing the nonsexual trait are located on the X chromosome, then the recessive nonsexual trait will occur more often in males than in females. Because there is only one X chromosome in males, a single recessive gene on this X chromosome will produce the recessive trait. However, in females which have two X chromosomes, both alleles must be recessive. This is much less likely than receiving just one recessive gene. Therefore, X-linked recessive traits occur much more often in males than in females. If the gene producing the nonsexual trait is located on the Y chromosome, the trait will occur only in males.

In summary, linked traits are divided into three categories: Y-linked traits (holandric traits), X-linked traits, also called sex-linked traits, and autosomal-linked traits.

Y-Linked Traits

Since in humans only males have a Y chromosome, traits controlled by alleles on this chromosome will show up only in males. Hairy ear rims apparently is such a trait. However, confirmation of this is still in question. The trait involves the presence of long, stiff hairs on the rim of the ears. Though present in all populations, the trait is more common in people from India.

X-Linked Inheritance

In contrast to Y-linked inheritance, there are over 100 definite human traits definitely known to be X-linked. Some of these are: (1) certain types of color blindness, (2) night blindness, (3) hemophilia or bleeder's disease, and (4) at least four types of muscular distrophy, the most common one being Duchenne's muscular dystrophy. The most common sex-linked abnormality is probably the red-green type of color blindness, which causes confusion in distinguishing between red and green.

Let C stand for the dominant allele for normal vision and c for the recessive allele for red-green color blindness. About 8% of males in the United States have this type of color blindness, while only 0.4% of females have it. The possible genotypes for a female are X^CX^C, X^CX^c, and X^cX^c, and those for a male are X^CY and X^cY. That is, the Y chromosome does not have a gene locus for this trait. Thus, a Punnett square can be used to solve various genetic problems involving red-green color blindness. For example, mating between a normal male and a normal female who has one gene for color blindness produces 50% of all males with this trait but no females with it, see table 11.9.

Table 11.9

Male Sperm	Female Ova	
	X^C	X^c
X^C	X^CX^C normal female	X^CX^c normal female
Y	X^CY normal male	X^cY color blind male

▶ **Table 11.10**

Male Sperm	Female Ova	
	X^H	X^h
X^h	X^HX^h normal female	X^hX^h female with hemophilia
Y	X^HY normal male	X^hY male with hemophilia

There are several kinds of hemophilia, but the most common type, *hemophilia A*, accounts for about 75% of the cases. These diseases are due to a recessive allele, designated as h, which results in a lack of one of the factors necessary for blood clotting. Table 11.10 shows the possible children from a female carrier of h and a male with hemophilia. Hemophilia in females is rare because the individual must be homozygous to express the disease, and thus must receive one recessive allele from each parent. The chance of a female carrier mating with a male with this disease is low because such males seldom survive through reproductive age or they reproduce at a very low rate. As a result, there are very few cases of hemophilia in females, while one male in every 10,000 is born with this disease.

Gene Interaction

There are situations where one or more pairs of alleles will modify, suppress, or mask the action of one or more other pairs of alleles in the cell. When this interaction involves the gene loci controlling sex, the resulting traits may be of two types: sex-limited and sex-influenced traits. Gene interaction involving nonsexual gene loci is called epistasis.

Sex-Limited Traits

A trait is said to be sex-limited if it appears only in one sex. The pair (or pairs) of alleles occurs in both sexes but it is expressed only in one. Many secondary sex characteristics probably are determined by several pairs of alleles. Though we may not know the details of their inheritance patterns, we know they are sex-limited. For example, a beard and a larger larynx occur only in males while developed breasts and wider hips usually occur only in females. Another example is precocious puberty. At least one form of this condition is the result of a dominant allele located on one of the autosomes. Four-year-old heterozygous males with this gene develop secondary sex characteristics and undergo an adolescent growth spurt. Since the ends of the long bones fuse at an early age, the affected boys are much taller for a time than their contemporaries, but they soon stop growing and as adults are short men. The same abnormal gene may be present in females but is not expressed.

Sex-Influenced Traits

Sex-influenced traits are traits that are determined by genes located on autosomes but are expressed to a greater degree in one sex than in the other. The sex of the individual apparently determines certain alleles to be dominant in one sex but recessive in the other. Thus, when both parents are heterozygous for a sex-influenced trait, there will be a 3:1 ratio in the sex in which the allele is dominant and a 1:3 ratio in the sex in which it is recessive. The following traits are more common in males than in females: (1) pattern baldness, (2) hairlip, (3) some forms of cleft palate, and (4) some allergies. For example, consider the inheritance of pattern baldness. Let B^1 represent the allele for pattern baldness which is dominant over B, the allele for normal hair, in males but is recessive in females. If a normal female possessing this recessive allele (B^1) mates with a normal male, then the possible children will be: (a) female children: BB and BB^1 and both will have normal hair, and (b) male children: BB which will have normal hair and BB^1 which will develop baldness. The son of a bald father may not inherit baldness; he may get the B rather than the B^1 gene. Conversely, the son of a nonbald father may develop baldness; he could get the B^1 gene from his heterozygous mother.

Epistasis

Epistasis is the phenomenon of alleles at one gene locus modifiying the expression of alleles at other gene loci. If we consider two independent gene loci, for example, A and B, there are at least three types of interaction: (1) dominant epistasis, A suppresses B, (2) recessive epistasis, aa represses B, and (3) complementary gene interaction.

Recessive Epistasis, for Example, Classical Albinism.   The double recessive pair of genes that produce albinism blocks the normal production of the pigment melanin. Skin color is determined by the amount of melanin in the skin cells. The allele pair for classical albinism blocks several other pairs of alleles, namely those that determine eye color, hair color, and skin color. Typically, the albino has white hair, white skin, and pink eyes which result from the reflection of light from the unmasked blood in the capillaries of the eye. The albino may have children with some hair, skin, and eye color because the genes for these traits were present in the albino parent but masked by the double recessive alleles for albinism.

Tearout #1

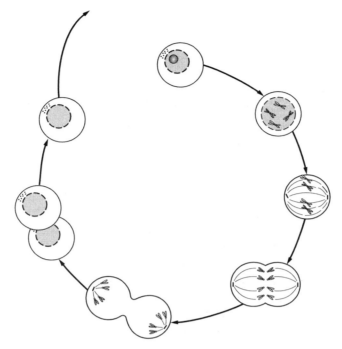

FIGURE 11.3

Directions:

1. Using figure 11.3 as a reference, fill in all labels.

Tearout #2

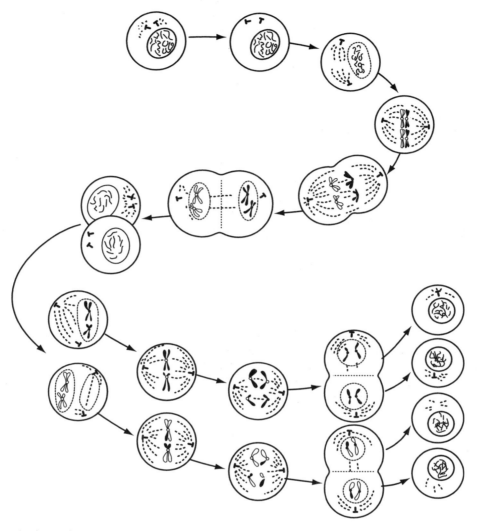

FIGURE 11.5

Directions:

1. Using figure 11.5 as a reference, fill in all labels.

Tearout #3

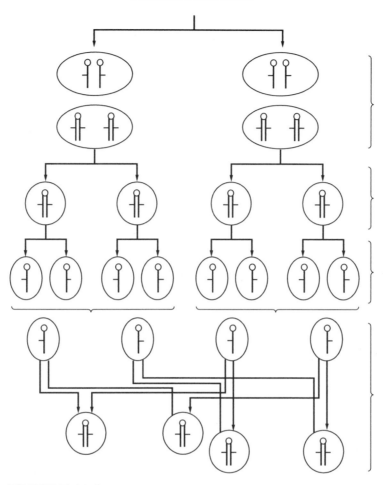

FIGURE 11.6

Directions:

1. Using figure 11.6 as a reference, fill in all labels.
2. State Mendel's first law in terms of Tt × Tt.

Epigenesis: Developmental Individuation

Epigenesis in Plants and Animals

Epigenetic development is one of the core ideas of the modern synthesis theory of evolution. Theodosius Dobzhansky, one of the major formulators of the modern synthesis, summarizes epigenesis as follows:

> Development is then a sequence of interactions between the successive states of the organism and of its environment. A fertilized egg [or a single-celled nonsexually reproducing organism] is a highly organized system. This system is so contrived that the built-in response of the living body to the environments which the species usually encounters in its habitats is embryonic development, maturation, and aging. In normal environments, development follows a predictable path. . . . However, the trajectory through time which the development of an organism may follow is not irrevocably fixed. This is because the environment of any species is more or less variable, and because a living body is a time-binding machine which carries the record of its life experiences within itself. Every response of the organism to its environment is determined by genes, jointly with the sequence of the environments which this organism had encountered during the individual's lifetime. . . . Ultimately, any character is "hereditary," because the potentiality of its development lies necessarily within the inherited norm of reaction of the organism. But any character is also "environmental," because the development can take place only within a certain range of environments, and life and development involve transformation of a part of the environment into the living body.[1]

Epigensis is the developmental process of making via a sequence of stages an integrated whole system, e.g., an organism, where: (1) Each stage is determined (programed) by a set of directions (for example, a genetic program) and (2) Each state is influenced by: (a) incoming components for it, (b) what occurred in the immediate previous stage, (c) communication among components of each stage (for example, cell-to-cell dialogue) and (d) communication between each stage and the environment.

Epigenesis reinforces three aspects of the process perspective: (1) Each organism is interrelated to its environment which, of course, includes other organisms. (2) Each organism is a story jointly

---

[1]Dobzhansky,T. and J. B. Birdsell, 1957. "On Methods of Evolutionary Biology and Anthropology." *Am. Scientists*, 45: pp. 382–383.

determined by inherited potentials, the genes, and its environment; so time is irreversible because organisms are forordained to grow old, not the reverse. (3) As biologists, we are forced to rely on subjective insights of developing organisms which exist independent of our knowing them. We may not comprehend such objects the way Aristotle and medieval philosophy would have us believe, but the subject-object duality remains. These entities with varying degrees of individuality are not illusions emerging into "quasi-existence" from pure consciousness as some physicists would have us believe.

Even prior to the proposal and acceptance of the cell theory, biologists carefully observed the gross structure changes that occur in developing embryos. The cell theory enabled embryologists (logy means "study," therefore embryology means "study of embryos") to analyze these structural changes in terms of multiplication of cells and the production of different types of cells. This cellular analysis led to the following descriptive theory of development. A zygote divides many times to produce a mass of identical cells. During a process called differentiation these cells separate into different types. Each cell type further differentiates into various types of tissue. For example, in animals there are three primitive cell types called *germ layers*. The layers are named according to their place of origin: (1) ectoderm (ecto means "outer"), (2) endoderm (endo means "inner") and (3) mesoderm (meso means "between" and refers to the fact that mesoderm lies between the ectoderm and the endoderm). The ectoderm differentiates into more specialized cells such as primitive nerve cells, which mature into the various types of nerve cells. Likewise the mesoderm produces primitive muscle cells that differentiate into the three types of muscle cells. The above statements exemplify the idea that development consists of cell differentiation which occurs in a sequence of stages that go from the general to the more specific.

The synthesis of genetics with the cell theory provided a new dimension to the analysis of development. The genetic material contains information that "programs" various stages of cell differentiation. These programs represent "potentials" any cell has to develop more specific traits in response to a certain type of environment. In other words, differentiation results from the interaction between a cell with a particular set of "genetic potentials" and a particular environment that can activate some of these genetic programs. The more recent science of "experimental embryology" verified that these types of interactions do occur in embryos.

The synthesis of embryology with cell biology, biochemistry, molecular biology, and genetics has produced the discipline called "developmental biology." In this framework all development is thought of as consisting of four component processes: (1) determination, (2) differentiation, (3) growth, and (4) morphogenesis. Determination is the process by which the potential of a cell or of a part of an embryo becomes restricted to form only a particular type of cell or organ. Differentiation is the manifestation of determination and so is the sequence of events where new properties in a single cell or in a group of cells appear. Growth is somewhat ambiguous since it could mean any increase in size or mass of an organism. In developmental biology, growth refers to a permanent enlargement of a part, e.g., cells, organ, or whole organism, as distinct from random or cyclic fluctuation in size. Cell growth involving increase of cytoplasm may occur before cell determination. After cell determination, the pattern of growth in cells, tissues or organs is one of the visible outcomes of differentiation. Finally, morphogenesis—the generation of a form—is the process of a part of an organism taking on a new shape. Morphogenesis, of course, involves differentiation and growth and usually is associated with massive cell movements in animals or different rates of growth in plants.

Development of Flowering Plants

Flowering plants provide a good first approach to visualizing the four component processes of development. In various sections of a developing plant, one readily may distinguish four regions: (1) region of cell division producing partially determined cells; (2) region where few or no cells are dividing but all cells are growing longer; (3) region of cell differentiation, and (4) region of organ morphogenesis where cells are dividing, growing, differentiating, and becoming organized into a specific plant organ such as a leaf or flower. Plants also nicely illustrate the different factors which influence cell determination: intrinsic factors, cell-to-cell communication, and environmental factors.

Intrinsic Partial Cell Determination

Root-Shoot Polarity.   Developing and adult plants show polarity in that one end of the plant grows up as a shoot (the part of the plant that includes stem, branches and leaves) and the other end grows down as a root, see figure 12.1. The polarity of shoots and roots is a built-in property of the component tissues. This is shown in regeneration experiments. If one cuts a stem into pieces and suspends them in a moist atmosphere, the lower end of each piece will develop roots and the upper end will develop shoots. Likewise, if one plants pieces of root in moist sand, the lower end of each piece will develop roots and the upper end will develop shoots. This polarity of development occurs even if the roots are planted upside down or sideways. If one cuts a piece of stem or root into smaller pieces and induces regeneration, the same polarity prevails: upper end produces shoots and lower end produces roots. This polarity may be traced to the original zygote which has its nucleus and dense cytoplasm at one end and a vacuole filled with water at the other end. The zygote divides to produce a larger basal cell and a smaller terminal cell. The basal cell gives rise to the root end of the

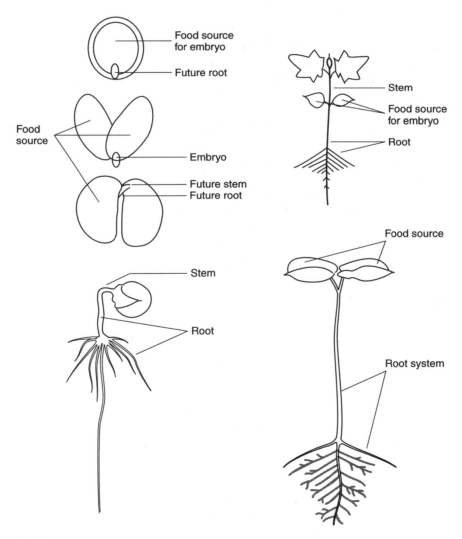

FIGURE 12.1

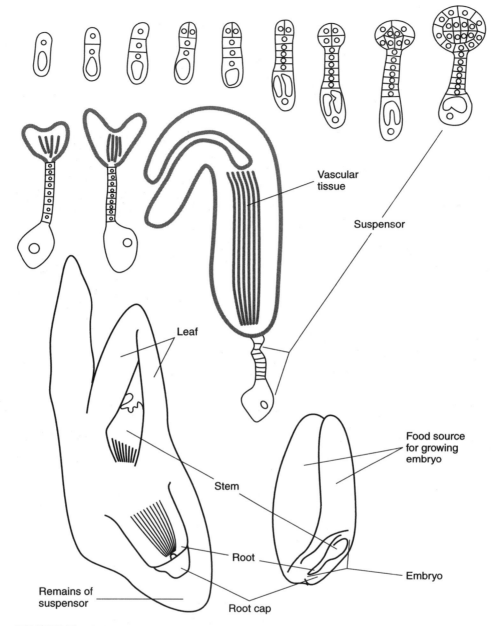

Vascular tissue

Suspensor

Leaf

Food source for growing embryo

Stem

Root

Embryo

Remains of suspensor

Root cap

FIGURE 12.2

embryo, whereas the terminal cell gives rise to the shoot end, see figure 12.2. Thus, these initial cells have identical genetic information but are determined to develop into different organs.

   Totipotency of Plant Cells.   Note that the basal cell and terminal cell referred to above are determined to develop into different types of organs, but as yet cell differentiation has not occurred. The early development of an embryo into a root end and a shoot end exemplifies the distinction between cell determination and differentiation. In general, cell determination in plants which leads to a particular pattern of differentiation is never complete. Under certain circumstances it is possible to cause differentiated plant cells to start dividing again, producing daughter cells that are undifferentiated. With careful treatment in tissue culture, any one of these undifferentiated cells may be induced to produce a new plant. Thus, the first division of a zygote produces partial cell determination. The basal and terminal cells and all subsequent daughter cells retain totipotency—under the right conditions each cell can regenerate a new plant.

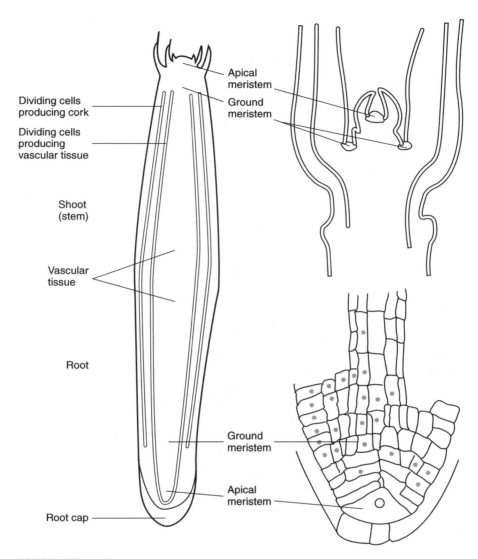

Apical
meristem

Ground
meristem

Dividing cells
producing cork

Dividing cells
producing
vascular tissue

Shoot
(stem)

Vascular
tissue

Root

Ground
meristem

Apical
meristem

Root cap

FIGURE 12.3

Self-Determining Plant Organs

The term *primordial* refers to cells that are partially determined but not yet differentiated in a specific way. The polarity resulting from the first division of the zygote is only the first of a series of partial cell determinations. The primordial root cells and shoot cells divide into daughter cells of two types: meristem cells which continue to divide, see figure 12.3, and cells that after progressive determination, differentiate into specific cell types. In roots, these further cell determinations probably are influenced only by intrinsic factors; however, cell-to-cell communications bring about the coordination of differentiation of different primordial cells to produce the organized tissues of a root. In shoots, cell-to-cell communications influence some of the further cell determinations as well as coordinating tissue differentiation. This coordination provided by cellular communication means that the developing organ, in this case roots and shoots, are self-determining—environmental factors influence but do not help guide the differentiation.

Developing Root.   The developing root, perhaps the simplest land plant organ, spatially illustrates the four component processes of development, see figure 12.4. Root meristem cells continue to divide to produce primordial root cap cells on the underside and primordial root tissue on the above

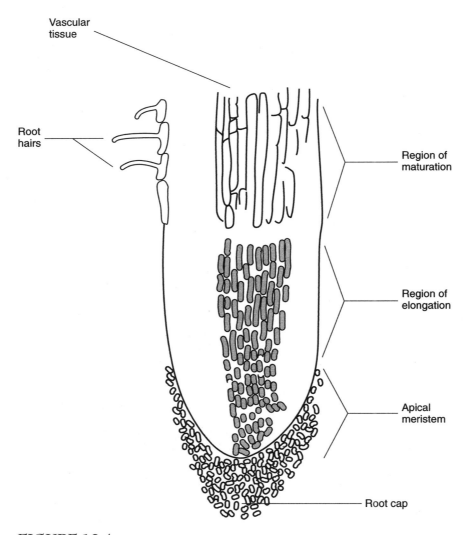

Vascular
tissue

Root
hairs

Region of
maturation

Region of
elongation

Apical
meristem

Root cap

FIGURE 12.4

side. The meristem represents growth in terms of cell division and individual cell increase in cyto-plasm. The area above the root meristem consists of cells partially determined to become specific root tissues. These cells are in the process of elongation by accumulating water in vacuoles; thus, this area represents cell determination and individual cell growth. Above the area of growth, cells begin to show specialized structural features which illustrate cell differentiation. The developing root as a whole represents morphogenesis.

1. Cell Growth by Vacuolation.   Meristematic cells and their daughter cells are cube or rectangular in shape with a thin primary cell wall. Some of these cells continue to divide whereas others are pushed upwards and begin to enlarge by taking in water. The primary cell wall expands resulting in a decrease of intracellular water pressure. This leads to more water entering the cell that produces more cell wall expansion, more decrease of intracellular water pressure, more inflow of water, and so on. The expansion of the primary cell wall is regulated so that instead of the cell expanding equally in all directions to form a sphere, the cell elongates. Water is taken into small vacuoles which then coalesce to form a large vacuole, see figure 12.5. While cell elongation is occurring, the growing vac-uoles push cytoplasm to the periphery of the cell against the cell wall; however, some cytoplasmic strands criss-cross the single vacuole of an elongated cell. Cells in the early stages of vacuolation may continue to divide, but the more mature elongated cells lose this ability.

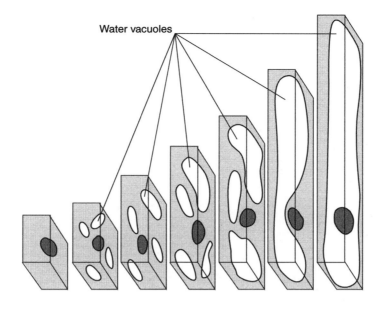

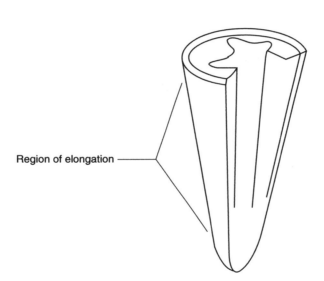

FIGURE 12.5

2. Cell Differentiation.   The biochemical and functional changes of a differentiating cell are associ-ated with changes in cell wall structure. Secondary cell wall is usually deposited between the outer cell membrane and the primary cell wall, see figure 12.6. The pattern of secondary wall deposition is related to the function of the differentiated cells. Some columns of cells become hollow, dead tubes, called xylem, which conduct water absorbed from the soil up to stems and leaves of the plant. Other columns of cells, called phloem, become living sieve tubes which conduct substances from one part of a plant to another. Still other cells become something like connective tissue in animals—some of these are called parenchyma cells and others are fiber cells that provide mechanical support. The out-ermost cells of the root differentiate into epidermis analogous to skin cells in animals, and some of these cells develop root hairs, as shown in figure 12.4. Eventually some of the parenchyma cells dif-ferentiate into lateral meristem, see figure 12.3, that divide so that the root expands in width while the apical meristem enables the root to expand in length.

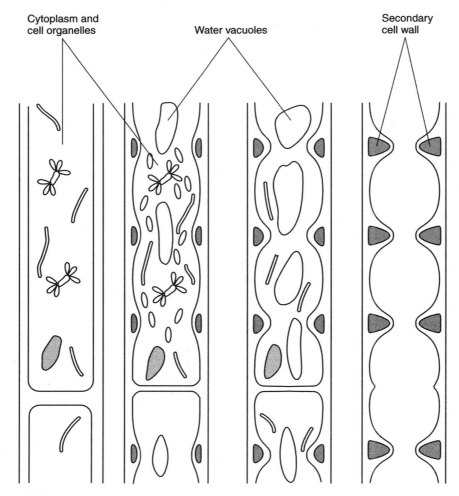

FIGURE 12.6

Cell-to-Cell Communication.   Plant cell membranes contain receptor sites that form a specific key-lock type interaction with substances from other cells or with foreign plant cells. A more widespread form of cell-to-cell communication occurs via transport of substances among cells. Xylem and phloem tubes transport fluids, sometimes over great distances, from one part of the plant to another. Plasmodesmata (singular, plasmodesma) are fine cytoplasmic channels which pass through intervening cell walls, see figure 12.7, and allow substances to move among cells in a tissue. The membrane of one cell is continuous with the membrane of its neighbors through these channels. As a result, instead of a plant being just a collection of individual cells, each flowering organism is an interconnected commune of living, self-sustaining units. Most of the plasmodesmata are formed at the time of cell division.

Flowering plants produce five types of hormones. These hormones influence many plant functions, and in particular coordinate cell differentiation so that instead of having a collection of many different types of cells, there is an organized organ. Plasmodesmata probably regulate the passage of substances—in particular, hormones—from one cell to another.

Developing Organ-Environment Interaction

Flowering Shoot Apex.   The shoot apex is, like the root, a self-determining whole, but the details of development are more complex than root development. The shoot apex produces cone-shaped aggregates of cells that are leaf primordia or bud primordia. A *bud* is a shoot meristem that produces new leaves and is enclosed by older leaves. Like the shoot apex, morphogenesis of a leaf or a bud is self-determining—influenced by genetic programs and cell-to-cell communication and relatively

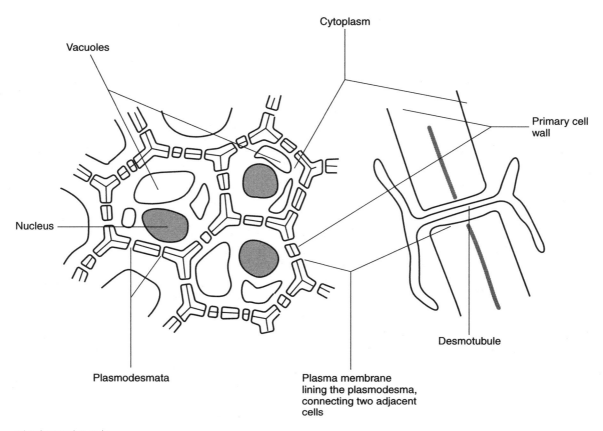

FIGURE 12.7

independent of environmental factors. Environmental factors can induce the shoot apex to undergo a basic change in structure and become a flowering shoot apex whose further development does depend on interactions with the environment.

Epigenesis of Flowers.   The development of flowers is too complex to describe in this introductory treatment of plant development. Flowers are mentioned in order to contrast self-determination epigenesis with partial environment-determinating epigenesis. Environmental factors play a key role in flower development in contrast to the self-regulation of leaf and bud primordia, roots and shoots. A detailed illustration of partial environment determining epigenesis of male and female reproductive organs and of gender commitment is described toward the end of this chapter.

Development of Animals

Phases of Animal Development

Egg, A Very Special Type of Cell.   Fertilization is the union of gametes that triggers the egg to begin to develop into an adult organism. From the genetic viewpoint the combination of two sets of genetic material is the important event, but from the standpoint of the potentially new individual, the triggering of development is the important event. Actually these two events can be separated though they usually occur together. In 1899 Jacques Loeb was the first to demonstrate that eggs could begin to develop into an adult organism without fertilization by male gametes. This phenomenon is called parthenogenesis. Loeb caused sea urchin eggs to begin to develop by pricking them with a needle or by changing the salt concentration of the seawater in which they were being cultured. The announcement of artificial induction of egg development greatly upset some people. According to the science historian, Garland Allen:

> To laymen, the successful production of parthenogenetic organisms was both exciting and terrifying presaging the era in which life would be created in a test tube. In a less philosophical vein, contemporary wits noted that ever since Loeb's discovery about salt water stimulating parthenogenesis, maiden ladies had expressed grave doubts about ocean bathing; the less subtle observer noted that weekends at one seashore did indeed seem to be remarkably fertile.[2]

Male honeybees develop from unfertilized eggs, and every other day the female water flea, *Daphnia*, produces about a hundred live, fatherless young. However, natural parthenogenesis is rare in animals having backbones, though it sometimes happens in turkey eggs. Since the time of Jacques Loeb, a number of different types of eggs, such as mouse and rabbit eggs, have been artificially induced to begin development. Thus, the egg is a very special type of cell.

Production of Gametes. 1. Formation of Eggs.   The egg is the result of a complex process of differentiation consisting of three stages: (1) the formation of the cells called primordial germ cells (because later they form mature gametes), (2) growth, and (3) maturation of the germ cells.

In the embryo, primordial germ cells migrate to the ovary and begin to divide by mitosis to produce several primary oocytes. The human female is born with all the eggs she will ever produce—in fact, most of the primary oocytes never reach maturity. As the primary oocytes enlarge, they become surrounded by a single layer of ovary cells, called follicular cells, thus producing primary follicles, see figure 12.8a. The primary follicles remain dormant until after puberty, but then with the rise of follicle stimulating hormone (FSH) during the first phase of the menstrual cycle, these follicles begin to grow. During growth more follicular cells divide to produce protective coats around the developing primary oocyte, which changes in two ways. It acquires the cell machinery (RNA) for making proteins, and the egg cytoplasm accumulates yolk which contains fat droplets and granules of lipid, proteins and lipoprotein and pigment, as shown in figure 12.8a. Just before ovulation the oocytes undergo the first maturation division. This is a first meiotic division that distributes most of the cytoplasm to only one cell. During or just after ovulation, the second maturation division, the second meiotic division, begins to occur but does not go to completion. If fertilization occurs, this division is completed, and again most of the cytoplasm is distributed to only one daughter cell, the mature ovum, see figure 12.8a.

Animal eggs sometimes are classified according to the amount of yolk present: (1) little yolk, e.g., humans, sea cucumbers; (2) a lot of yolk, e.g., birds; and (3) intermediate amount of yolk, e.g., amphibians such as frogs. The latter two types of eggs have a top and a bottom. The bottom which has more yolk is called the *vegetative pole*, whereas the top portion is the *animal pole*.

2. Formation of Sperm.   The sperm cell is the end result of a process of differentiation analogous to that for eggs. Primordial germ cells migrate to the testes and begin to divide and grow to become spermatogonia. After puberty, spermatogonia begin to differentiate into spermatocytes. These undergo two divisions producing four spermatids of equal size. Two or three weeks later the spermatids have transformed into mature sperm cells, see figure 12.8b.

The mammalian sperm cell is specialized to survive the hazards of being in the environment of the vagina, to swim up the female reproductive tract to meet an ovum, to penetrate the ovum, to contribute genetic information, and to stimulate ovum development. The head region of the sperm contains half the amount of genetic material as contained in nongamete cells and specialized structures for penetrating the ovum. The neck leads to the midpiece, which contains mechanisms that energize the contractile motion of the tail region. Thus, the third and fourth region of a sperm cell provide the mechanism for it to swim up to an ovum.

Fertilization.   Fertilization is a complex sequence of events that implies an extensive coordination of differentiation of male and female gametes. The structure of each can best be understood in terms of the fertilization process. As the sperm contacts the outer surface of the egg coats, the granule at the

---

[2]Garland E. Allen, *Life Science in the Twentieth Century* (New York: John Wiley & Sons, Inc., 1975) p. 78.

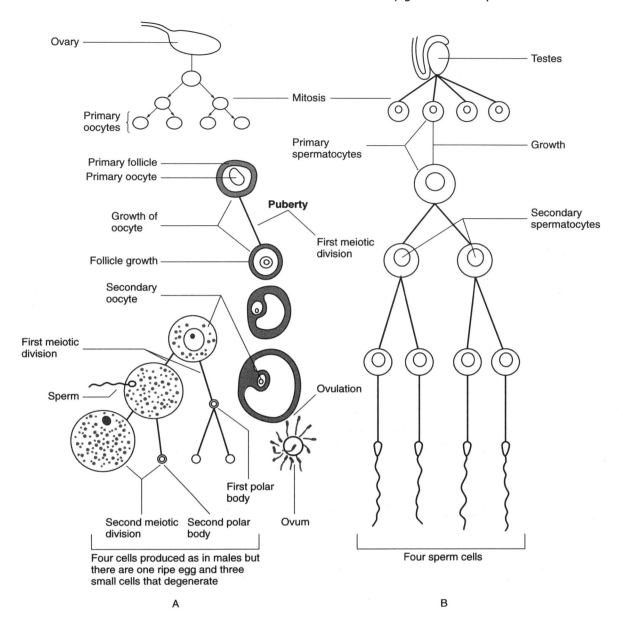

FIGURE 12.8

tip of the sperm head breaks down, releasing enzymes that digest away envelope material. As the sperm makes its way through the egg coat, membranes that once surrounded the enzyme-releasing granule in the sperm head elongate into tubules that contact the plasma membrane of the egg. These plasma and tubule membranes fuse, thus providing a pathway for the sperm head to enter the egg, see figure 12.9. This penetration brings on many changes. The egg contents become rearranged, the plasma membrane changes its permeability, and becomes resistant to penetration by another sperm cell. In the meantime, the egg completes its second division, the male and female nuclei fuse, and the resulting cell, now called the *zygote,* begins to develop. In summary, fertilization is the first phase of animal development that leads to the first stage, the zygote. The whole process of sperm penetrating an egg occurs in about 10 seconds.

    Cleavage.   The growth phase of the differentiating egg was in preparation for the zygote to undergo rapid mitotic division into 2, 4, 8, 16, 32, and 64 cells. This cell division, known as *cleavage,* produces progressively smaller cells known as blastomeres. The end result of cleavage is a mass of cells called a blastula. The cavity which eventually forms inside this mass of cells is called the blastocoel.

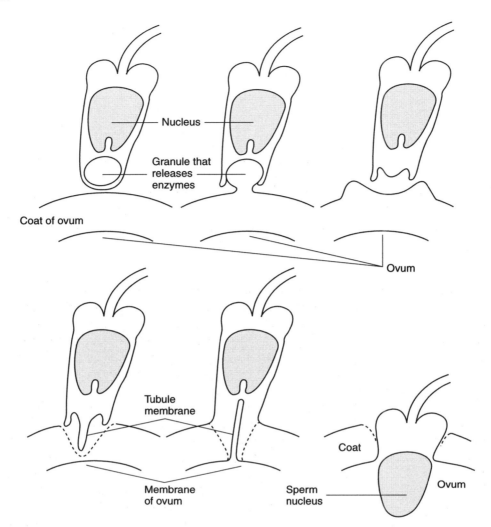

FIGURE 12.9

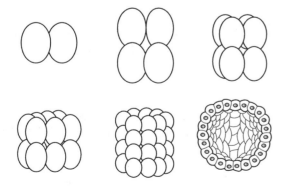

FIGURE 12.10

The pattern of cleavage and the shape of the blastula depend on the type of egg involved. In eggs with little yolk, cell division is uniform, see figure 12.10, whereas in eggs with intermediate or a lot of yolk, cell division is more rapid at the animal pole than at the vegetative pole, because the inert yolk slows the active division, see figure 12.11. The blastula of mammals such as humans is similar to that of sea urchins—both involve eggs with small amounts of yolk—except the mammalian blastula has an inner cell mass and the remaining cells form structures called the trophoblast, see figure 12.12.

Gastrulation. Gastrulation is a highly integrated movement of individual cells and groups of cells that transforms the blastula into a three-layered structure called a gastrula. The outer and inner cell layers are called ectoderm and endoderm respectively, and the layer of cells between them is the mesoderm. We do not know how or why cells begin to move during gastrulation. Some biologists theorize that discontinuities and changes in cell surfaces start and coordinate these cell movements. We are sure that gastrulation is a very important phase in development. If gastrulation is stopped, so also is all further development. Gastrulation is a critical period where disturbances rapidly bring about

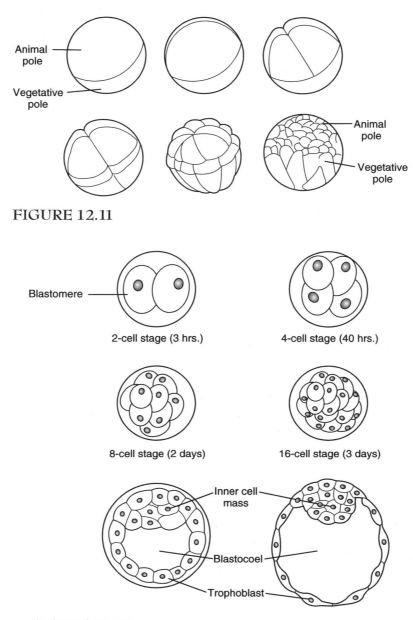

FIGURE 12.11

FIGURE 12.12

developmental malformations. Most important of all and in partial explanation of the above two facts, events occurring during gastrulation cause the partially determined cells of the blastula to become further restricted in their potentialities.

The cell determination and differentiation that begins during gastrulation is due to the new associations resulting from cell movements. The new contacts between cells after gastrulation allow these partially determined cells to exchange materials and/or form new types of intersurface boundaries that lead to further cell differentiation.

The pattern of gastrulation will depend on the pattern of cleavage, that in turn depends on the type of eggs involved. Gastrulation in sea urchins is the easiest to visualize. One group of cells in the blastula begins to move inward, invaginate, as when one presses one's finger into an inflated balloon. As this invagination continues, a group of cells come to lie within the blastocoel. These cells become the endoderm and enclose a second cavity, the archenteron. Other cells break away from the outer ectoderm layer and lodge between the endoderm and ectoderm, thus starting the mesoderm,

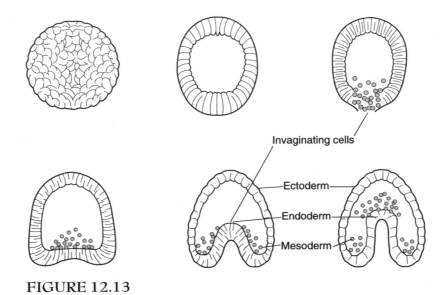

FIGURE 12.13

see figure 12.13. Gastrulation in mammals, and in human beings in particular, is complicated by the process of implantation.

Implantation.   Fertilization occurs in the upper one-third of the Fallopian tube, so the zygote has become a blastula by the time it reaches the uterus (4 to 7 days after fertilization). Trophoblast cells next to the inner cell mass cause the blastula to stick to the uterine wall where there is a blood supply underneath the uterine lining. Trophoblast cells then divide, invade the uterine lining, and form a multinucleated cell mass with no outer definite boundary with the uterus. This cell mass continues to penetrate among separated cells of the uterine wall, thus causing the blastocoel to sink into and become partially surrounded by the uterine wall. In the meantime, the cells of the inner cell mass facing away from the uterus have become the endoderm germ layer, and the layer just above it is the ectoderm germ layer, see figure 12.14a.

The invading trophoblasts form tissues which will form the embryonic membranes. Isolated cavities appear within the trophoblast involved with implantation. These cavities become filled with blood from broken vessels in the uterine wall. As a result, the blastocoel sinks further into the uterine wall so as to be completely covered by it. At this stage, implantation is complete, the woman is now pregnant, and the developing embryo has a source of nutrients from the maternal blood. The cavity below the endoderm, the blastocoel, is now called the primitive yolk sac, and the cavity above the ectoderm is called the amniotic cavity (or just amnion), see figure 12.14b.

Formation of Embryonic Membranes.   The embryo consists of the inner cell mass called the embryonic disc with the amnion above and yolk sac below, see figure 12.14b. Spaces between the embryo and the invading trophoblast cells fuse to form the chorionic cavity surrounded by the chorionic sac (or chorion), see figure 12.15. The developing embryo is attached to the chorion by a body stalk, see figure 12.15. In the meantime, the amnion expands laterally and begins to surround the embryo, see figure 12.16. Eventually the amnion will completely surround the embryo except for the body stalk which connects the embryo to the basal portion of the chorion—the part of the chorion facing the lining of the uterus, see figure 12.16. The basal chorion in conjunction with the body stalk becomes the umbilical cord, see figure 12.17.

Developmental Interactions

Cell Determination.   All cells in the developing embryo—except the gametes—start off with identical sets of genes and, therefore, they all have the same potential for development. However, after a while cell determination occurs; the cell becomes progressively restricted in its potential for becoming various types of cells. As cell determination progresses, the cell goes through a critical period.

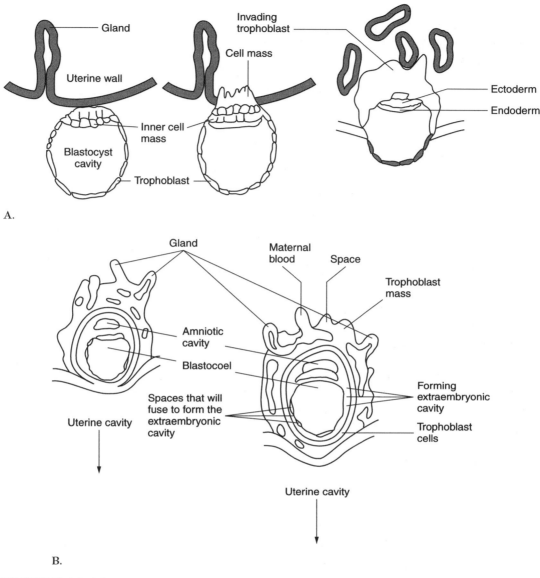

FIGURE 12.14

During the critical period, the cell is sensitive to environmental factors that may disturb its future differentiation; for example, some environmental factors may cause it to reverse its differentiation. After the critical period, cell determination is irreversible and so the subsequent pattern of differentiation also is irreversible. Note, this is a major difference between animal and plant development—in plants, cell determination usually is reversible.

Beginning of Biological Individuality.   From a biological point of view, when does the embryo become an individual organism? Cell determination and subsequent differentiation continually occur throughout the first three months of development and to a much lesser extent throughout one's whole life, so using this as a criterion of individuation there is no one time at which the organism irreversibly passes over to being an individual. On the other hand, we can use this criterion to suggest a time when the embryo is not an individual. A number of experiments or observations indicate that the early blastomeres are not irreversibly determined, so the blastula is not a biological individual, though it has the potential to become one, just as does the ovum.

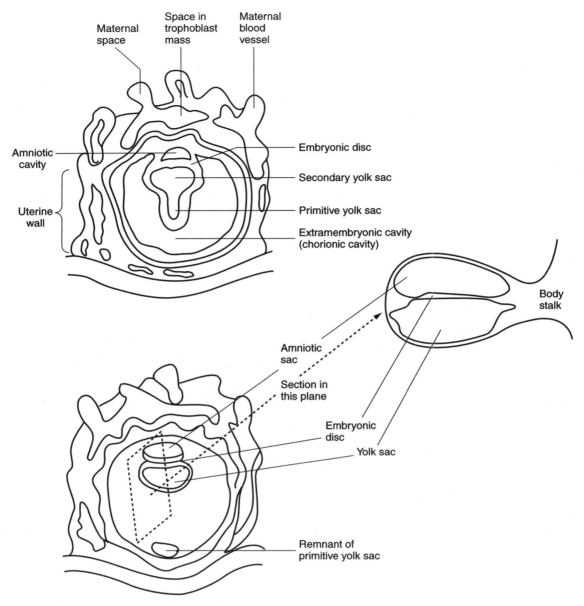

FIGURE 12.15

The results of some of these experiments are as follows: (1) If some of the blastomeres of a rabbit blastula are destroyed, the remaining cells still develop into a normal adult rabbit, (2) Special treatments can cause the blastomeres of sea urchin or amphibian blastulas to separate into individual cells. Any one of these cells nevertheless can, under the right circumstances, develop into a normal adult organism—this is analogous to the totipotency of plant cells, (3) Identical twins in humans as well as other mammals occur at the end of the first week by a division of the inner cell mass. The resulting two embryos have their individual amnionic sac, but they develop within one chorionic sac and eventually share a single placenta. (Note, each member of a pair of nonidentical twins, fraternal twins, will have its own placenta). Thus, a whole blastula may become several whole organisms, (4) When two mouse embryos in as late as the 8-cell stage are fused in the laboratory, they develop into a single blastocyst. If this blastocyst is transplanted into a mouse uterus, it develops into a single normal adult mouse.

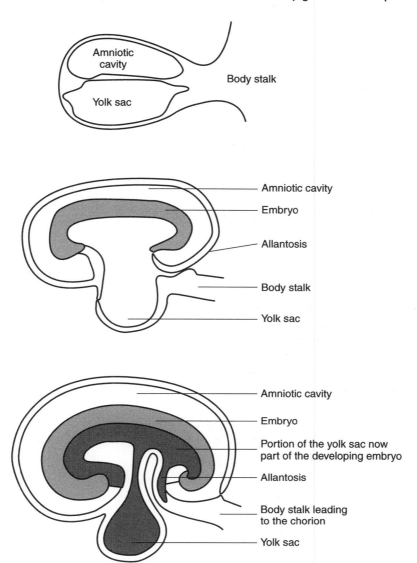

FIGURE 12.16

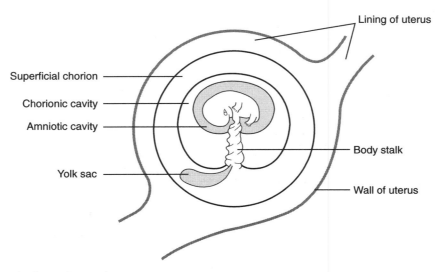

FIGURE 12.17

None of the above experiments produce normal individuals when performed on early gastrulas. Therefore, individuation begins to occur in the early stages of gastrulation.

Progressive Determination of Parts of the Embryo.   Cell determination is brought on (to some extent) by interactions with neighboring cells. Evidence for this comes from some quite interesting experiments using animals. It is possible to mark blastomeres with a dye and then observe what type of tissue they develop into. Using this technique it is possible to identify prospective epidermis (skin cells) and prospective neural tube cells in early amphibian gastrulas. When the upper part of an early gastrula was cut, rotated 180° and then replaced, the cells of each region came into new relations with cells of the lower gastrula. After the graft healed, the prospective epidermis became neural tissue like the other cells in its surroundings, and correspondingly the translocated, prospective neural cells became epidermal tissues. Apparently the cell determination had not begun in an early gastrula so that the new environment could bring about the changes.

Two closely related species of newts (a kind of salamander) have slightly different pigments, so embryonic cells from one can be visually distinguished from those of the other. A reciprocal graft was made between early gastrulas of these two species. A piece of prospective epidermis of one was transplanted into the prospective neural region of the other, and conversely, a piece of this latter region was transplanted into the prospective epidermis of the first gastrula. Just as in the previous experiments, the transplanted tissues developed in conformity with their surrounding cells. The transplanted cells were influenced by their new neighbors. If the same experiment was performed at the end of gastrulation, just two days later, the prospective epidermal cells transplanted to the neural region still differentiated as epidermis, and prospective neural tissue transplanted to the epidermal region still developed into nerve tissue. By late gastrula the cell determination in these cells is irreversible.

Human Embryonic Development

Rollo May, a well-known clinical psychologist and author, commented that several of his female patients who were college graduates, knowledgeable about methods of contraception, single, and successful in a career, nevertheless became pregnant—sometimes two or three times. This is puzzling because these women affirmed that they did not want to marry or have the responsibility of raising children, as these ties would interfere with their careers. Other psychologists and sociologists have noted that in some poverty areas young girls (14–15 years old) actively sought to become pregnant even though they could not support their offspring. One explanation that has been proposed for these findings is that some women unconsciously want to give birth to children because this gives them the feeling of personal meaning and self-worth. Before, during, and for some time after birth the woman is seemingly participating in a universal creative process. As such, some women feel a part of a process that is mysterious because it seemingly transcends the routine events of daily life. Indeed, the series of transformations beginning with the fertilization of an egg is a sequence of biological masterpieces uniting form and function. The developing embryo is an ongoing creation that occurs so rapidly that we can hardly fail to marvel at it. This section will describe the phases of development occurring between fertilization and birth.

Interactions between Mother and Child

Key Role of the Chorion and Placenta.   Epigenesis implies that any change in the environment of the embryo will tend to modify its normal development. Before implantation the young embryo is exposed to substances present in the female genital tract. Most of the genes in the blastomeres are dormant; so these undifferentiated cells merely reproduce themselves. Cell determination begins to occur in early gastrulation by selective activation of some sets of genes and selective repression of others. A foreign substance could alter cell determination and produce a developmental malformation. It is highly significant that during implantation the chorion is formed completely around the embryonic disc before the embryo enters the final stages of gastrulation (formation of the primitive streak and of the intraembryonic mesoderm). Gastrulation is a critical period, in which many developmental errors may occur. An error at this time is likely to influence many developmental steps and be manifested as major malformations. The chorion provides a protective buffer between the inner world of the embryo and that of the mother.

Throughout development, the chorion is the site of many interactions between the mother and the embryo. The trophoblast cells begin secreting Human Chorion Gonadotropin (HCG) the same day they begin invading the uterine wall. This hormone maintains the secretion of progesterone by the corpus luteum (a body of cells that remains in the ovary after an ovum has been expelled into the body cavity and quickly picked up by a tube [fallopian tube] that carries it to the uterus). If it were not for this interaction, in a few days the corpus luteum would stop secreting progesterone and spontaneous abortion would occur. Next, the chorion sprouts many primary villi that make contact with pools of maternal blood. Consequently, very early on, the embryo has a means of obtaining nutrients and of getting rid of waste material. At this stage, when the blastula is completely embedded in and surrounded by the tissues of the uterine wall, the chorion envelope is completely covered with primary villi. By the end of the third week, the embryo has developed its own functional circulatory system, see figure 12.18.

As the embryo grows, it begins to protrude into the cavity of the uterus so that eventually only 25 percent of the uterine surface is in contact with the chorion. By the eighth week the protruding chorionic surface begins to lose its villi, whereas the remaining 25 percent of the surface increases the number and size of its villi. In the meantime, the spaces into which the villi extend increase in size at the expense of the uterine wall tissue. More maternal blood vessels develop in these areas, see figure 12.19. By the twentieth week the enlarging spaces, the intervillous spaces, are kept separated from one another by septa which develop from the uterine wall. The 25 percent of the chorion closely associated with complementary developments of the uterine wall is called the placenta.

Several arteries propel blood in jet-like streams into each intervillous space of the placenta, where exchange of oxygen and carbon dioxide and other materials occurs across the villi into the fetal circulatory system. The oxygen-depleted blood slowly seeps down to the floor of these spaces where the blood is carried away by several veins, see figure 12.20. As the embryo increases in size, thus requiring more efficient nutrient and waste exchange, the intervillous spaces increase in size and more maternal arteries develop to supply them with blood. The villi grow larger, branch and rebranch in order to fill the enlarged space. As a result, the placenta continues to occupy 25% of the uterine surface. The placenta and the uterine surface increase at the same rate, which is just enough to keep the embryo adequately supplied with nutrients. Thus, the formation and growth of the placenta is due to the cooperative interaction between the mother and the embryo.

The placenta carries out three major activities: (1) metabolism, (2) transfer of materials, and (3) secretion of hormones. The placenta makes glycogen, cholesterol, and fatty acids and probably helps supply nutrients and energy to the young embryo. Many different materials can pass across the placenta, usually by simple diffusion, but in some cases by active processes requiring energy. The materials that can cross are: (1) various gases including oxygen, carbon dioxide, and carbon monoxide; (2) waste products such as urea and uric acid; (3) nutrients such as sugars, amino acids, and vitamins; (4) most drugs; (5) steroid hormones such as estrogen and androgen; (6) proteins such as antibodies; and (7) viruses, such as rubella virus, which causes measles in adults and may produce congenital malformations when the infected adult is a pregnant woman. Materials that cannot cross the placenta are maternal and fetal blood cells and bacteria. The placenta synthesizes and secretes protein hormones such as human chorionic gonadotrophic (HCG) and the steroid hormones, estrogen and progesterone. These latter maintain the pregnancy after the corpus luteum degenerates at about the twelfth week.

Stages in Human Embryonic Development

The period of human development before birth may conveniently be divided into the following processes and corresponding developmental stages: (1) fertilization → zygote; (2) cleavage → morula

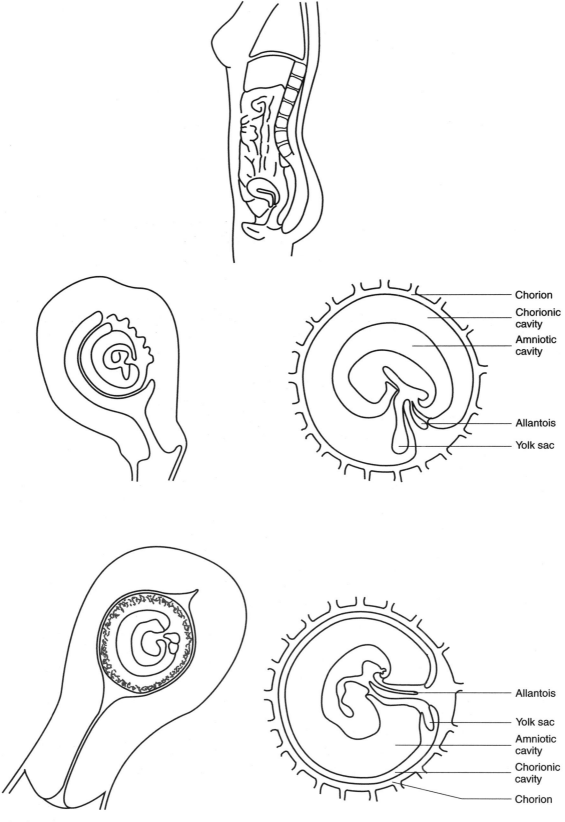

Chorion

Chorionic cavity

Amniotic cavity

Allantois

Yolk sac

Allantois

Yolk sac

Amniotic cavity

Chorionic cavity

Chorion

FIGURE 12.18

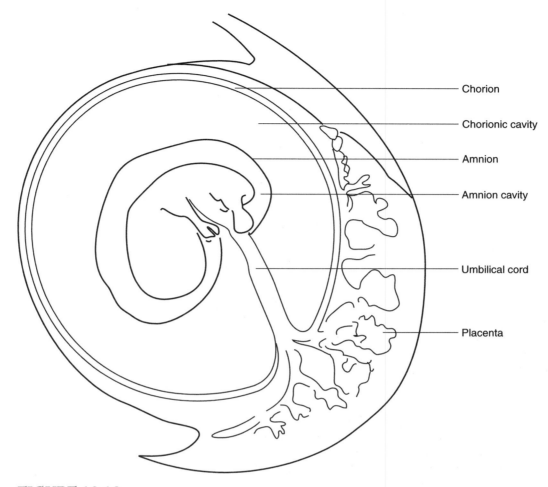

Chorion

Chorionic cavity

Amnion

Amnion cavity

Umbilical cord

Placenta

FIGURE 12.19

(16 cell, solid ball of cells) → blastula with fluid-filled cavity, the blastocyst; (3) gastrulation → gastrula, and (4) organogenesis → organ systems. The developing individual is called an embryo from the second week when the inner cell mass is converted into the two-layered embryonic disc to the end of the seventh week, after which it is called a fetus. Conceptus refers to the embryo or fetus and its membranes.

The period of development before birth is called prenatal and development after birth is postnatal. Prenatal development generally lasts nine months or 266 days.

The zygote starts out about the size of a period on a printed page, and by the end of the first month, though a great deal of cell differentiation has occurred, it still is smaller than a pea. By the end of the seventh week the embryo definitely looks like a human form but still is only about 1½ inches long. By the tenth week the fetus has doubled in size. The thirteenth to sixteenth week begins a period of rapid growth which then slows down in subsequent stages. Overall, the prenatal period is divided into three periods, called trimesters, each consisting of three months.

Birth

The usually accepted view of initiating birth is as follows. The steady, high level of progesterone secreted during pregnancy is thought to inhibit contraction of the muscles of the uterine wall. At term, this inhibitory action of progesterone is withdrawn, the uterine wall begins to contract, and labor culminating in birth begins. More recent studies with sheep indicate that the fetus, not the mother, controls labor. A potent group of substances called prostaglandins initiates the birth process.

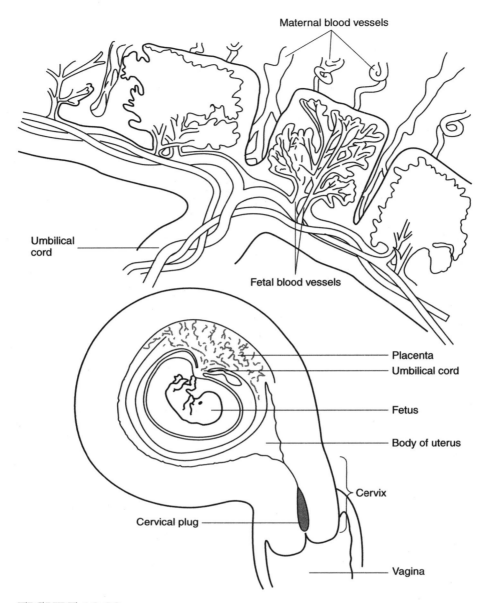

FIGURE 12.20

In the later stages of labor, the posterior lobe of the pituitary gland secretes a hormone called oxytocin which increases the strength and tone of uterine muscles. A positive feedback mechanism takes over in the last stages of labor. As the uterus contracts, sense receptors there send impulses to the hypothalamus which stimulates the posterior pituitary gland to secrete more oxytocin which causes greater uterine contraction followed by still greater amounts of oxytocin and so on until birth is complete.

Childbirth (parturition) takes place in three stages: (1) initiation of the birth process and dilation of the cervix, (2) actual delivery of the baby, and (3) delivery of the placenta, the "afterbirth," see figure 12.21. Any one of three signs indicates the onset of the first stage of parturition. Powerful contractions of the muscles of the uterus start to occur usually at intervals of 15 to 20 minute, each contraction lasting about 30 seconds. Since these contractions often are painful, they are called labor pains. The contractions steadily increase in frequency, intensity, and duration until they occur about

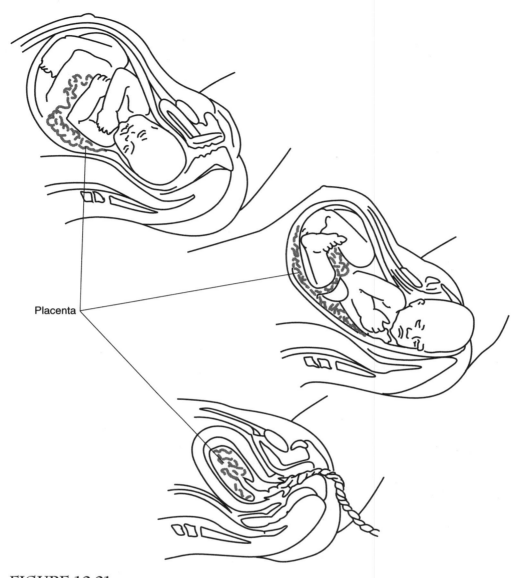

Placenta

FIGURE 12.21

every 3 to 4 minute and last up to 1 minute. These contractions push the baby up against the opening of the cervix, causing it to dilate from about 3.5 mm (0.13 in.) before labor to 100 mm (4 in.) at the end of this first stage of labor. This stage often lasts up to 16 hours for a first baby, but this is variable. Some women pass through it in less than an hour and others remain in labor for 1 or 2 days.

Another indication of the onset of the first stage of labor is the expulsion of the mucous plug from the cervix. During pregnancy this plug functions as a barrier between the vagina and uterus against invasion by bacteria and other undesirable matter. A third physiological announcement of onset of labor is the rupture of the membrane surrounding the fetus. Since the fetus is suspended in a fluid, the rupture of this membrane causes fluid to pass out from the vagina; so the process is called the "breaking water."

The second stage of parturition lasts from the time the cervix is fully dilated to about 10 cm (4 in.) until the fetus is expelled. This expulsion takes about 2 hrs. for first babies and 1 hour in subsequent deliveries. Delivery of the placenta about 15 minutes after delivery of the baby constitutes the third stage of parturition.

The Human Reproductive System

Mammals have adapted their reproductive systems so that the embryo can form and develop within the female until it is able to survive on its own. Let us look at how humans carry out reproduction which exemplifies the mammalian reproductive process.

Males

The male reproductive system consists of: (1) two gonads called *testes;* (2) the penis; (3) a set of tubes leading from the testes through the penis to the outside; and (4) the male accessory glands, see figure 12.22. The testes are oval bodies 4 to 5 cm (1.6 to 2.0 in.) long that produce male gametes, called spermatozoa or sperm cells, and sex hormones. The penis is the organ by which the male deposits his gametes into the female, and consists of the glans[3] or tip; the urethra, or tube through which semen travels; and three bodies of spongy tissue that become engorged with blood during erection. The set of tubes "connects" the testes to the penis, and the male accessory glands assist in storing and transporting sperm cells.

The testes produce and secrete small amounts of estrogen, the female sex hormone, and androgens, the male sex hormones, of which the major type is called testosterone. The testes develop from tissue in the abdominal cavity, but at the beginning of the seventh month of development of a fetus, they descend into a pouch, the scrotum, outside the body cavity. The temperature here is somewhat lower than that of the body cavity. The sperm cells develop and survive longer at this lower temperature. If for some reason one or both of the testes do not descend into the scrotum, the individual is likely to become infertile because the sperm cells do not survive very long at body temperature. This happens in about 3% of all male births.

Sperm cells produced in the testes pass into a long tube—about 6 m (20 ft.) long—called the epididymis. This tube is folded back on itself many times so that it is packed into a small space just on top of each testis. Sperm cells undergo final maturation as they pass from the head region to the tail region of the epididymis. These cells are stored in the tail region sometimes for several months. The tail region is continuous with the vas deferens, a tube that passes up and behind the urinary bladder where it enters the urethra just below the bladder. A blind outpocketing of each vas deferens just before it enters the urethra is called the seminal vesicle, which is one of the male accessory glands. The last portion of each vas deferens from which the seminal vesicle arises is called the ejaculatory duct.

How do sperm cells rapidly travel from their place of storage in the epididymis, up and through the penis into the female during copulation? Considering the size of these cells (smaller than red blood cells), this is a long journey. Once the sperm cells reach the seminal vesicle, they are carried along in seminal fluid, which is composed of secretions from the three male accessory glands. These are the prostate gland, which lies just below the bladder and surrounds the urethra emerging from it, the seminal vesicle, and the bulbourethral gland (Cowper's gland), an outpocketing of the urethra at the base of the penis. Together the seminal fluid and sperm cells are called *semen.*

Sperm cells have a long wirelike tail emerging from the midpiece, see figure 12.23. This tail undulates and thereby propels the sperm along in the seminal fluid. Still it would take the sperm a relatively long time to swim through the penis to the outside. However, the walls of the epididymis, vas deferens, and seminal vesicles contain layers of smooth muscles. These muscles contract forcefully during the final phase of copulation leading to male orgasm and so propel the semen out through the penis in a matter of seconds.

Females

The female reproductive system consists of: (1) two gonads, the *ovaries;* (2) a set of tubes which at one end catches the female gametes, ova or egg cells, coming from the ovaries and at the other end opens

---

[3]At birth the glans is covered with a foreskin called the prepuce. This is the skin which is removed when a male is circumcised.

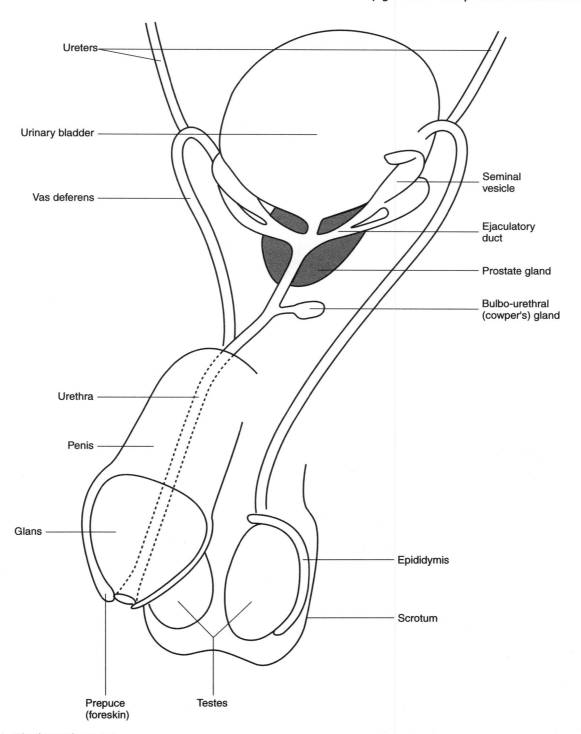

Ureters

Urinary bladder

Vas deferens

Seminal vesicle

Ejaculatory duct

Prostate gland

Bulbo-urethral (cowper's) gland

Urethra

Penis

Glans

Epididymis

Scrotum

Prepuce (foreskin)

Testes

FIGURE 12.22

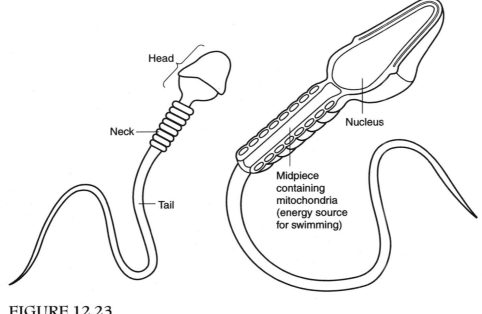

Head

Neck

Tail

Nucleus

Midpiece
containing
mitochondria
(energy source
for swimming)

FIGURE 12.23

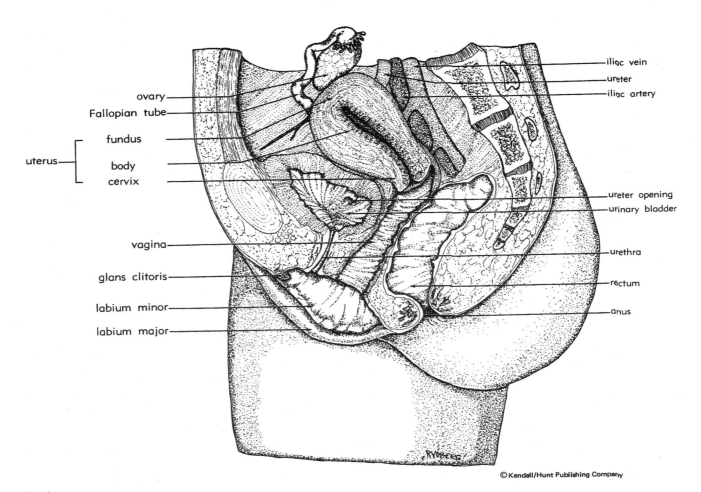

ovary

Fallopian tube

fundus

uterus

body

cervix

vagina

glans clitoris

labium minor

labium major

iliac vein

ureter

iliac artery

ureter opening

urinary bladder

urethra

rectum

anus

© Kendall/Hunt Publishing Company

FIGURE 12.24

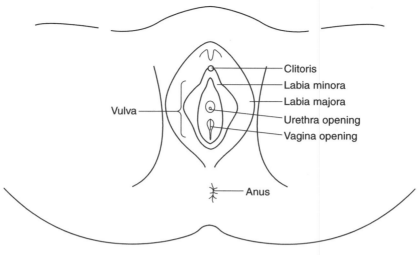

FIGURE 12.25

to the outside so as to receive the male sperm cells; and (3) the external genitalia, figures 12.24 and 12.25.

The ovaries are ovoid-shaped glands about 3.8 cm (1.5 in.) long lying within the body on each side of the pelvic cavity, see figure 12.24. The ovaries, like the testes, produce the sex hormones, the androgens and estrogen, but, of course, the androgens are secreted only in small amounts whereas estrogen is predominant. In the embryo, primitive sex cells (derived from the lining of the yolk sac) migrate to the developing ovaries and become young sex cells. Each of these becomes surrounded by cells from the ovary. The spherical layer of cells containing an immature female gamete is called a primary follicle. At birth and during childhood before puberty, there are about 1 million primary follicles per ovary. After puberty, each month several of these primary follicles begin to mature, but usually only one persists to maturity and releases the almost mature ovum, the female gamete. The ovum is swept into one of a pair of oviducts (or Fallopian tubes) and if it unites with a sperm cell there, it completes its development. The union of a sperm cell and an almost mature ovum is called fertilization. The entrance of the sperm cell into an ovum activates the final stages of maturation of the ovum. The sperm cell also activates the fertilized ovum, now a zygote, to begin dividing so that by the time it reaches the uterus, it will be able to attach itself to the uterine lining. This process of attachment, called implantation, leads to the embryo becoming something like a parasite in that it obtains all its nourishment from its host, the mother. If the ovum is not fertilized in the Fallopian tube, it does not complete its maturation but instead usually starts to degenerate by the time it reaches the uterus.

The set of tubes that transports the ovum consists not only of a pair of oviducts (or Fallopian tubes) but also the uterus and the vagina. Each of the Fallopian tubes consists of an open end with delicate, fingerlike projections near each ovary and a tubular portion that leads to the upper part of the body of the uterus. In monkeys and rabbits, the fingerlike projections have been observed to move to and fro over the surface of the ovary at the time an ovum (egg cell) is released. This movement, which probably also occurs in humans, actively collects the ovum as it ruptures from the ovary. Each Fallopian tube transfers sperm cells upward and ova and/or fertilized ova (zygotes) downward to the uterus. The cells lining the Fallopian tubes maintain an environment suitable for survival and development of a zygote as it passes through to the uterus. If an ovum is fertilized while passing down one of the Fallopian tubes, it lodges in the uterine wall. If an ovum is not fertilized by the time it reaches the uterus, it passes out of the uterus through the vagina to the outside.

The uterus is a hollow, muscular organ consisting of two parts: an upper pear-shaped portion, the body, and a lower cylindrical portion, the cervix.

The cervix is a tubelike structure with a thick wall consisting primarily of connective tissue. Glands in the lining of the cervix secrete mucous material having a gel-like, semisolid consistency.

This material temporarily traps and so accumulates sperm cells that have been deposited in the vagina during intercourse. Phagocytes (cells that engulf bacteria, debris, and other cells) in the uterus and Fallopian tubes attack sperm cells there, whereas sperm cells accumulated in the mucus of the cervix are protected from this action. Because of this, sperm cells survive less than a day in the uterus or Fallopian tubes, but remain alive and fertile in the cervix for up to 48 hours. During this time, sperm cells continuously released into the body of the uterus swim up to each Fallopian tube. This ensures a high probability of fertilization should an ovum pass through one of the tubes.

The vagina is the receptacle for the penis during copulation. Usually the vagina in virgins is partially covered by a thin mucous membrane known as the hymen. The hymen usually is broken after the first copulation, but it may be accidentally broken by other, nonsexual activities, particularly if it is poorly formed in the first place. Before puberty, the vagina has a relatively thin lining. After puberty, two important modifications occur which are due to the increased levels of estrogen in the blood. The vagina develops the same kind of lining as that found in the mouth. This tougher lining protects against the possible damage due to the mechanical rubbing of the penis against the vaginal walls during intercourse. The outermost cells of this lining accumulate glycogen, a carbohydrate, which then is converted into lactic acid by bacteria living in the vagina. As a result, the acidity in this area increases and thereby helps to prevent invasion of the vaginal walls by harmful types of bacteria.

The external genitalia (the vulva) in the female consists of labia majora and labia minora, which surround the vestibule, see figure 12.25. Within the vestibule the urethra and, just below it, the vagina open to the outside. Just above the vestibule in association with the labia minor is the clitoris. The greater vestibular glands (Bartholin's glands) within the labia majora open on either side of the vagina. Biologists have speculated that these glands help to lubricate the vagina in preparation for sexual intercourse. It is now known that the amount of mucoid secretion is insignificant (usually about a drop) and discharge occurs too late after sexual arousal for it to be of aid as a lubricant. The function of the glands is unknown.

Epigenesis of the Human Reproductive System

The adult reproductive system consists of specialized tubes which carry gametes produced in gonads to the outside. Many of these structures in the embryo are taken over from other systems and readapted to genital functions in a secondary and relatively late phase of their development. Components of the primitive urinary system are candidates for these genital readaptations because the urinary and genital systems begin to develop at about the same time and in close association with one another. For example, a pair of tubes in one of the embryonic urinary systems becomes a component in the male reproductive system.

Genetic information received at conception initiates genital readaptation in a seven-week-old embryo. Some structures, such as primitive gonads that begin to form in the 5th week, can differentiate either into testes or ovaries. Of the two pairs of primitive tubes which form in the 5th week, one can differentiate into components of a male and the other into components of a female reproductive system. Thus, between 4 and 7 weeks, the embryo is potentially bisexual. At the beginning of the 7th week, genetic information initiates the epigenetic development of the individual to be exclusively male or female. This process may be analyzed into six major stages: (1) genetic sex, (2) gonadal sex, (3) fetal hormonal sex, (4) reproductive organs, (5) brain sex, and (6) postpubertal hormonal sex.

Genetic Sex

All cells in humans, except gametes, contain 23 pairs of chromosomes of which one pair, the sex chromosomes, are known to initiate epigenetic development of sex. When these chromosomes are viewed under a microscope, the pair in females resembles two Xs and the pair in males resembles an X and a Y. The other 22 pairs of chromosomes are called autosomes. Gametes produced from meiosis (the process of forming gametes) contain half the amount of DNA of a parent cell as a result of receiving only one chromosome from each of the 23 pairs. Thus, female gametes always possess 22 autosomes and 1 X chromosome, while male gametes possess 22 autosomes and either an X or a Y chromosome. A zygote, therefore, receives 22 autosomes from each gamete, but also receives an X chromosome from the ovum and either an X or a Y from the sperm cell. An XX zygote is a genetic female, and an XY zygote is a genetic male. This is the genetic sex of the individual.

Gonadal Sex

The gonads develop in association with the primitive kidneys. Early in the fourth week a mass of cells differentiate into a pair of ducts, called the pronephrose, which empty into the cloaca, see figure 12.26a and b. This primitive kidney is never functional in human embryos and begins to degenerate toward the end of the fourth week. As this is happening, another large mass of cells differentiates into the mesonephrose, the adult kidney of fishes and amphibians. Tubes of the mesonephros connect with the remains of the pronephros which now are called the mesonephric ducts, see figure 12.26c. At the beginning of the 5th week, an outgrowth at the base of the mesonephric duct begins to differentiate into the metanephrose, the adult kidney, as shown in figure 12.27a and b. Also at this time

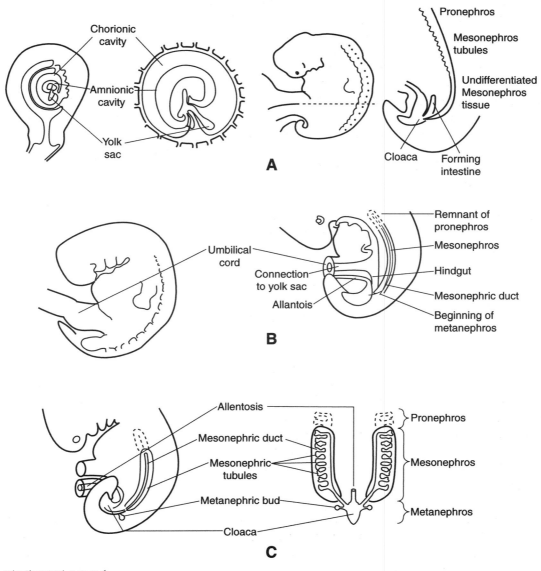

FIGURE 12.26

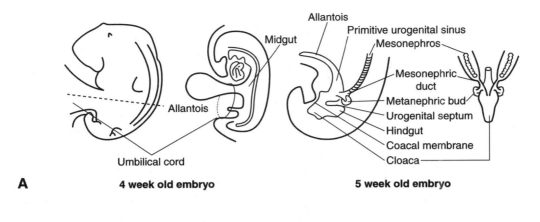

A   **4 week old embryo**   **5 week old embryo**

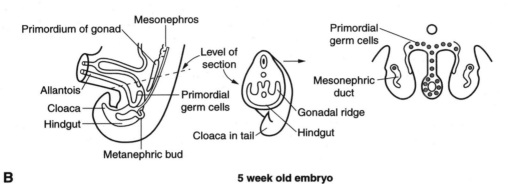

B   **5 week old embryo**

C   **6 week old embryo**

FIGURE 12.27

primordial germ cells in the wall of the yolk sac begin to migrate toward the gonads, see figure 12.27c. By the 6th week these cells have reached the gonads which have differentiated into an outer cortex and an inner medulla. The gonads are designated as "indifferent" at this stage because the cortex can differentiate into ovaries and the medulla can differentiate into testes. In the 7th week, the presence of a Y chromosome initiates the development of testes. This, in turn, overwhelms and prevents the later differentiation of primordial ovary cells. In genetic females, the primordial ovary cells differentiate first and convert the gonad into an ovary. Occasionally, the timing is disturbed so that a genetic male or female may have one or both gonads that are part testes and part ovary. Still, more rarely, a fetus may have gonads which develop into testes on one side of the body and an ovary on the other. This condition produces what is known as a true hermaphrodite—one-half the body will exhibit male reproductive organs and the other half, female reproductive organs.

Fetal Hormonal Sex

The normal testes begin to secrete two hormones in the 8-week-old fetus: androgen and a substance called MIS (mullerian inhibiting substance). The normal ovaries do not produce estrogen until the fetus is about 12 weeks old. Thus, a genetic male becomes a testes male which results in the *presence* of androgen and MIS, and a genetic female becomes an ovary female which results in the lack of androgen and MIS. This dimorphic hormonal condition has a profound influence on all subsequent stages of development by means of a mechanism known as the Adam principle. According to this principle, both male and female features tend to differentiate as females. The female body plan is altered—"masculinized"—by the presence of specific hormones during the critical periods of determination of cells associated with the reproductive system. Androgen induces cells to differentiate according to a male program, and the hormone called MIS causes partially differentiated female tubes, e.g., fallopian tubes, uterus, and vagina, to degenerate. If MIS is absent or deficient in a genetic male fetus, the individual will develop male reproductive organs and develop a vagina and uterus, see figure 12.28. If both androgen and MIS are absent, the genetic male will develop a female body.

There is a rare abnormality which dramatically exemplifies this important point. In this inherited abnormality, none of the body cells respond to androgen. Therefore, at eight weeks a genetic male fetus with a pair of testes will begin circulating androgen in the blood, but none of the body cells will respond to it. Consequently, the fetus will, on its own, develop some female sex organs just as if no androgen were present. The individual will have no uterus (or a very small one) and no ovaries, and, of course, will be infertile. However, she will have a vagina and a clitoris, and as an adult she will be able to enjoy sexual intercourse with males.

According to the same Adam principle, if a genetic female fetus is exposed to androgen during critical periods of development, the female organs will be present but masculinized. In particular, the clitoris will be enlarged to look like a penis and the labia major may unite to form an empty scrotum, see figure 12.29. This condition occurs when the mother or fetus has overactive adrenal glands (glands on top of the kidney). In both males and females, adrenal glands secrete small amounts of androgen. On the other hand, if there is a deficient level of androgen in genetic males during the critical periods of development, the male organs will be underdeveloped. In particular, the baby will have a micropenis—a penis so small that it is like a clitoris and, as such, cannot function in coitus, see figure 12.30.

Thus, according to the Adam principle, genetic females lack androgen and MIS and therefore develop female reproductive organs. Conversely, genetic males have androgen and MIS and therefore develop male reproductive organs. In summary, the Adam principle consists of four interrelated ideas:

1. Both the genetic, gonadal, and fetal hormonal male and the genetic, gonadal, and fetal hormonal female have an epigenetic program that directs the 6–8-weeks-old fetus to develop a female sexual reproductive body plan.
2. The developing female sexual reproductive body plan will start to develop a male sexual reproductive body plan by the presence of androgen and MIS during the critical period between 5 weeks and 12 weeks after conception.
3. Androgen induces the cells developing the male or female sexual reproductive body plan to differentiate according to a male program.
4. MIS causes partially differentiated female tubes to degenerate.

Reproductive Organs

Development of the Urinary System.   An extension from the yolk sac and the developing digestive tube connect to the cloaca. On each side of the fetus a succession of three distinct kidneys—pronephrose, mesonephrose, and metanephrose—develop and connect to a duct that also empties into the cloaca. The pronephrose—6 to 10 pairs of tubules—connect to a pair of primary ducts that

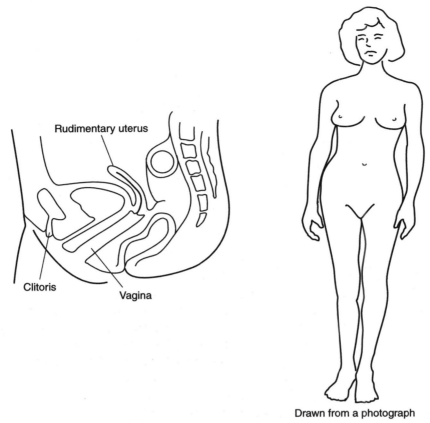

Rudimentary uterus

Clitoris

Vagina

Drawn from a photograph

Genetic, gonadal, androgenic "male"
with female body (cells do not respond to androgen)

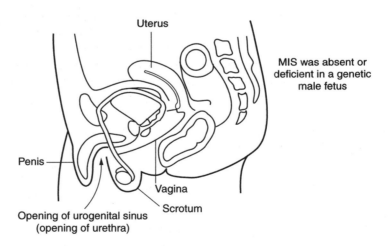

Uterus

MIS was absent or
deficient in a genetic
male fetus

Penis

Vagina

Scrotum

Opening of urogenital sinus
(opening of urethra)

FIGURE 12.28

empty into the cloaca. The pronephrose degenerates by the end of the fourth week, but as they degenerate, tubules of the mesonephrose differentiate and connect to the primary ducts which now are called mesonephric ducts, see figure 12.26. During the sixth week, a bud from the base of the mesonephric duct grows upward collecting mesoderm cells around its tip, see figure 12.27c. During the seventh week the cloaca becomes divided into a front section, the urogenital sinus, and the back section, the rectum, which is the last portion of the digestive tube, see figure 12.31. By the eighth week, due to growth of the cloaca and bud, the bud separates from the mesonephric duct

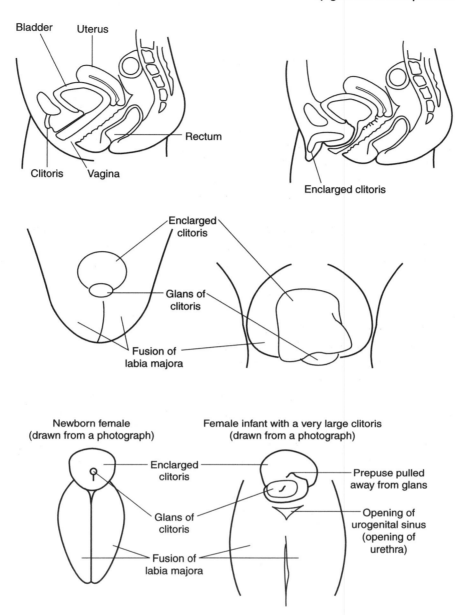

Bladder   Uterus

Rectum

Clitoris   Vagina

Enclarged clitoris

Enclarged clitoris

Glans of clitoris

Fusion of labia majora

Newborn female
(drawn from a photograph)

Female infant with a very large clitoris
(drawn from a photograph)

Enclarged clitoris

Prepuse pulled away from glans

Glans of clitoris

Opening of urogenital sinus (opening of urethra)

Fusion of labia majora

FIGURE 12.29

and empties into the portion of the urogenital sinus that will become the urinary bladder. The mesoderm associated with the bud begins to differentiate into metanephrose tissue which will be the outer portion of the adult kidney. The upper part of the bud becomes the core portion of the kidney and the remainder becomes the ureter, the tube that carries urine from the kidney to the bladder, see figure 12.31.

Internal Reproductive Tubes.   At the same time as the kidneys are differentiating, a pair of ducts called Mullerian ducts form close and parallel to the mesonephric ducts. By the ninth week, the lower ends of the Mullerian ducts have fused into a single tube that connects with the urogenital sinus, see figure 12.32. The Mullerian ducts are partially determined to develop into female internal reproductive tubes, and the mesonephric ducts can be induced to differentiate into male structures. Thus, between six and eight weeks, the embryo—no matter what its genetic or gonadal sex—is bisexual with respect to internal reproductive tubes.

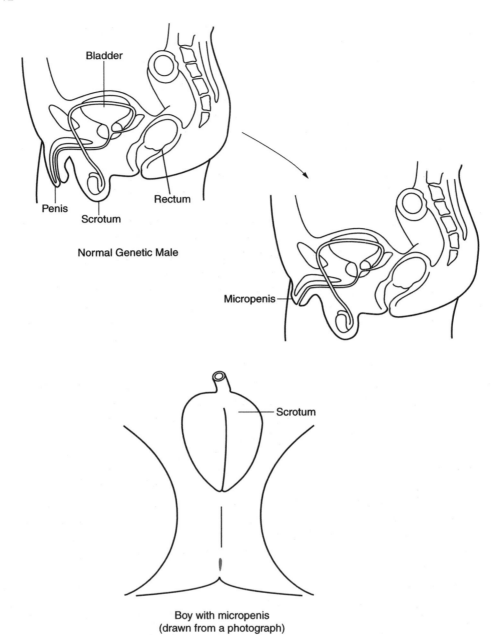

Bladder

Penis

Scrotum

Rectum

Normal Genetic Male

Micropenis

Scrotum

Boy with micropenis
(drawn from a photograph)

FIGURE 12.30

After the eighth week, the Mullerian ducts of genetic males begin to degenerate, while the mesonephric ducts begin to differentiate into male sex tubes. Correspondingly, after the eighth week, the Mullerian ducts of genetic females continue their development of female internal sex organs, while the mesonephric ducts begin to degenerate, see figure 12.32. At this point, none of these developmental changes are irreversible. If a genetic female embryo is exposed to androgen, the development of female sex organs will be arrested and male organs will begin to differentiate. Likewise, if a genetic male embryo is deprived of androgen or is exposed to high levels of estrogen (which antagonizes the action of androgen), the development of male sex organs will be arrested and female organs will begin to differentiate. The last two days of the third month of pregnancy constitute a critical period. If androgen is present during these two days, the developing reproductive system is irreversibly determined to have male features (no matter what the genetic sex of the individual). If androgen is not present, the developing reproductive system is irreversibly determined to have female features.

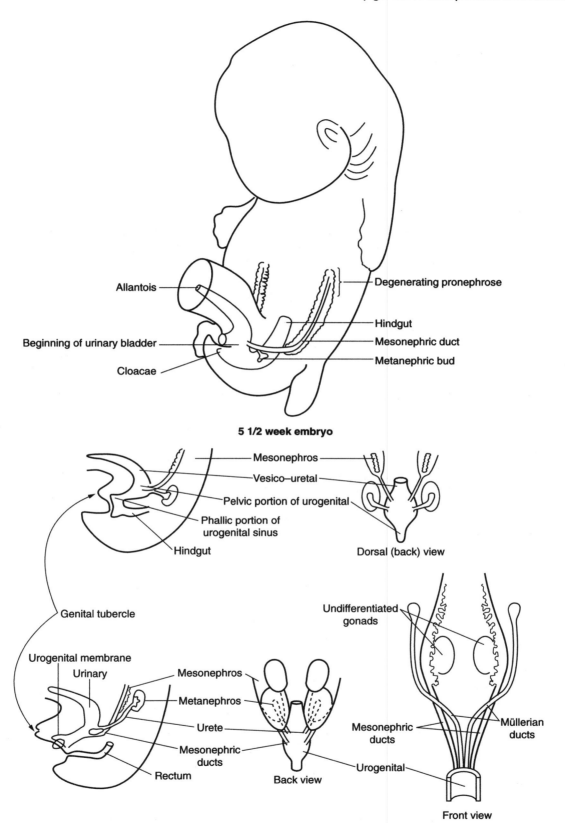

Allantois

Degenerating pronephrose

Beginning of urinary bladder

Hindgut

Cloacae

Mesonephric duct

Metanephric bud

5 1/2 week embryo

Mesonephros

Vesico–uretal

Pelvic portion of urogenital

Phallic portion of
urogenital sinus

Hindgut

Dorsal (back) view

Genital tubercle

Undifferentiated
gonads

Urogenital membrane

Urinary

Mesonephros

Metanephros

Urete

Mesonephric
ducts

Rectum

Mesonephric
ducts

Müllerian
ducts

Urogenital

Back view

Front view

FIGURE 12.31

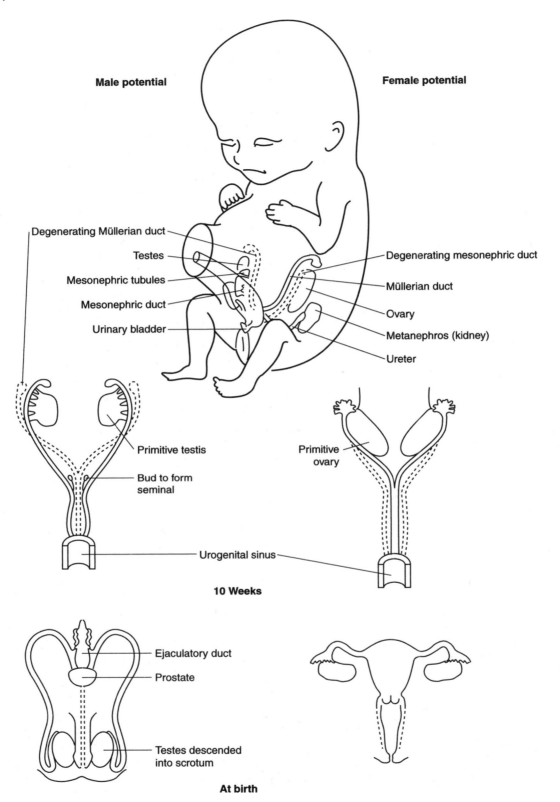

Male potential

Female potential

Degenerating Müllerian duct

Testes

Mesonephric tubules

Mesonephric duct

Urinary bladder

Degenerating mesonephric duct

Müllerian duct

Ovary

Metanephros (kidney)

Ureter

Primitive testis

Bud to form
seminal

Primitive
ovary

Urogenital sinus

10 Weeks

Ejaculatory duct

Prostate

Testes descended
into scrotum

At birth

FIGURE 12.32

External Sex Organs.   At seven weeks both genetic male and female embryos have the same primitive external sex tissue consisting of: (1) the genital tubercle, (2) the genital groove, (3) the genital fold, and (4) the genital swelling. Later in the development of both genetic males and females, the genital tubercle becomes the phallus, the genital fold becomes the urethrolabial fold, and the genital swelling becomes the labioscrotal swelling, as shown in figures 12.33 and 12.34 and table 12.1. The phallus becomes the penis in males and the clitoris in females. In males, the urethra formed from the urethrolabial fold enclosing the genital groove runs through the penis. In females, the much shorter urethra opens to the outside just below the clitoris and above the vaginal opening in the vestibule derived from the genital groove. On each side of the vestibule are the labia minora derived from the urethrolabial fold. The labioscrotal swelling becomes the scrotum in males and the labia majora in females. Just as with the internal sex organs, the external sex organs are irreversibly determined only after the last two days of the third month of pregnancy. Androgen induces male external sex organs to develop, and lack of androgen allows the built-in female program in the body to express itself.

Brain Sex

The presence or absence of androgen determines the type of neural circuitry that is laid down. In normal males, the hypothalamus secretes releasing factors at a more or less constant rate. This, in turn, stimulates the pituitary gland to maintain a steady-state level of LH and FSH in the blood. In normal females, the hypothalamus cyclically secretes the releasing factors which, in adults, regulate the menstrual cycle.

Postpubertal Hormonal Sex

After the onset of puberty, the hypothalamus secretes higher levels of releasing factors which stimulate the pituitary gland to secrete higher levels of LH and FSH. In males the LH (also known as interstitial-cell-stimulating-hormone (ICSH) stimulates the testes to secrete androgen and FSH stimulates the testes to produce sperm cells. In females, the cyclical release of LH and FSH regulates the maturation and release of usually one ovum per month and regulates the other changes that occur during a menstrual cycle which includes higher levels of estrogen in the blood. The androgen in males and estrogen in females stimulate the development and maintenance of the secondary sex characteristics.

Thus, the epigenesis of the human reproductive system may be divided into seven stages of biological sexuality as summarized in table 2.2.

► Table 12.1   Parallel Development of External Sex Organs

Undifferentiated Structure Present in Both Genetic Males and Females	First Step of Differentiation Present in Both Genetic Males and Females	Structure Present at Birth in Either Genetic Males or Genetic Females	
Genital Tubercle	Phallus	Female	Clitoris
		Male	Penis
Genital Groove	————————	Female	Vestibule
		Male	Cavity of urethra in penis
Genital Fold	Urethro-Labial Fold	Female	Labia minora
		Male	Urethra of penis
Genital Swelling	Labioscrotal Swelling	Female	Labia majora
		Male	Scrotum

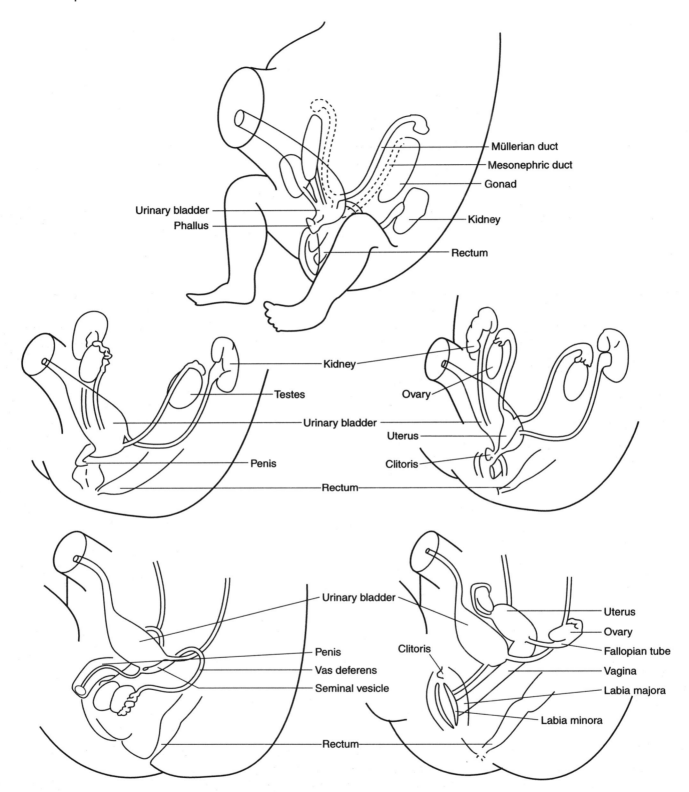

FIGURE 12.33

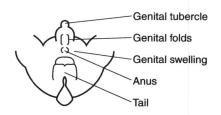

Bisexual potential in 7 week embryo

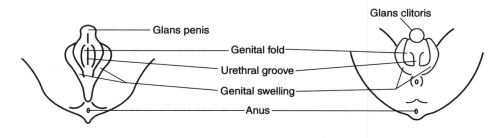

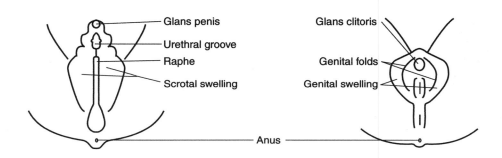

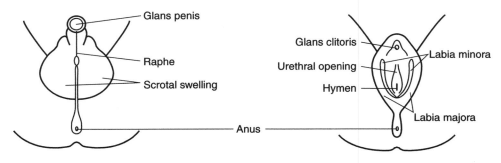

FIGURE 12.34

▶ **Table 12.2 Seven Stages of Biological Sexuality**

Stage	Male	Female
1. Genetic Sex	XY chromosomes	XX chromosomes
2. Gonadal Sex	testes	ovaries
3. Fetal Hormonal Sex	androgen	no androgen
4. Internal Sex Organs	mesonephric ducts	müllerian ducts
5. External Sex Organs	penis	clitoris
6. Brain Sex	hypothalamus is not cyclic in influencing the pituitary gland no menstrual cycle	hypothalamus is cyclic in influencing the pituitary gland menstrual cycle
7. Postpubertal Hormonal Sex	androgen, which leads to male secondary sex characteristics	estrogen, which leads to female secondary sex characteristics

Tearout #1

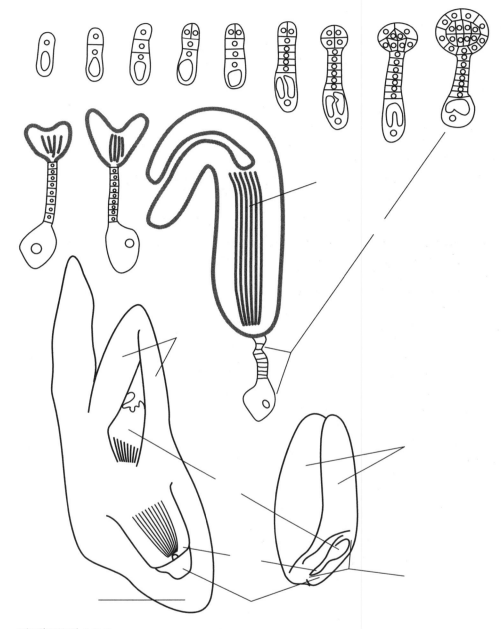

FIGURE 12.2

Directions:

1. Using figure 12.2 as a reference, fill in all labels.

Tearout #2

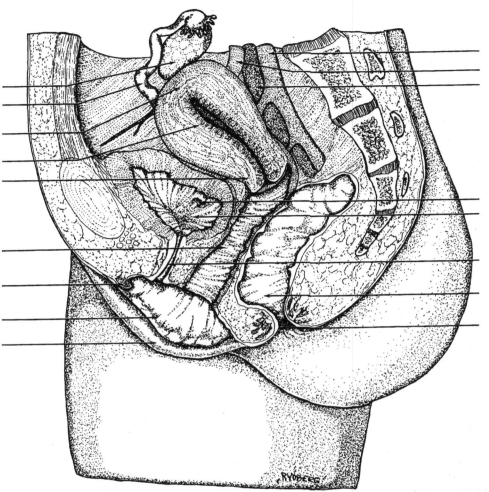

FIGURE 12.24

Directions:

1. Using figure 12.24 as a reference, fill in all labels.

Tearout #3

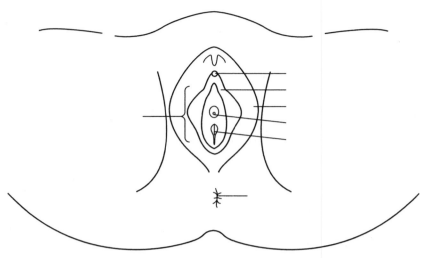

FIGURE 12.25

Directions:

1. Using figure 12.25 as a reference, fill in all labels.

2. Summarize the Adam principle with respect to external genitalia.

Biological Theory of Evolution

Theory of Evolution

Biological mechanism describes individuals adapting via negative feedback regulation to factors in the environment that tend to disrupt an organism. Eventually aging sets in and the individual dies. Thus, because of aging, life in individuals always leads to death. However, if we focus on the life pattern of a species, life leads to rebirth. Individuals transcend death by participating in the species life cycle. The species life cycle goes on only in a suitable environment with characteristics that allow individual members of the species to adapt to it. If, over a period of many thousands of years, the accumulation of many gradual changes results in a drastically different environment, individuals will not be able to adapt and/or participate in reproduction. Then the species becomes extinct.

There is evidence that 3.5 billion years ago, bacteria-like primitive cells lived and were adapted to an atmosphere of water vapor, carbon dioxide, nitrogen, methane, and other substances but no oxygen. Five hundred million years later, the earth's atmosphere contained oxygen, which is toxic to bacteria totally adapted to a nonoxygen atmosphere. This drastic change in the atmosphere would cause all primitive species to become extinct. However, today bacteria are found virtually everywhere on earth. How can this be? In some manner, populations of primitive bacteria were transformed into new species. That is, a population over many thousands of generations acquired so many new traits that it can be called a new species. This transformation, called *evolution*, is the core of the biological theory of evolution.

Evolution consists of two themes: (1) continual variations in a species over a long period of time can lead to the "creation" of one or more new species, and (2) all of the complex organisms that exist now or ever have existed according to fossil records have evolved from bacteria that were the first forms of life on earth. In his book, *The Origin of Species,* Charles Darwin (1809–1882) described in great detail both of these themes. Even though the idea of evolution had been proposed by many thinkers since the beginning of the 19th century, Darwin's theory was the first to be convincing to the scientific community and has in modified form come to be one of the unifying principles of 20th century biology. The major reason for Darwin's success was that not only did he give ample evidence for the historical fact that evolution has occurred, he also provided a convincing naturalistic mechanism for how creation of a new species can occur.

Darwin's Insight

Suppose one believed, as did many biologists in the eighteenth and nineteenth centuries, that all humans descended from a single couple, Adam and Eve. Wouldn't this belief seem incongruous in view of the present great diversity among humans? For example, suppose Adam and Eve had oriental features—yellow skin color, slanted eyes, short stature. With this couple as our most ancient ancestors, how do we explain the current existence of people with very different features—skin color ranging from light to very dark, stature ranging from pygmies to 8 feet tall giants, diversity of eye outline and color, etc.? Apparently, a single homogenous population of humans in some manner gave rise over many thousands of generations to very diverse offspring. Biologists also realize that many other species show the same sort of diversity as found among humans.

Furthermore, one may group many different species into a single category according to some unifying features; for example, primates include humans, gorilla, orangutan, and chimpanzee, and the category Felis includes bobcats, cougar, and domestic cat. These higher level classification categories show the same underlying unity plus diversity that is what is found in virtually all species. This, at the very least, suggests that the mechanisms that produce diversity in a species also produce diversity among species classified in some higher level category. That is to say, perhaps these similar species descended from a single population just as the diverse races probably descended from Adam's family (if one believes in Adam) or from some primitive single, human population.

There is a hierarchy among organisms according to complexity of structure. Carolus Linnaeus (1707–1778) devised a classification scheme which at the same time represented the hierarchy of structural-functional complexity among organisms. Linnaeus did not believe even in the possibility of evolution, but any portion of his classification scheme showed categories of organisms with variations of an underlying unity of structure-function, for example, vertebrates exhibiting many variations of homeostatic body plan built around a circulatory system. No matter where we look among organisms and no matter how broad our vision is, we see unifying structure-function patterns with many variations and elaborations. If several species could emerge from a single ancestor population, given enough time perhaps all living things could have emerged from a single primitive life form.

These are the kinds of ideas people talked about in the early nineteenth century (except they did not use the notion of population). Then Charles Darwin (1809–1882) made a number of very thought-provoking observations. While on a five year voyage on the ship the *Beagle,* to various parts of the world as an unpaid naturalist, Darwin observed some mind-boggling things, particularly around the Galapagos Islands off the coast of Ecuador. The different islands supported different sets of species, but several of these species (mocking-thrush, finch, tortoise, and various plants), though different, were closely related and occupied similar environments on the islands. These species also were similar to but different from those occurring on the mainland. Darwin asked himself, "Why would God create slightly different but similar species suited for each island and the mainland? Could not natural interactions better explain these facts?" Darwin also was aware of Charles Lyell's (1797–1875) book, *Principles of Geology,* which brought together evidence that the earth was very much older than the generally accepted age of 6000 years. Darwin also noted that breeders could "select" various traits in plants and animals by planned breeding, a selection he called "artificial selection." Darwin was astounded at the number of breeds of pigeons such artificial selection could produce. Could not different natural environments cause the same sort of selection?

Still, from what was described earlier, selection enables populations of organisms to adapt to their environments, so perhaps several primitive (created?) species over the years have become much more diversified, even to the point of producing new species. One day in October, 1838, Darwin happened to read *An Essay on the Principles of Population* by Thomas Malthus (1766–1834). According to Malthus any population of animals or plants can produce offspring in a geometric ratio (2, 4, 8, 16, 32, 64, . . .), while food production increases only in an arithmetic ratio (2, 4, 6, 8, 10, 12, . . .). All populations produce more individuals than the environment can possibly support. That's it! From a sea of randomly inherited changes among life forms and random changes in the environment (which geologists had shown had occurred), nature selects those traits that make a plant or an animal better able to survive and then reproduce itself. By "natural selection" a population accumulates small advantageous changes. After a long period of time some groups of organisms that did not inherit these new traits became extinct. The resulting new population represents a transformation from one species to another. The history of life is the progressive transformation of a primitive form of life into many diverse species that continue to diversify as they adapt to new environments.

During his voyage on the Beagle, Darwin had the creative insight that somehow species evolved. This novel idea was radically different from the collective perception of the times which was determined by the integration of a religious tradition stemming from Moses and the book of Genesis and a philosophical tradition originated by Socrates, Plato, and Aristotle. According to the traditional consensus, the world consists of things created and sustained in existence by God. Each thing is an individual substance that is what it is by possessing or participating in an eternal form. In particular, each organism has or participates in an eternal form and the classification category, species, is the population of organisms each having or participating in the same form. Therefore, a species can never transform to a new species. Thus, Darwin's perception of evolving species was absurd because impossible according to the normal perception of the world.

Fear of ridicule would cause most people to retreat from such a radical insight. Others may arrogantly proclaim it without justifying it or describing how a species could transform. Darwin stayed open to the insight, collected evidence that evolution did occur, and pieced together ideas from diverse disciplines for a theory of how evolution could occur. Though Darwin suggests in his autobiography that the idea of natural selection occurred to him in a sudden flash after reading Malthus, Stephen Gould (1980) argues that this was only one of the final ideas for formulating his theory of evolution by natural selection.

> The theory of natural selection arose neither as a workmanlike induction from nature's facts, nor as a mysterious bolt from Darwin's subconscious, triggered by an accidental reading of Malthus. It emerged instead as the result of a conscious and productive search, proceeding in a ramifying but ordered manner, and utilizing both the facts of natural history and an astonishingly broad range of insights from disparate disciplines far from his own. . . . Darwin was continually proposing, testing, and abandoning hypotheses, . . . He began with a fanciful theory involving the idea that new species arise with a prefixed life span, and worked his way gradually, if fitfully, towards an idea of extinction by competition in a world of struggle. [Gould (1980) pp. 64–65.]

As may occur to any one of us, Darwin had a transpersonal insight. It was transpersonal in that he did not plan for it to happen though his previous studies plus the *Beagle* voyage did dispose him to have such an insight. Also the insight was nonrational rather than the rational knowing associated with ego consciousness which would judge it as irrational according to its consensual, objective understanding of the world. Darwin's response to the insight illustrates *impeccability*.[1] Being at the point of no pity, he did not retreat from the insight out of fear nor embrace the idea out of a romantic fascination with it or out of an egocentric desire to be different; rather he remained open to the possibility of its validity. This choice of openness propelled him from a state of ORDER, the conventional view of nature, to a state of CHAOS, an apparently absurd view of nature. After several years of struggle, he formulated a theory that was reasonable and testable according to conventional scientific standards. In this manner his theory incorporated conventional wisdom into a transformed view of nature; i.e., a NEW ORDER emerged from the chaos. As described in Chapters 1 and 2, this process of ORDER → CHAOS → NEW ORDER is the core of universal evolution of which species transformation is one example.

---

[1]Don Juan explains impeccability as the way of life that leads to personal transformation. ". . . to all appearances, having the assemblage point shift [transpersonal experience] is the first thing that actually happens to a [warrior] apprentice,' he replied. So, it is only natural for an apprentice to assume that this is the first principle of [becoming a warrior or mystic]. But it is not. Ruthlessness [being at the point of no pity] is the first principle. . ." (Castaneda, C. ,1987, *The Power of Silence*. New York: Pocket Books/Washington Square Press, p. 118.)

Ruthlessness is the standpoint that enables one to view any situation with respect to actualizing new possibilities. The warrior is ruthless with himself and then with everyone else. This refusal to pity himself or anyone else is what enables the warrior to see possibilities. As Don Juan explains: "A warrior cannot complain or regret anything. His life is an endless challenge, and challenges cannot possibly be good or bad. Challenges are simply challenges. . . . The basic difference between an ordinary man and a warrior is that a warrior takes everything as a challenge . . . while an ordinary man takes everything either as a blessing or as a curse." (Castaneda, C., 1974, *Tales of Power*. New York: Pocket Books/Washington Square Press, pp. 105–106.)

Darwin's theory of how one species may gradually transform over thousands of generations to a new species may be summarized by the following six statements. (1) Each species tends to produce more individuals than will be able to survive and reproduce, and therefore, organisms compete for resources. (2) All species continually produce some offspring with new traits. (3) Some new traits have positive survival value because they enable organisms to more successfully compete for resources than organisms without these traits. (4) Traits with positive survival value are passed on to future generations and gradually increase in frequency due to natural selection. (5) Natural selection is: organisms with new positive survival traits are better able to survive and reproduce more off-spring than organisms without these new traits; that is, operating on a particular species (population) the environment (nature) prefers (selects) one group of organisms over another. (6) As a population of organisms accumulates new naturally-selected traits, it is gradually transformed first into a subspecies and eventually into a new species. Though individual interactions drive natural selection, it is the population over thousands of generations that evolves. An individual organism does not evolve; so in this respect the conventional view still is correct.

Overcoming Greek-Medieval Philosophy

Two books published in the 19th century announced the emergence of biology as an experimental science: Darwin's *Origin of Species* and *An Introduction to the Study of Experimental Medicine* by Claude Bernard (1813–1878). Bernard was explicitly opposed to Scholasticism (the Medieval philosophy that used ideas such as forms, essences, substances) whereas

> Darwin was no philosopher. He seems to have been quite unaware that his objective approach to the facts of life had implicit philosophical premises. He had no world view, no cosmic ontology, or rather he took the one he had for granted without introspective identification.[2]

Nevertheless, Darwin's theory more directly attacked three cornerstone ideas of Medieval philosophy of nature: (1) Aristotle's theory of matter and form, (2) knowledge of a thing by abstraction of its species, and (3) philosophical teleology, also see Chapter 1.

These ideas are closely related. Every material object is composed of nonautonomous "parts," matter and form. Matter is something like the quantum mechanics notion of a tendency of an atom represented by a probability function to express itself in a particular way, which would be the form, during a measurement. In more concrete terms, matter defined as "pure potential" is analogous to marble which has the potential to take on whatever form a sculptor imposes on it in creating a statue. This form, then, is called a species which is "individualized" by matter when present in the statue. We know the statue by abstracting the *species* which then becomes a universal concept in our mind that perfectly represents the statue. In this manner we have absolutely true, objective knowledge of the nature of the statue and of the purpose given to it by its creator which in this case is to express some aspect of the beauty of things. Furthermore, knowledge of natures (forms) leads us to understand natural laws which are the foundations of morality. Material things can gain or lose forms (undergo substantial changes) and change in accidental ways, but the forms are spiritual and eternal. Thus, in this perspective we are absolutely sure that evolution cannot occur. We know a horse by abstracting its form, horesness; we know its purpose by noting its innate potentials which direct it to achieve specific goals, eat grass, copulate, and so on.

The key ideas of Darwin's proposed mechanism for evolution contradicted all of these philosophical notions. In convincing his peers of the validity of a naturalistic understanding of evolution, Darwin was one of the midwives for the birth of biology as an experimental science.

Darwin's proposed mechanism of evolution is based on five ideas that shift mechanistic science toward a process perspective.

---

[2]Ibid., p. 50.

1. Population replaces the idea of substance. An organism is an aggregate of properties, and a species is an aggregate of organisms with sufficiently many properties in common to be put into one category. Taxonomists provide an operational definition of "sufficiently many properties."

2. The properties of an organism are determined by genetic information interacting with patterns in the environment. This directly contrasts with the Scholastic (e.g. Thomistic philosophy) and mechanistic views that properties of an organism are determined by potentials totally within the individual.

3. The world is at least partially probabilistic as a result of random mutations and other statistical processes.

4. Natural selection is a teleological principle in that by providing order and direction to evolution, natural selection gives each species a purpose, which is to be adapted to a particular environment in order that it (the population) can survive. A *surviving population* may be defined operationally by means of statistical properties of the species (age pyramids, survivorship curves, and fertility). However, the concept of adaptation implies a tautology: if a species is adapted to an environment, it is surviving, and conversely, if a species is surviving, it is adapted to an environment. Nevertheless, this tautology represents a subjective insight about the mutuality of species and its environment. Thus, mechanistic analysis must be modified to accommodate necessary subjective insights that supplement operational definitions.

5. The gradual emergence of a new species over time is like any historical process; it is irreversible; thus, time is irreversible.

Overview of the Process of Biological Evolution

There are six key interrelated ideas necessary for understanding the process of biological evolution: (1) continual emergence of new traits, (2) some new traits have positive survival value, (3) competition, (4) natural selection, (5) species adaptation, and (6) species transformation.

Continual Emergence of New Traits with Positive Survival Value.   Reproduction provides the species the opportunity to try out, so to speak, variations of the species life pattern. These variations can be brought about by changes in the genetic information of a species, and indeed, random changes of DNA structure do occur from time to time. When these changes, called *mutations,* occur in the germ cells of individuals participating in reproduction, they are passed on to some individuals in the next generation. Usually these mutations disrupt the species life pattern and therefore are said to have negative survival value. However, occasionally, a mutation may have positive survival value. For example, a change in DNA structure may lead to the synthesis of an altered enzyme that is now more efficient than the one that usually is produced. Alternatively, mutations in DNA could lead to the production of a new enzyme that could be very useful to an organism in a particular environment. For example, a mutation in a primitive bacterium (exposed to oxygen for the first time) could lead to production of an enzyme that enabled the organism to use oxygen in some simple way to gain energy. Such an organism would have a better chance of surviving in an atmosphere that is beginning to accumulate oxygen.

However, what is one innovative mutant among millions of conservatives? Random change would cause this innovation to be lost from the species if it were not for two other interrelated processes: competition and natural selection.

Competition.   All species reproduce more offspring than the environment can support. Therefore, individual organisms compete for resources. However, individual survival is only important to the survival of the species when the organism successfully participates in producing offspring capable of reproduction. The "winner" of biological competition, then, is that organism which produces the greatest number of prolific offspring. Obviously, in order to win in this competition, the organism must be able to survive long enough to participate in reproduction. Some traits help an organism win by making it better able to survive in a particular environment than some other organisms. However, other traits facilitate winning by increasing the efficiency of reproduction.

Natural Selection.   The continual competition among organisms implies that usually an innovative mutant will reproduce more offspring than the conservative organisms. A particular environment selects the innovative mutant to win against any conservative organism. This process is called "natural selection" because any particular environment is thought of as representing nature. For example, before the industrial revolution, various species of moths (e.g., Biston betalaria) in England were uniformly peppered light grey similar to the lichen that covered trees in the area. The camouflage helped these moths escape predatory birds. The occasional dark colored moth or its unfortunate offspring in each generation did not survive to produce more dark colored moths. However, in the middle of the nineteenth century, the vegetation in industrial regions was blackened by soot and other pollution. In this new environment, the dark colored moths were camouflaged and the light grey moths were easy prey to predator birds. Thus, the pre-nineteenth century environment selected light grey moths to survive, whereas the environment of industrialized England selected dark colored moths to survive. Thus, natural selection may be defined as the process where a particular environment usually interacts with an innovative mutant so that it produces more offspring that become fertile, sexually mature adults than noninnovative individuals.

Species Adaptation.   Many natural selections in each generation will result in the next generation having more innovative mutants—individuals with mutations with positive survival value. Eventually—perhaps after many thousands of generations—virtually all individuals of the species will have this new trait. Since the new trait makes the species more adapted to a particular environment, the overall process of acquiring this new trait is called species adaptation. Species adaptation is the increase of a mutation with positive survival value from a few individuals in one generation to virtually all individuals of a subpopulation of the species at some later generation. Since species adaptations may occur again and again, each adaptation may also be represented as the following cycle: ORDER = species life pattern, followed by CHAOS = environment becoming less suitable for a species plus accumulation of a naturally selected mutation, followed by NEW ORDER = species life pattern modified by a new trait.

Species Transformation.   Actually, the introduction of a new trait into the species life pattern is more complex than indicated in the definition of species adaptation. Usually the original positive mutation must occur in several individuals in the same generation. Also, the new trait may be improved upon by a sequence of other mutations. Then, too, any new trait must fit in with the pattern of traits that individuals of the species already possess. Moreover, the new trait may be expressed in different ways in correspondingly different environments. However, as a first approximation, we may think of a species adding one new trait, then another, then still another, and so on, each time by the process called species adaptation. Eventually the species will have so many new traits that its life pattern is different enough to be called a new species life pattern. This process of accumulating many new traits is a species transformation which is the change of one species into a new species as a result of the accumulation of many species adaptations. Evolution is a continuous process in which many species transformations occur. Therefore, a species transformation may be represented as a cycle analogous to species adaptation as follows: ORDER = species life pattern, followed by CHAOS = environment becoming less suitable for a species plus accumulation of species adaptations, followed by NEW ORDER = new species.

The process of evolution may be summarized by the following six statements: (1) Some of the offspring of each generation of a species have mutations. (2) Mutations express traits which have either positive or negative survival value with respect to a particular environment. (3) Organisms continually compete for resources; the winners are those who survive long enough to reproduce the greater number of prolific offspring. (4) In a process called natural selection, a particular environment causes organisms with mutations with positive survival value to reproduce a greater number of fertile offspring than reproducing organisms without these mutations. (5) Many thousands of natural selections occurring over many thousands of generations result in a species adaptation which is the increase of a few individuals in one generation to virtually all individuals of the species at some later generation. (6) When a subpopulation of a species accumulates many new traits as a result of many species adaptations, it becomes sufficiently different to be classified as a new species in the process called species transformation.

In the 20th century, Darwin's mechanism of evolution was modified into two versions: (1) mutationalism and (2) the modern synthesis (in some books misnamed as Neo-Darwinism) which is a synthesis of Neo-Darwinism and Mutualism and draws evidence from all subdisciplines in biology.

There are many current biologists who remain committed to a primarily mechanistic rather than a process perspective, and in so doing they hold an updated version of mutationalism involving molecular biology. In both the semimechanistic and process versions of evolution, a mutation is a random change in a segment of DNA which may be expressed as a new trait in an adult organism. Mutationalism holds: (1) Any particular mutation will survive and be a step in evolution if there happens to be (by chance) an environment that is suitable for the trait produced by the mutation. Therefore, evolution totally is the result of random processes and thus has no intrinsic purpose or direction. (2) Since a mutation may produce a new trait in a particular organism, evolution happens in individuals. In contrast, the modern synthesis holds:

> [1] . . . [evolution] has been oriented or directed toward achieving and maintaining adaptive relationships between populations of organisms and their whole environments. . . . The mechanism of orientation, the nonrandom element in this extraordinarily complex history, has been natural selection. . . .[3]

> [2] Evolution does not happen in individuals, but in *populations.* . . . The population is a continuing, integrated unit. . . The integration of populations is not within but among discrete objects. . . . The natural unit of evolution, then, is a continuing population, a lineage, the members of which exchange genes through the generations.[4]

Systems Perspective of Evolution

Aspects of process science as well as modifications of the Neo-Darwinian theory of evolution and its extension to "a grand evolutionary synthesis" applicable to all levels of organization in the universe have led to what has come to be known as "the systems perspective of evolution," see Chapter 2.

Mechanistic Versus Process Perspective

Mechanistic

1. Science studies material substances that can change.
2. a. Science studies material substances that during time intervals interact and change position.
   b. Mass, momentum, and force are the fundamental concepts of physics.
3. Time is reversible.
4. a. Entropy and the trend toward disorder are secondary theoretical ideas necessary for practical applications of science.
   b. Deterministic order pervades the universe.

Process

1. Science studies statistical properties of populations of things or events.
2. a. Science studies "really existing" fields that contain energy and transfer electromagnetic waves.
   b. Energy is the fundamental concept in physics.
3. Time is irreversible.
4. a. Entropy and the trend toward disorder are fundamental ideas in science.
   b. Partial indeterminism pervades the universe due to the intrinsic probability of microscopic phenomenon ranging from subatomic to molecular events.

---

[3]George Gaylord Simpson, *This View of Life* (New York: Harcourt, Brace & World, Inc., 1964), p. 24.
[4]Ibid., pp. 71–72.

5. All material things, events (changes, inter-actions) and space and time are made up of ultimate subunits which are continuous (infinitely divisible).

6. Any system in nature is made up of autonomous parts or events.

7. Autonomous entities making up any system change by interactions that are determined by laws of nature.

8. Science forbids any appeal to teleological explanations.

9. Science either is positivistic or approximates objectively true knowledge of reality.

10. Science seeks a single, logically consistent, objectively true model that at least approximates reality.

11. Science advances by making logical dichotomies.

5. Subatomic particle-events which make up all reality are discontinuous (not infinitely divisible).

6. Material things making up nature are not totally autonomous.
   a. Bell's theorem: all subatomic particles are simultaneously interrelated.
   b. Genes within an individual are interrelated.
   c. Each organism is interrelated to its environment.
   d. All species are interrelated by the process of evolution.

7. Reality is dynamic in that it consists of irreversible processes rather than things that interact.

8. Science permits certain types of teleological explanations.
   a. Natural selection that directs adaptation and the process of evolution.
   b. Systems not in equilibrium sometimes "evolve" to attractor states.

9. Science either is positivistic or involves some form of subjectivity depending on the context of experimentation.
   a. Measurement of macroscopic events must be relative to the observer's coordinate system.
   b. Measurement of subatomic events must involve probability calculus and only can be intuitively understood by means of nonliteral metaphors.
   c. Observation-experimentation with organisms requires subjective insight into patterns in individuals (basis of classification of organisms) and adaptation of an organism to its environment.

10. Science acknowledges:
   a. There may be many relatively true models which may be contrary or even contradictory because: (1) no one model can ever approximate all aspects of reality; (2) each model represents a different aspect of reality.
   b. There are fundamental phenomena that may be appropriately described by nonliteral metaphors and myths, e.g., subatomic events, evolution, open systems far from equilibrium moving to an attractor state.

11. Science sometimes advances by recognizing complementary dualities, e.g., nature-nurture.

12. Systems are predetermined by internal, static potentials that are expressed in theoretically predictable ways. In practice, the potential may not be totally predictable, but this is due to human limitations rather than to the structure of reality.

12. a. The randomness of microscopic events sometimes produces new, unpredicted, ordered structures.

   b. An organism is an unfolding story that is only partially predictable because the details of the plot are determined by interactions between internal potentials of the organism and a randomly changing environment.

13. Science emphasizes mathematical-physical models of nature.

13. Many models of science employ nonmathematical models of nature, e.g., developmental models and the theory of evolution.

14. The mechanistic perspective does not include the process perspective.

14. The process perspective includes the mechanistic perspective.

Short Version

Mechanistic	Process
1. Study material things that change.	1. Study statistical properties of populations.
2. Interactions between things produce motion.	2. Fields contain energy and transmit electromagnetic waves.
3. Time is reversible.	3. Time is irreversible.
4. Ultimate measurable reality is continuous.	4. Ultimate measurable reality is discontinuous.
5. Total determinism.	5. Partial indeterminism.
6. Systems are predetermined by internal potentials that are expressed in predictable ways.	6. Systems have random aspects that sometimes produce unpredictable new structures.
7. Science seeks a single objectively valid model of nature.	7. Science acknowledges the need for diverse models of nature each appropriate for some aspect of nature.
8. Science emphasizes mathematical-physical models.	8. One can describe some aspects of nature with nonmathematical models.
9. Science approximates objectively true knowledge of reality.	9. Science involves some form of subjectivity depending on the context of experimentation.
10. Science proposes logical dichotomies, where each is the opposite or the inverse of the other, but (1) both are patterns that are *not* coded versions of one another and (2) when both are members of a time sequence, time is reversible.	10. Besides using logical dichotomies Science proposes complementary pairs in which (1) each is a pattern which is a coded version of the other and (2) when both are members of a time sequence, time is irreversible; e.g., stimulus vs response.
11. No teleological explanations.	11. Teleological explanations allowed for some aspects of nature, e.g., natural selection *guiding* adaptations to an environment.
12. Any system is made up of autonomous parts.	12. In some systems the parts are mutually dependent.
13. Understand systems in terms of reversible interactions among parts.	13. Understand systems in terms of irreversible processes among parts, e.g., stimulus-response processes.
14. The mechanistic perspective does not include the process perspective.	14. The process perspective includes the mechanistic perspective as a special case.

Process Aspects of Biological Evolution

The scientific knowledge of the history of life on earth adds three ideas to the process perspective.

1. The material things making up the world are not totally autonomous. a. Each species is interrelated to its environment which includes some other species; i.e., each species is adapted to its environment. b. All life forms, past, present, and future, are interrelated by the historical process of evolution. c. An individual organism is a unified whole determined to be what it is by its inherited genes. (1) The genes determine traits only within a certain range of environments; genes are not autonomous with respect to environment. (2) The genes interact with each other to produce adaptation to an environment; genes are not autonomous to one another.

> . . . no gene has a fitness in isolation from other genes. The development of an organism is brought about by all the genes which it has acting in concert. Adaptedness of individuals and species is always a function of the entire constellation of genes which the organism carries.[5]

Thus, mechanistic analysis must be modified to accommodate interrelatedness.

2. The classification of organisms is one type of evidence for the theory of evolution which then interprets and modifies classification schemes to represent historical relationships. BUT the process of classifying things demands a subjective insight into a pattern of traits existing in an objective individual organism.

> All systematic categories (phylum, class, order, family, genus, species) are universal concepts abstracted from the individuals. As concepts, they exist only in the mind, but we find the basis for these concepts in nature. . . . [according to Schindewolf] Only the individuals have true reality, tangible objectivity, which can be perceived by the senses. But a single individual never represents the whole species concept. . . . [thus though some taxonomists adopt the philosophical concepts of matter/form and species, we are not compelled to do so] *Population* [italics mine] is a taxonomically empty concept and has no meaning so long as the species to which it belongs is unknown. But the species can only be recognized by morphological comparison of several species of the same group and this comparison is always made on the individual. The individual is the primary object of research of the taxonomist.[6]

Thus, mechanistic analysis must be modified to accommodate our subjective perceptions of patterns in individuals in objective reality.

3. The evolution of life is a story; it has a plot. We have and must continue to use the guidelines of the scientific method to objectify this plot. Also, by knowing the plot so far with ever greater detail, we can influence and thus partially control the evolution of our species. Nevertheless, we are drawn to participate in nature rather than only using IT for our benefit. Moreover, this story lends itself to being described by mythology. As Thomas Berry notes:

> Our scientific understanding of the universe, when recounted as story, takes on the role formerly fulfilled by the mythic stories of creation. Our naturalists are no longer simply romanticists or transcendentalists in their interpretative vision; they have absorbed scientific data into their writings. A new intimacy with the universe has begun within the context of our scientific tradition. . . . Science [especially evolution] is providing some of our most powerful poetic references and metaphoric expressions.[7]

---

[5]Dobzhansky, T. and J. B. Birdsell, 1957. "On Methods of Evolutionary Biology and Anthropology." *Am. Scientist*, 45: 381–400.

[6]Borgmeier, T., 1957. "Basic Questions of Systematics." *Systematic Zoology*, 6: 53–69.

[7]Thomas Berry, *The Dream of the Earth* (San Francisco: Sierra Club Books, 1988), p. 16.

<u>Tearout #1</u>

Directions:

Using the six ideas listed in the section on the overview of the process of biological evolution, describe the evolution of a hypothetical or real business that over a few years transforms to a new business.

Core Mutualities of Nonhuman Individuation

Ethic of Mutuality

Ever since the classical Greeks (600 B.C.) sought rational comprehension independent of any traditional, mythological, or religious view of the world, Western cultures have sought for *the* method for obtaining objective certitude. In the 17th century the method used to propose and validate Newton's theory of motion became accepted as the only way of obtaining objective truth. This way of rationally comprehending objective reality became known as the scientific method or mechanistic analysis, and today a somewhat modified version of it is sometimes referred to as hypothesis testing. Many scientists still believe that this method is the only way of obtaining true knowledge. However, some scientists such as Jacques Monod fully admit that they are totally committed to hypothesis testing, which Monod called objective knowing as a result of an ethical choice rather than as a result of objective knowing itself.

In an analogous manner I make the ethical value judgment that though the Cosmos is ultimately rationally incomprehensible, that is, mysterious, it does have ultimate meaning. This amounts to saying that the Cosmos is partially intelligible and partially mysterious. This, in turn, implies that we can relate to the Cosmos in two ways: (1) control of nature based on rational understanding and (2) participation in nature based on love and reverence, that is, we love and revere nature as a result of choosing not to claim that we can totally comprehend her via rational models. This choice, of course, comes from our realization that we cannot comprehend her but rather must, by an act of faith, go out to her, participate in her. This is analogous to any two people loving one another. Love is a reciprocal relationship. If either lover claims to totally comprehend the other or attempts to dominate the other, then the reciprocity and therefore the love relationship is destroyed. To put this positively, love reciprocity implies that while the lovers know some things about one another, each replaces control and any claim to total objective knowledge of the other with faith in the value of the other. This leads to their mutual participation in one another, that is, mutual empathetic knowing of one another.

Thus, the ethical choice I propose is that we not only love one another, but each of us loves all aspects of the Cosmos. However, *love* is such an ambiguous term and the admonition to love all aspects of the Cosmos appears to be paradoxical. For example, must we love destructive and the personally painful aspects of the Cosmos? Must we love the events in the Cosmos such as rape and murder that we judge to be evil? Alternatively, I propose that we focus on the reciprocal nature of a true love relationship and rephrase the admonition about love to state: each of us must enter into mutual relationships with all aspects of the Cosmos that confront us. This mutuality does not negate rational comprehension or some degree of control, but rather mutuality subordinates these to a higher level system that is the relationship. That is, the individual having some rational understanding and control over some aspect of the Cosmos, is incorporated into the mutual relationship. Yes, one must also enter into mutuality with destructive, painful, and even evil events in the Cosmos because only by doing so can one transcend them and possibly help others transcend them. The hard truth, which is discussed in the next section dealing with creativity, is that chaos, which may be the manifestation of evil events, is the occasion for transcending to higher levels of consciousness; or put more concretely, chaos is the occasion for creating solutions to rationally unsolvable problems. In summary, I propose to incorporate the ethic of objective knowledge including scientific, objective knowledge, which is the core of the Enlightenment, into an ethic of mutuality.

Mutualities of the Biosphere

Ecology Implies Thing-Process Mutuality

Ecology may be described as the ". . . myriad of ways in which organisms . . . interact with, influence, and are in turn influenced by their natural surroundings."[1] The core idea here is *interaction between an organismal unit and its environment*. Organismal unit includes all types of organisms that may be studied as an individual, a family unit, a population, a species, or a community of species. The environment of an organismal unit includes other organismal units as well as nonliving aspects. Ecology draws upon all subdisciplines of biology as well as all disciplines dealing with physical interactions, physics, chemistry, geography, climatology, and geology. Thus, in studying all types of interactions ecology is all-inclusive like physics, which describes all types of motion. However, unlike physics which reduces all motion to quantitative descriptions, ecology attempts an integration of the diverse ways of knowing the myriad types of interactions. Hence, while physics, the exemplar of scientific knowing, reduces any whole phenomenon into interactions among autonomous parts, ecology attempts to understand parts in terms of a previous intuitive grasp of a whole. In particular, ecology analyzes all of our experiences of nature into autonomous organismal units interacting with an autonomous network of environmental factors. Then, ecology brings in the idea of complementarity: environment implies organismal unit(s) and organismal unit(s) implies environment. The environmental network is caused to be what it is by organismal units interacting with it, and any organismal unit is caused to be what it is by an environmental network interacting with it. Thus, the parts, organismal units and environmental network, are understood in terms of the unified whole which is the network of interactions in nature.

Story Perspective of the Biosphere

The unified whole = network of interactions in nature is neither merely a "thing," an entity called the biosphere, nor is it merely a "process," a sequence of interactions. Rather, "the unified whole of nature" is a thing-process. That is to say, at any moment in time, the whole of nature is a unified whole system characterized by a nonquantifiable mutuality among nonautonomous parts. This idea of nonquantifiable mutuality among nonautonomous parts follows from a story perspective of the earth.

---

[1]Eric R. Pianka, *Evolutionary Ecology* (New York: Harper & Row, Pub., 4th ed., 1988), p. 1.

At some unique place-time in our earth's unfolding, physical-chemical interactions led to a stimulus-response process between an organism and its environment. This story does not presume the irrational claim that somehow less complex physical-chemical interactions generated the higher order stimulus-response processes of living things. Rather our interpretation of our experiences leads us to accept that a complex network of interactions at some place-time became a system that began a dialogue between this network system which we call a living organism and its environment. In other words, we accept the emergence of life in the same way as we accept the unpredicted and not totally comprehended but reasonable-plausible "turn of events" in any story we create or know about. The emergence of life is something like a mother having a verbal monologue with her baby until one day, the baby talks back (of course, from the moment of first interaction between mother and baby, there is a feeling dialogue going on between them). Though this "story perspective" does not explain why "life emerged from nonlife," it does provide a framework for how any particular organismic dialogue is related to other dialogues as well as the network of physical-chemical interactions in nature. The "story perspective" is called evolution and the interrelation of all dialogues and physical-chemical interactions in nature is called ecology.

Monologues vs. Dialogues.   A primitive classification scheme of the "thingness perspective" distinguishes between nonliving and living things. Mechanistic science reduced all things to physical-chemical processes. I propose to reintroduce the distinctions between nonlife and life without resorting to vitalism or to the naturalistic (Platonic) classification schemes of the 19th century. The distinguishing characteristic of life is stimulus-response processes always associated with epigenetic development and sometimes involved with the reproduction of a species. The stimulus-response processes (S-R processes) are dialogues between an organism and its environment. The organismic dialogues always consist of physical-chemical processes that are monologues.

Physical-Chemical Monologues.   Force exemplifies a physical-chemical monologue in that force is analogous to a monologue communication understood as the transfer of information from a source to a receptor. The source is the giver and the receptor is the receiver and the meaning of this communication is the information exchanged in the context of the place where this exchange occurs. Of course the information given is equal to the information received. In like manner, contact force is the directional interaction at a point in space involving the transfer of momentum from one body to another. That is, force is the monologue communication involving the "giving" of information by one body, the source, to another body, the "receptor" at a point in space. The "meaning" of this force monologue is *momentum = information* which is lost by the source-body and gained by the receiver-body. In modern mechanistic science contact force is the starting concept for defining all other types of physical-chemical interactions in nature. Thus, all other physical-chemical interactions are complex monologues that are elaborations of this elementary concept of contact force understood as a simple monologue at a point in space.

Organismic Dialogue.   Any organism (a system that is said to be alive) exhibits stimulus-response processes. A stimulus is a physical-chemical communication between the environment as source and the organism as receptor. The meaning of this communication is some directly or indirectly observable event which itself directly or indirectly activates a pattern in the organism to produce a response. The stimulus is like a key inserted into a receptor of a car engine. The key does not "cause" the car engine to start but rather leads to activation of the starter motor which causes the car engine to start. The response is a physical-chemical communication between the organism as source and the environment as receptor. Though the stimulus communication is distinct from the response communication, they both are linked in a complementary way analogous to cause-effect. The response is a necessary consequence to the stimulus. If there was no stimulus, there is no response, and conversely, if there is a response, there must have been a stimulus. That is, stimulus implies the idea of response and response implies the idea of stimulus. Stimulus and response mutually imply one another just as cause and effect imply one another. This complementary mutuality between stimulus and response communications is a dialogue between an organism and its environment. In a manner of speaking, the environment "talks" to the organism, and it "talks" back.

A computerized robot is something like an organism in that some factor, for example, light, may activate a computer program in the robot causing it to carry out some mechanized activity; i.e., light stimulates the robot to respond. However, organismic stimulus-response processes have two other aspects unique to living things. First, the ongoing dialogue between the environment and an organism influences it in two fundamental ways. The environment communicating with genetic programs of the organism partially determines phases in its development—we call this epigenetic development. For example, corn seeds with the identical genetic programs will develop into short plants in one type of environment, medium height in another environment, and tall in still a third environment. Furthermore, the environment may stimulate an organism to participate in reproduction of other organisms just like itself. Second, organismic dialogues are aspects of an unfolding story.

Organism-Environment Mutuality.   The "story perspective" leads to what we may call organism-environment complementary dialogue. From the "story perspective" it is misleading to say that a primitive organism was created from nonliving matter just like a vase may be created from clay. The "thing" that has emerged is not the primitive organism but rather a network of monologues (physical-chemical events) has been radically modified by the emergence of a dialogue between this network and a more individualized system we designate as an organism. The story of nature radically changed when some physical-chemical communications led to organismic responses. A new kind of polarity emerged. An organism in communicating with the (nonliving) environment modifies it, but the environment does not respond to the organism. In contrast, the environment in communicating with an organism not only modifies it but also elicits a response from it. The organism and environment are radically different from one another and at the same time imply one another. The organism as well as the environment each can be described as having an intrinsic structure, but each is related to the other by means of the idea of an organism adapting to and therefore surviving in a suitable environment. Organismal adaptation and suitability of an environment are complementary ideas.

A business analogy may help describe this fundamental ecological complementarity between any organismal unit and its environmental. If a business is making a profit for an extended period of time, then two things are simultaneously true. First of all, the people running the business are doing what is necessary to be successful. In particular, they are making the necessary adjustments to changes in the market, such as changing their product or service to meet the needs of their customers. Secondly, the market is such as to enable a well-run business to survive. The most well-run, aggressive refrigerator company could not survive if it depended on selling refrigerators to native Eskimos living in igloos in Alaska. In like manner, if a species is surviving, that is, successfully reproducing itself for several generations, then: (1) the species is adapted to a particular environment and (2) the particular environment is suitable to support the species. Species adaptation means that there are a sufficient number of individual organisms that are adapted to the environment in such a way that they can survive long enough to mate and produce fertile offspring. Of course, these offspring must be adapted to the environment in a similar way as their parents. Adaptation to and suitability of a particular environment are complementary ideas. Each idea implies the other. Just as there is no UP without a DOWN or LOSS without a GAIN, so also there is no "adaptation to" without a "suitability of" an environment. Just as no one—not even God—can make a square circle, no organismal unit can be adapted to a totally unsuitable environment. Adaptation to and suitability of a particular environment are distinct component ideas of a unified whole pattern called organism-environment relationship.

Mutuality of Evolution of Life and Ecology

Evolution of life really is the evolution of the primitive organism-environment dialogue to more complex dialogues. As soon as there were more than one organism, the environment of any one organism always consisted of a living as well as a nonliving aspect; that is, now the living aspect of the environment could respond to an organism. Thus, we have dialogues between each organism and its physical-chemical surroundings and dialogues among organisms.

According to the current biological perspective, a species is a population of organisms having sufficient "family resemblances" to be classified in the category designated as a particular species.

When there are two or more species, then the environment of any one species consists of the physical environment plus all other species. If there are two species A and B, then A may be affected by B and simultaneously B may be affected by A. In other words, as soon as there are two species, there is a network of relationships among all aspects of the environment. An evolutionary process leading to the emergence of a new species leads to a transformation of this environmental network of relationships. Thus, evolution may be described by the following sequence of network transformations: a network without component organisms is transformed into a network with one component primordial species. The network then is transformed into a network with one new species, and so on to the current network on earth consisting of the physical aspects of the environment plus millions of different species.

The following analogy may make these somewhat abstract ideas more understandable. The emergence of a family in society may be thought of as the following sequence of network transformations involving a male, M, and a female, F. The nonmarried M and F marry to form the couple, MF, which then have a baby, thereby changing the couple into a family. Suppose the marriage is an ideal, stereotypic one that produces a stable family unit. When M and F marry each other, each drastically changes his/her social interactions, for example, he no longer dates women, and likewise, she no longer dates men. When the couple has a baby, the network of interactions between M and F and between MF and the rest of society drastically changes. Likewise, society itself changes and all interactions within it not directly involving MF are indirectly affected by the emergence of this family. As the baby grows up, M and F also must change. Moreover, if F changes, the offspring and M also are affected and must make some adaptations to these changes. If M changes, the offspring and F are affected and must make adaptations. Thus, each family emerges from a network of actual and potential relationships among nonmarried persons. Also, the family is itself a particular network of actual and potential relationships that continually change as each of its members undergo developmental-evolutionary changes.

Evolution is the process by which a purely physical network of relationships led to the emergence of primordial life forms, thereby producing a primitive living network embedded in the physical network. This in turn led to increasingly more complex living networks that influence and are influenced by the physical network. Ecology may be defined as ". . . the study of the relations between organisms and the totality of the physical and biological factors affecting them or influenced by them."[2] In other words, ecology is the study of the dynamics of the result of evolution that produced living networks embedded in the physical network. Thus, evolution implies ecology.

There is a wealth of evidence that the physical environment has drastically changed since the appearance of primordial life forms on earth. For example, at one time the earth's atmosphere did not have oxygen and now it does. Modern ecology has demonstrated that except for species that are becoming extinct, life forms "fit in" with the network of interrelationships in an environment; that is, each species contributes to the stability of an ecological community. Diverse species interrelate to form communities that are able to survive and reproduce themselves in a current physical environment. The continual shifting of the structural patterns of communities is due to the sum of the mutual influences of each organism and its environment. Therefore, all current species must have made drastic changes in order to "fit in" with one another to form communities that survive with some pattern that is continually adjusted to a changing environment. For example, instead of the primordial bacteria that emerged 3.5 billion years ago in an atmosphere without oxygen, there now are many species of bacteria that utilize oxygen and coexist with species that do not use it. Thus, in studying the interrelationships of living and nonliving aspects of our world, ecology points to the necessity of evolution. In fact, Darwin proposed the theory of evolution in order to explain the ecological relations he observed in nature. Ecology implies evolution, and since evolution and ecology imply one another, there is the global *evolution-ecology complementarity*.

---

[2]Ibid., p. 4.

Eros Drive Mutuality

The ecology-evolution complementarity is related to Eros drive mutuality (see Chapter 1 for description of Eros drives). Ecology is Order manifested as organism-environment complementrity. Eros-chaos disrupts this order and thereby opens up new life potentials. The process of evolution is Eros-order that actuates and integrates some of these new potentials into a new order, a new organism-environment complementarity. Thus, Eros drive complementarity is: Eros-chaos that leads to the emergence of new possibilities and then Evolution = Eros-order that leads to a New Ecology. The overall process is: Logos which is a species-environment complementarity, Eros-chaos which disturbs this complementarity to produce a chaotic Logos that opens up new possibilities, and then Eros-order actuates and organizes some of these possibilities to produce a New Logos, a new organism-environment complementarity.

Mutualities of Biological Evolution

The modern synthesis theory of biological evolution is, in effect, the biological version of the Buddhist's idea of dependent co-origination. (See *www.zenpribor.org*; THEMES, Zen Postmodernism dependent co-origination.) Like the physicists of the Enlightenment (1500–1700), as discussed in chapters 1 and 3, biologists in the late 19th century abandoned the Aristotelian ideas (incorporated into Medieval thought by Thomas Aquinas) of being, substance, and causation. After the acceptance of Darwin's theory of evolution, biologists—like the physicists—adopted nominalism as the basis for classifying different types of entities. The Linnaeus classification system was kept but with a major shift in perspective. A species no longer was thought of as a being or substance that had a nature (form or essence) that determined it to be what it is. Rather, a species is a population of organisms that empirically are seen to have enough similarities to be classified in the same category. Further empirical observations and inductions may lead to some organisms being reclassified. Indeed the classification scheme may be radically changed as happened in the 1950s. R. H. Whitaker proposed a 5-kingdom system to replace the scheme that classified life forms into animals and plants. Individual organisms including humans were no longer considered to be caused to exist by a creator God operating through biological reproduction. Likewise any final causality or teleological explanation was no longer applicable for a biological description of life forms. That is, the empirical science of biology does not indicate that any life form, including humans, has a purpose or intrinsic meaning. This shift in perspective provided the framework for describing how the biosphere has dramatically changed, that is, evolved over time. In particular, biological evolution describes how some species become extinct and other species change over thousands of generations to produce one or more new species.

If we believe that biological evolution has occurred and continues to occur and if we also abandon the notion that a creator God causes evolution, how does one explain this process? The Buddhist explanation of evolution is that there is no being that created the biosphere and continues to cause it to evolve. Rather, Emptiness nondifferent from absolute Nothingness manifests both Being and non-Being. With respect to biological evolution, Emptiness manifests both the extinction of some species and the emergence of new species. What is more, since all things in the universe are interrelated, the extinction of some species is related to the emergence of some new species. That is, a species becoming extinct and a new species that is emerging each is that it is—its "suchness"—not because of some cause or some essence or some purpose. Each is that it is as a manifestation of absolute Nothingness. Likewise the relation of the species becoming extinct to the new species that is emerging is a manifestation of Emptiness. Emptiness manifests both the chaos of extinction and the order of emergence. Neither stands alone. In general, neither is permanent in that chaos leads to order and order leads to chaos. Thus, Eros-chaos, Eros-order and the mutuality of Eros-chaos and Eros-order are manifestations of Emptiness. Likewise Emptiness manifesting the mutuality of ecology and evolution is the biological version of dependent co-origination. A modern contemplative Catholic, Bernadette Roberts, in three books, gives a Christian interpretation of this same insight: *The Experience of No-Self, The Path to No-Self,* and *What Is Self?*

More specifically, component organisms dying continually disrupt the order of a stable ecosystem. The ecosystem order is reestablished by reproduction of the life-pattern of each component species.

However, reproduction involves both Eros-order and Eros-chaos. Eros-order leads to reemergence of species life patterns and to the stabilized structure of the ecosystem. This structure involves a mutuality between each organismic unit (the individual organism, the species, the species community if there is one, and the ecosystem) and the environment. Eros-chaos of reproduction is the randomness of the process that produces diverse outcomes. Most outcomes are instances of the species life-pattern. But there always are outcomes—offspring—which are instances of some degree of departure from the species life-pattern. These deviant offspring are called mutants. Each mutant results from one or more of its genes changing structure. This change in gene structure is called a mutation.

The organism-environment mutuality is a dynamic process called natural selection which determines whether a species is stable or becoming extinct or evolving. If the environment is not changing and there is no competition among component species of the ecosystem, then the organism-environment mutuality selects offspring that have the species life-pattern to survive. The offspring that deviate from the species life-pattern will be less likely to survive. Of course survival itself is somewhat random. There always will be some normal organisms that do not survive and some mutants that do survive to reproduce a mutant life-pattern. The more deviant the mutant, the less likely it will survive. The offspring of a non-radical mutant again have a probability of not surviving that is greater than that of a normal organism.

If there is competition, the random aspect of reproduction generates evolution. Most of the mutant offspring will have a less probability of surviving than that of normal offspring. One or more mutants may have genetic traits that give them a slight advantage in competing for the resources of the environment or for being able to reproduce. If there is only one such mutant and it by chance happens not to survive to reproduce itself, then no evolution will occur. However, if there are several such mutants and some of them survive to reproduce themselves, then their offspring will have a greater chance of surviving than normal offspring. As a result, each new generation will produce a slightly greater percent of mutants and a correspondingly slightly less percent of normal organisms. Over many thousands of generations the more adaptive mutant organisms will have totally replaced the less adaptive normal organisms. The species will have undergone what is called a species adaptation. Likewise if there is no competition but a small change in the environment, then again mutants will arise that are slightly more adapted to the changed environment than the normal organisms. After many thousands of generations, there will be a species adaptation. After hundreds of thousands of generations a continually changing environment plus competition among species will produce many species adaptations. At some point in time the accumulation of species adaptations will make the evolving species so different from the species before it started to evolve that it will be classified as a different species. The accumulation of species adaptations that led to this reclassification of the "original species" is called species transformation.

Organism-environment mutuality is neither Good nor Evil. It is not caused to be nor does it have any intrinsic meaning. It is "such that it is." The mutuality of randomness and organism-environment mutuality generates evolution which generates ecology. That is, random events disturb a particular organism-environment mutuality. The random aspect of reproduction produces mutants some of which may exhibit greater organism-environment mutuality than normal organisms. We say these mutants are more adapted than the normal organisms. The increase in percentage of mutants over hundreds of thousands of generations is the process of evolution that produces the accumulation of species adaptations that lead to a species transformation. The new species represents a new organism-environment mutuality that in turn represents a new ecology. Another way of stating this is: old ecology plus randomness leads to evolution involving randomness that leads to a new ecology which incorporates a modified old ecology. Biologists tend to say that evolution results from or is driven by randomness. This is equivalent to saying order comes from chaos or being comes from nonbeing, i.e., new ecology comes from chaotic old ecology. Randomness, a concept taken from the mathematical theory of probability, is elevated to an absolute principle of reality. Buddhists would say that Emptiness manifesting as dependent co-origination is what "drives" evolution. The perspective of this book is that INDIVIDUATION, which is absolute Nothingness, generates evolution. INDIVIDUATION also may be taken to be only Emptiness or it also may be taken to be Christ or Allah or God or Tao or some other interpretation.

Universal Harmony vs Transformations

The perspective of this text is that ecology is the expression of INDIVIDUATION representing absolute Reality, which produces harmony among diverse things and diverse potentials. Ecology of the universe includes randomness that produces disharmonies, i.e., a random event is equivalent to a disharmonious event. However, the ecology of the universe transcends randomness as a result of implying a harmonious story. That is, not only does ecology and evolution imply one another, as was described earlier in this chapter, but also the disharmonious events that appear are transcended by emergent creativity that is evolution. Random events are aspects of a "mini-story plot." Some changes in the harmony of a portion of the universe produce disharmony, but these random events open up many new, nonharmonious possibilities. By a process analogous to natural selection in the biological theory of evolution, some of the random possibilities come together to form a New Harmony that includes a modified version of the Old Harmony. The network of "mini-story plots" occurring over time results in a GRAND STORY in which the cosmos expresses a continually emerging, harmonious hierarchy of harmonies.

Birth exemplifies *the* most fundamental feature of the unfolding story of the universe. Disharmonies occur and when taken to be autonomous events are *evil*, but when these evil events are understood in relation to the unfolding story, they are *transcended* by virtue of being an aspect of a mini-plot which most fundamentally may be represented as: Harmony ⇒ Disharmony ⇒ New Harmony that includes a radically modified "Old Harmony." I find it convenient to represent this mini-plot metaphorically as LIFE ⇒ DEATH ⇒ REBIRTH. With respect to this particular mini-plot, the New Harmony is a *Holarchy* consisting of two levels of organization, the higher level new harmony and the lower level of the modified old harmony. In general, the unfolding story of the universe is a many-leveled Holarchy. From the perspective of this book which is consistent with a similar idea of quantum mechanics, there is no "lowest level Harmony" or "fundamental building blocks of matter." (This paradox of the no lowest level starting point of the unfolding story of the universe is *transcended* though not rationally comprehended by saying that the emerging cosmos Holarchy simultaneously is descent of the One into the Many and the ascent of the Many to the One. These ideas of Plato and Plotinus are described by Wilber in: Wilber, K. (1995). *Sex, Ecology, Spirituality* (*the Spirit of Evolution*). Boston: Shambhala Pub. Inc., pp. 319–344.

Life.   Before birth the baby is in paradise. No matter how uncomfortable the mother is because of very hot or very cold weather or because of lack of food or water, so long as she is not severely ill, the baby will be in total comfort. It literally is taking a nice warm bath in a superbly controlled environmental room. Thousands of biological servants (mechanisms) take care of all its needs. Food is brought to it and wastes are carried away without the fetus even helping in these processes. The fetus sleeps when it wants to and moves around if it so desires. The womb experience is the ultimate leisure vacation that provides escape from the hard realities of survival. It is Heaven on earth where all is growth, development, and pleasure without any pain.

Death.   Birth is death! After what seems like an eternity in which all is One, True, Good, and Beautiful, in a matter of hours, the baby is literally thrown out of paradise; its life line to mother is torn off and destroyed, and it is given the nonnegotiable command: "You're on your own, kid!" There is only one other life catastrophe that will be so dramatic and so totally nonnegotiable, irreversible, and nonmodifiable so as to produce an ultimate disruption of life; that event, of course, is a person's death.

Rebirth.   The birth trauma would quickly lead to biological death if an equally dramatic series of events were not immediately set in motion. After the "fall from Heaven," the baby (usually) is reborn into a world of dualities. Initially these dualities consist of pleasure-pain and Nature-Society, which complement one another so as to sustain the baby. Pain expressed in anger or fear leads to the baby being fed or changed or comforted by a human "mother" (male or female) who represents the nurturing of human society. Instincts and emotions representing Mother Nature take care of some needs but for the most part, nature "pushes" the infant and a mother toward bonding with one another. Once bonding is established the baby's life is sufficiently stabilized to begin to burst forth, like a bud in the first warm day of spring, into a human psychic flowering. The infant now is an embryonic ego which will spontaneously create the next most central duality of humankind: I versus everything else!

Modern Biological Theory of Evolution as a Creative Process

Evolution Begins with Disharmonies

As indicated earlier in this chapter, Ecology and Evolution imply one another. We can see this mutuality in relation to the idea of harmony in nature. Evolutionary creativity begins when a particular ecology expresses disharmonies. The Species–Environment complementarity represents harmony in nature. When the environment permanently changes so that a species is no longer adapted to or not sufficiently adapted to its environment, then harmony is partially destroyed. This disharmony in an ecology (an ecosystem) starts a process—evolution—that will reestablish harmony in nature. Thus, a stable ecosystem that has survived changes in the environment implies that evolution must have occurred; i.e., ecology implies evolution. At the same time evolution always leads to a species-environment complementarity, which is harmony in the ecosystem; i.e., evolution implies ecology.

Adaptation Creativity

One aspect of the modern theory of evolution is that species or higher level organismal units are continually adapting to a changing environment. This adaptation creativity has five aspects.

Emergence of a New Semistable Ecology.   Natural selection of a mutation with positive survival value leads to a species adaptation. This is a new semistable ecology involving a species adapted in a new way to survive in its environment.

Differentiation That Leads to Greater Diversity.   The newly adapted species represents the expression of a potential that was present between the nonadapted species and its environment. This statement is just another way of saying that the world may be thought of as a probabilistic world; i.e., at any moment, it may be thought of as having an infinite, nondenumerable number of possible expressions. (The systems scientist who is a reductionist would replace the phrase "may be thought of" with "is.") Thus, the emergence of a new adaptation is analogous to differentiation that occurs during epigenesis. That is, a naturally selected mutation of DNA is expressed, and the environment in which it is expressed modifies its expression. In a certain sense we may say that the ecology is differentiating as a result of one of its component species evolving to a new adaptation. As is true in epigenesis, differentiation leads to diversity.

Diversity Leads to Conflict.   The new adaptation resulting from a new trait in each individual of a newly adapted species may not be compatible with one or more other traits present in that individual. Likewise, a new adaptation that involves a change in behavior may not be compatible with behaviors of nonadapted individuals. Also, a newly adapted species may disrupt the organism-environment complementarity of one or more other component species of the ecology. In all of these cases, diversity produces conflict that, of course, produces disharmony. As before, this new disharmony starts more evolutionary changes.

Conflict Leads to Cooperative Individuality.   The idea of cooperative individuality will be described in greater detail in the discussion of a trans-systems theory of creativity. For the time being, we may note that the conflict that initiates the "more evolutionary changes" involves natural selection of one or more other mutations in the species. This leads to the conflicting traits being complementary to, i.e., fitting into, one another, rather than conflicting with one another. Alternatively, mutual species adaptations of the newly adapted species and one or more other species in the same ecosystem lead to nonconflicting pairs of species in a modified ecology. This same phenomenon describes-explains the emergence of or modification of ecological niches.

Cooperative Individuality Leads to Hierarchal Incorporation Producing a New Ecology.   The newly emerged species adaptation and the other traits that were modified to be compatible with, that is, complementary to, one another are thus incorporated into a new species pattern. Likewise, the species modified by a new species adaptation and the other species in the same environment that "co-evolved" to adapt to it are incorporated into a new stabilized ecology. This new ecology = new holarchy is higher in the sense that it has incorporated the old pattern plus the evolved new pattern in such a way that leads to harmony. It has transcended conflict meaning that the conflict between traits or species was not resolved by one winning or dominating the other, but rather, the conflicting factors developed mutuality. Thus, there is a hierarchy in which a higher pattern incorporates a modified lower pattern.

Transformation Creativity

One aspect of the modern theory of evolution is that there is the ongoing process of emergence of new species. In effect, the biosphere not only is continually being modified, it is transformed each time a new species emerges. Many evolutionary thinkers take the gross reductionist view that the emergence of a new species results from the accumulation of many species adaptations. Systems scientists reject this gross reductionist view. They attempt to describe-explain the emergence of a new species as analogous to the emergence of life from nonlife. A probabilistic interpretation of thermodynamics and information theory is the foundation for these theories of the emergence of new species.

Evolution as Narrative

The process leading to the emergence of a new stable ecology is a story. The evolutionary story may involve only adaptation creativity or both adaptation and transformation creativity. In either case, the story has the overall plot of Harmony that breaks down into Disharmony from which emerges a New Harmony that includes modified aspects of the old harmony. With respect to humans' analytical understanding of this story, the plot is represented as Order that breaks down into Chaos from which emerges a New Order that includes a modified old order. Metaphorically speaking the story is represented as a Life, Death, Rebirth process.

Trans-Systems Theory of Creativity

Transcending Reductionism

A trans-systems theory of creativity combines ideas from systems science with the four interconnected styles of creative learning described in the end of Chapter 3. One aspect of this synthesis is the necessity of reductionism. Because of the limitations of language, all of us are compelled to be gross or subtle reductionists. If we believe that any of our rational models in some degree represent patterns that "really are there" in reality, we reduce our existential knowing stemming from direct experiences to a set of concepts. This is *destructive nihilism*. Like Nietzsche, many postmodern thinkers today believe that all rational models merely superimpose meaning onto REALITY that in ITSELF has no ultimate, transcendental meaning. This way of avoiding reductionism is inauthentic nihilism. Creative scientific thinking, Chapter 3, proposes that the use of conceptual language forces one to reduce existential knowing to distorted representations of REALITY. By realizing this, which is authentic nihilism, we can create these necessary illusions and yet remain connected to the REALITY from which they emerge. Creating illusions is a fundamental aspect of the human condition. We need them. Science provides one type of illusion, art another type, philosophy and religion another, and politics and ethics still another.

This realization combined with the systems theory of creativity provides the foundation for all creativity. All new systems come from the REALITY which we can never totally understand via language models. As David Bohm and F. D. Peat (1987, p. 265) proclaim in several places in their book, *Science, Order, and Creativity:* "Whatever we say the totality is, it isn't—it is also more than we say and different from what we say." REALITY manifests Itself as continuous, evolutionary change. As some systems "fall," new systems will rise. We cannot afford to cling to any system or idea, not even our own existence. However, we can choose to participate in this universal evolution by acknowledging all creations as impermanent illusions that nevertheless have meaning as a result of arising from (evolving from) eternal REALITY. This Zen disposition enables us to mirror the creativity of the universe. In this state new insights always will emerge in us.

The above insight may be described in terms of the scientific method in relation to the paradoxes in nature that modern science has disclosed. One of the characteristics of the scientific method is "scientific public consensus of truth" which includes guidelines for validating scientific theories, see Chapter 3. These guidelines work well for what Thomas Kuhn calls ordinary science (in his book, *The Structure of Scientific Revolution*), but they do not at all describe how humans formulate a scientific theory in the first place. More to the point, they do not describe how scientists resolve conflicting paradigms during a time of crisis in ordinary science. This points to a fundamental flaw in the mechanistic rational model for describing scientific knowing, namely, being totally rational, it does not allow for creativity which is nonrational. Einstein imagining himself riding a light beam and then proposing the special theory of relativity is not a rational process; rather it is like a gap experience of an avant-garde artist leading him to create a work of art. Mechanistic thinking could no more generate innovative transformations in science than it could in modern art.

Some positivistic scientists make a "fundamentalist leap of faith" that maintains that a mechanistic scientific model eventually will resolve all paradoxes. Einstein, himself, made this kind of commitment in spite of his gap experiences. Einstein could not accept the fundamental discontinuity and corresponding probability of nature postulated by quantum mechanics. This contradicted his firm belief in the fundamental continuity and corresponding determinism of nature postulated by the general theory of relativity. He believed that the probabilistic, process perspective of quantum mechanics was a temporary viewpoint needed until a more fundamental mechanistic theory could be created. Today many scientists are resigned to the paradoxes generated by quantum mechanics and do not seek to resolve them. The paradoxes themselves imply a higher level of understanding which holds that there must be a diversity of theories each relatively valid for some contexts.

This higher understanding is the natural science version of Heidegger's philosophy. The scientist neither discovers Logos Order in nature (as believed by Greek and Medieval philosophers) nor does he superimpose a self-generated Logos onto nature in order to comprehend it (as Nietzsche proposed). Rather nature (BEING) discloses itself to the scientist and partially determines the way he must create theories to understand it; i.e., ". . . *being* determines *thinking* which in turn interprets *being* [creates theories of nature]."[3] At the same time no scientific theory totally captures or even approximates the meaning of nature because nature simultaneously is *knowable* and *mysterious*; i.e., Heidegger translates a maxim from Parmenides as "There is a reciprocal bond between apprehension and being."[4] By this he means that being forces one to conceive what it [*Being*] is, but no conceiving, no idea can ever represent being. All conceiving of being leads one into glorious failure; *failure* because any possible conception is in principle inadequate but *glorious* because each new heroic conceiving leads one to passionately experience a new aspect of being.

The Creative Process in Terms of Seven Characteristics

Disharmonies

1st Characteristic.   Creativity begins with a particular ecology that expresses disharmonies.

A particular ecology refers to a whole system expressing a particular level of organization. Some aspects of this level of organization may be known at least in a vague way. Other aspects may not be known even in a vague way; or one may have incorrect opinions about the level of organization. Experimentation will reveal whether the system is at a higher or a lower level than one thought. If the whole system is oneself, then one may discover that he/she is at a higher or a lower level of consciousness than was initially believed to be the case.

---

[3]Heidegger, M. 1959. *An Introduction to Metaphysics.* New Haven: Yale Uni. Press, 144.
[4]Ibid., p. 145.

The whole system is a unified "thing-process;" that is, one can point to a thing that is continuously in the process of changing, such as, developing or evolving. The process of developing or evolving may be represented by a story that uses metaphors. In one's continual reevaluation of the story, the metaphors may have to be removed or added to or replaced by new metaphors. Likewise, the plot of the story may have to be modified. The story should be such that it can be represented by (transformed into) a rational model. The transformation to rationality involves converting metaphors (really metaphorical concepts, see Chapter 3) into pure concepts that can be logically defined, and whenever possible operationally defined, for example, see operational terms described in Chapter 3 as one of the four characteristics of the scientific method. Some concepts such as those used in philosophy or mystical disciplines, for example, Yoga meditation, may be "operationally defined in terms of one's personal experiences. This is one situation where a spiritual guide or teacher may direct a person to create for himself/herself these kinds of operational definitions. The story which uses well-constructed operational definitions now can be related to criteria for validity. One may set up his own criteria or he/she may accept the criteria given by some community or some spiritual guide or teacher.

A whole system may express disharmonies many different ways. Expected outcomes occur "as expected," but have unexpected distorting effects. Alternatively, expected outcomes do not occur or unexpected outcomes occur along with the expected ones. Over time a seemingly harmonious system expresses internal conflicts that were always there but not observed; or the system evolves or develops aspects that produce internal conflicts. For example, puberty is a time when a teenager develops internal conflicts which "push" him/her to create a new self-identity. Sometimes a system is maintained in order as a result of one part of the system dominating other parts, for example, husband dominating his wife or one ethnic group dominating another ethnic group. When the dominated or repressed aspect rebels or one way or another expresses negative symptoms stemming from being dominated or repressed, then the once stable system expresses disharmonies. Sometimes systems express disharmonies as a result of some aspect not being adequately differentiated in some way. For example, a manager or a parent may have a well-differentiated ability to make rational decisions but the individual is unable to be intimate or to be open to mindful creative dialogue.

Transformation Creativity

Transformation Creativity involves the same five sequence of ideas as occurs in adaptation creativity. These ideas are the 2nd through 6th characteristics of universal creativity.

2nd Characteristic.   There is the paradoxical emergence of a new semistable ecology.

Paradoxical emergence refers to the emergence of an ecology with a new set of properties which seemingly came from nothing because they were not present in the old ecology. For example, the emergence of life from nonliving matter is a circular (Uroboros) paradox, see Chapter 2. We can appreciate that this can happen, but we cannot understand how it happens. Thus, instead of trying to explain how "paradoxical emergence" occurs, we say that this emergence is a creative process that transcends the circular (Uroboros) paradox, see Chapter 2. The process is nonrational, that is, cannot be rationally described, and is not predictable—creativity always produces surprises. However, we can describe aspects of the process that then enable us to dispose ourselves to participating in the creative process; i.e., we may dispose ourselves to allowing the creative process to occur.

The newly emerged ecology is unstable. If it is not further modified or able to differentiate, it will disappear. The new ecology is like a newborn baby; if the baby is not taken care of while it further differentiates, it will die. A lot of new systems emerge from time to time, but only those systems that are sufficiently stable to further differentiate or are protected long enough for them to further differentiate will survive.

3rd Characteristic.   In becoming more stable via differentiation the new system produces internal diversity and expresses diversity with respect to its environment.

For example, when the young child created for itself a masculine or feminine gender identity, it simultaneously created internal and external diversity. Internally, a male child will begin to differentiate its masculine aspects and leave undifferentiated its feminine aspects, and in so doing the child will progressively differentiate itself from female children. The same dynamics applies to a female child. Thus, the 3rd characteristic of creativity is differentiation that produces diversity.

4th Characteristic. Diversity produces nonconflicting and conflicting aspects in the newly emerged ecology.

Some new aspects resulting from differentiation may be totally independent of one another. For example, the differentiation of secondary sex characteristics that occur during puberty produce some independent traits such as increased height and production of sex cells (sperm in males, eggs in females). Other traits produced at this time may complement one another, for example, characteristic fat deposit distribution and developing breasts in women. Overall each aspect in any pair interaction may enhance one or both members of the pair, but the stability and effectiveness of neither member is required for the stability of the other.

On the other hand, differentiation may produce aspects that conflict with one another. For example, when a baby emotionally bonded with its mother differentiates some degree of self-consciousness, it comes into conflict with its mother in some circumstances. When a hydrogen molecule breaks into two hydrogen atoms and each then loses an electron, the resulting two positive hydrogen ions repel one another. Likewise when an oxygen molecule breaks into two oxygen atoms and each atom picks up two electrons, the resulting two negative oxygen ions repel one another. In general, conflict leads to instability; we say the differentiating new ecology goes into chaos. Sometimes the instability is overcome by one aspect dominating and or repressing the other aspect. For a time, the mother may dominate the differentiating child. The differentiating masculine ego in boys will dominate and repress their feminine aspects. Alternatively, the unstable system disappears or reverts back to its original structure. The two negative oxygen ions become one neutral oxygen molecule and the two positive hydrogen ions regain electrons and then become one neutral hydrogen molecule.

5th Characteristic. Conflict is transcended as a result of the two differentiated aspects "developing" Cooperative Individuality.

For example, the conflict between the mother and the baby differentiating feeling self-consciousness usually leads to the mother and child developing a new way of bonding with one another. Likewise the chaos in the population of conflicting positive hydrogen ions and conflicting negative oxygen ions is transcended by two hydrogen ions chemically bonding to one oxygen ion to form a neutral, stable water molecule. We say there is cooperation among all components of the same order of differentiation because a higher order pattern directs the interactions (orchestrates the interactions) among the components. The new mother-child bonding directs how the baby child and mother relate to one another. Likewise, the chemical bonds between oxygen and two hydrogen ions direct how oxygen and hydrogen relate to one another. We also say that the components of the new higher order pattern have greater individuality because by being a component of a higher pattern, each has a reduced "degree of freedom." That is to say, each has reduced possibilities since a potential for a particular higher order has been selected over potentials for other possible higher order patterns. These components are more determined (by the higher order pattern of which they are a member) and therefore are less random. Being less random means they have greater individuality.

6th Characteristic. Cooperative individuality produces a new order via Hierarchal Incorporation.

Thus, via cooperative individuality the mother and child have been incorporated into the higher level pattern of a new mother-child relationship. Likewise, the two hydrogen ions and one oxygen ion have been incorporated into the higher level pattern of a water molecule. In general, lower-level patterns are modified so as to interact to form a higher-level pattern. Some properties of this new pattern are emergent properties. In particular: (1) These properties only can be described in terms of the whole higher order pattern. (2) These properties cannot be understood in terms of (reduced to) interactions among components; that is, the properties cannot be understood in a gross, mechanistic way. (3) These properties cannot (or could not) be predicted to occur from knowing the components of the new pattern. (4) The properties are the expression of imminent potentials of a set of apparently autonomous entities that interact in a context that actualizes the imminent potentials, see Chapter 2. One can know the existence of these imminent potentials only as a result of knowing their manifestation that shows up as emergent properties. One can know the context for the expression of these immanent potentials (and thus count on their expression when the appropriate context is set up as we do in chemistry, for example), but one never can know the how-why the context leads to their expression. Knowing the how-why implies a mechanistic description, and the expression of emergent properties is a nonmechanistic and indeed, a nonrational process.

Creativity as Narrative

7th Characteristic.   The emerging of a new ecology is a STORY.

Narrative Perspective of the Emergence of Life. In this perspective the cosmos may be viewed as an unfolding story of an ecology of monologues becoming progressively more complex until the emergence of life dialogues. A life dialogue is a stimulus-response process where the idea of stimulus implies the idea of *response* and conversely. This complementarity may be represented by the Uroboros, a snake eating its own tail, and thus, the emergence of a life dialogue is a *Uroboros Paradox.* We see the same sort of paradox in philosophy with respect to the "thing-process" or "thing-event" complementarity. A thing has "meaning," that is, is knowable, only in terms of a process, i.e., an observable event which is an interaction between two things. Likewise, process has "meaning" only in relation to things. In like manner, the Stimulus-Response Uroboros Paradox may be stated as follows: A stimulus has occurred only if a response has occurred, but a response has occurred only if a stimulus has occurred.

Levels of Creativity in the Human Biosphere.   There are four levels of creativity in the human biosphere punctuated by two radical discontinuities. The first level is the continuation of the Big Bang origin of the cosmos that produced a hierarchy of nonlife systems on earth, i.e., subatomic particles, atoms, molecules, and macromolecular systems. This hierarchy of nonliving systems provided the appropriate ecology for the emergence of Life, the first radical discontinuity. The second level results from creativity that produced the diverse ecologies that include nonconscious, living organisms. The third level results from creativity that produced conscious, feeling hierarchical societies, for example, conscious animal pair bonding, family units, and societies organized by feeling communications. The emergence of self-consciousness is the second radical discontinuity that led to the fourth level of creativity which produced self-conscious, language knowing that in turn, produced human, self-conscious, hierarchical societies with shared vision, for example, culture.

The Grand Story of the Human Biosphere

The perspective of this text is that ecology is the expression of INDIVIDUATION (= absolute Reality = Creator God or Emptiness or Brahman) which produces harmony among diverse things and diverse potentials. Ecology of the universe includes randomness which produces disharmonies, that is, a random event is equivalent to a disharmonious event. However, the ecology of the universe transcends randomness as a result of implying a harmonious story. That is, not only does ecology and evolution imply one another, but the disharmonious events that appear are transcended by emergent creativity which is evolution. Random events are aspects of a "mini-story plot." Some changes in the harmony of a portion of the universe produce disharmony, i.e., random events which in turn open up many new, nonharmonious possibilities. By a process analogous to natural selection in the biological theory of evolution, some of the random possibilities come together to form a New Harmony that includes a modified version of the Old Harmony. The network of "mini-story plots" occurring over time results in a GRAND STORY in which the cosmos expresses a continually emerging, harmonious hierarchy of harmonies. That is to say, this "continually emerging, harmonious hierarchy of things-events may be summarized by the following "acts.""

I. **ACT 1: THE COSMOS WITH A NONHUMAN BIOSPHERE**
   A. INDIVIDUATION of Matter
      1. Big Bang
      2. Galaxies Y solar systems including one which contains EARTH
      3. Hierarchal harmonies of matter found on EARTH, e.g., energy fields $\Rightarrow$ elementary particles $\Rightarrow$ subatomic particles $\Rightarrow$ atoms $\Rightarrow$ molecules $\Rightarrow$ macromolecules
   B. INDIVIDUATION of Life
      1. Hierarchal harmonies of matter incorporated into higher harmony of primitive living systems (bacteria)
      2. Hierarchal harmonies of life, for example, bacteria $\Rightarrow$ cells with a nucleus $\Rightarrow$ multicellular plants and animals make up the nonconscious biosphere which evolves to the conscious Biosphere as a result of the emergence of mammals and birds expressing (1) feeling awareness and (2) conscious communications sometimes leading to bonding, the basis of conscious social communities .
   C. INDIVIDUATION of harmonious hierarchy thus far:
      1. Matter $\Rightarrow$ Body (living systems) $\Rightarrow$ Heart (consciousness in mammals and birds expressed in feelings)
      2. Hierarchy of integrity-survival of systems
         a. Matter systems (for example, atoms and molecules) subordinated to Individual Organisms
         b. Individual Organism subordinated to Species
         c. Species subordinated to Ecological Community
         d. Ecological Community subordinated to the evolution of the Conscious Biosphere
II. **ACT 2: TRANSITION**
   A. Emergence of Hominids
   B. Primacy of Hominid Family Units; there now is the potential for the emergence of humans that as they differentiate MIND they will gradually break out of the INDIVIDUATION of harmonious hierarchy of the Conscious Biosphere
III. **ACT 3: EMERGENCE AND DIFFERENTIATION OF *MIND***
   A. Implications
      1. MIND progressively overshadows INDIVIDUATION
      2. Progressive Death of conscious biosphere ecology
      3. Progressive expression of INDIVIDUATION via human passion
      4. Progressive unfolding of possibility of rebirth of INDIVIDUATION as human ecology
   B. Stages:
      1. In terms of differentiation of patriarchal societies
      2. In terms of differentiation of rationality culminating in mechanistic science
      3. In terms of differentiation of personality defined in terms of ego control consciousness that produces objective knowing which suppresses participatory consciousness that leads to subjective insights
IV. **ACT 4: TRANSITION**
   A. Emergence of Process Perspective
   B. Emergence of Systems Perspective
   C. Avant Garde Modernist Art and Nihilism
   D. 1960s Revolution $\Rightarrow$ Postmodernism
   E. 1996–20??: Moment of Truth for Human Biosphere
      1. Ecological double binds produce great disharmony of Human Biosphere
      2. MIND approach to ecological double binds leads to:
         a. Nihilism = a doctrine or belief that conditions in the social organization are so bad as to make destruction desirable or inevitable for its own sake independent of any constructive program or possibility.
         b. Scientific humanism = the perspective that all human problems sooner or later can be solved by the application of scientific critical thinking that employs some version of the scientific method; most problems can be solved sooner rather than later. This perspective

ignores or denies the validity of ecological double binds or believes that new science/technology will lead to transcending these double binds.

3. INDIVIDUATION approach to ecological double binds:

   a. Regress to a social-cultural perspective in which MIND did not overshadow INDIVIDUATION (or did so to a very much less degree than now)

   b. Transform to a social-cultural perspective which transcends ecological double binds.

      1) Rebirth of ecology as a harmonious human biosphere

      2) Higher powers of INDIVIDUATION incorporates MIND powers

      3) First step: ethic of *Mutuality*

Tearout #1

Directions:

The student is to describe his/her puberty transformation in relation to the seven characteristics of the creative process.